COMPARATIVE DOSIMETRY OF RADON IN MINES AND HOMES

Panel on Dosimetric Assumptions Affecting
the Application of Radon Risk Estimates

Board on Radiation Effects Research

Commission on Life Sciences

National Research Council

NATIONAL ACADEMY PRESS
Washington, D.C. 1991

NATIONAL ACADEMY PRESS • 2101 Constitution Avenue, N.W. • Washington, D.C. 20418

NOTICE: The project that is the subject of this report was approved by the Governing Board of the National Research Council, whose members are drawn from the councils of the National Academy of Sciences, the National Academy of Engineering, and the Institute of Medicine. The members of the panel responsible for the report were chosen for their special competences and with regard for appropriate balance.

This report has been reviewed by a group other than the authors according to procedures approved by a Report Review Committee consisting of members of the National Academy of Sciences, the National Academy of Engineering, and the Institute of Medicine.

This project was prepared under Grant No. X-816115-01-0 between the National Academy of Sciences and the U.S. Environmental Protection Agency.

Library of Congress Cataloging-in-Publication Data

Comparative dosimetry of radon in mines and homes / Panel on
  Dosimetric Assumptions Affecting the Application of Radon Risk
  Estimates, Board on Radiation Effects Research, Commission on Life
  Sciences, National Research Council.
    p.   cm.
  Includes bibliographical references and index.
  ISBN 0-309-04484-7
  1. Radon—Measurement. 2. Radiation dosimetry. 3. Indoor air
pollution. 4. Mine safety. I. National Research Council. Panel
on Dosimetric Assumptions Affecting the Application of Radon Risk
Estimates.
RA1247.R33C66   1991
628.5′35—dc20   91-10157
                                                             CIP

Cover: Mine photograph by Samet.

Copyright © 1991 by the National Academy of Sciences.

No part of this book may be reproduced by any mechanical, photographic, or electronic process, or in the form of a phonographic recording, nor may it be stored in a retrieval system, transmitted, or otherwise copied for public or private use, without written permission from the publisher, except for the purposes of official use by the U.S. government.

Printed in the United States of America

PANEL ON DOSIMETRIC ASSUMPTIONS AFFECTING THE APPLICATION OF RADON RISK ESTIMATES

JONATHAN M. SAMET (*Chairman*), University of New Mexico, Albuquerque, New Mexico
ROY E. ALBERT, University of Cincinnati, Cincinnati, Ohio
JOSEPH D. BRAIN, Harvard School of Public Health, Boston, Massachusetts
RAYMOND A. GUILMETTE, Inhalation Toxicology Research Institute, Albuquerque, New Mexico
PHILIP K. HOPKE, Clarkson University, Potsdam, New York
ANTHONY C. JAMES, Battelle Pacific Northwest Laboratories, Richland, Washington
DAVID G. KAUFMAN, University of North Carolina, Chapel Hill, North Carolina

Staff

RAYMOND D. COOPER, Study Director
WILLIAM H ELLETT, Senior Program Officer
RICHARD E. MORRIS, Editor, National Academy Press

Sponsor's Project Officers

NEAL S. NELSON, U. S. Environmental Protection Agency
ANITA L. SCHMIDT, U. S. Environmental Protection Agency

BOARD ON RADIATION EFFECTS RESEARCH

MORTIMER L. MENDELSOHN (*Chairman*), Lawrence Livermore National Laboratory, Livermore, California
ERIC J. HALL, Columbia University, New York, New York
DANIEL L. HARTL, Washington University School of Medicine, St. Louis, Missouri*
LEONARD S. LERMAN, Massachusetts Institute of Technology, Cambridge, Massachusetts
FREDERICK MOSTELLER, professor emeritus, Harvard University, Cambridge, Massachusetts
JOSEPH E. RALL, National Institutes of Health, Bethesda, Maryland
WARREN K. SINCLAIR, National Council on Radiation Protection and Measurements, Bethesda, Maryland
THOMAS S. TENFORDE, Battelle Pacific Northwest Laboratories, Richland, Washington
ARTHUR C. UPTON, New York University Medical Center, New York, New York

Ex officio

BRUCE M. ALBERTS, University of California, San Francisco, California

Staff

CHARLES W. EDINGTON, Director
RAYMOND D. COOPER, Senior Program Officer
WILLIAM H ELLETT, Senior Program Officer
CATHERINE S. BERKLEY, Administrative Associate
DORIS E. TAYLOR, Administrative Assistant

---

*Resigned October 1990.

COMMISSION ON LIFE SCIENCES

BRUCE M. ALBERTS (*Chairman*), University of California, San Francisco, California
BRUCE N. AMES, University of California, Berkeley, California
FRANCISCO J. AYALA, University of California, Irvine, California
J. MICHAEL BISHOP, University of California Medical Center, San Francisco, California
MICHAEL T. CLEGG, University of California, Riverside, California
GLENN A. CROSBY, Washington State University, Pullman, Washington
FREEMAN J. DYSON, Institute for Advanced Study, Princeton, New Jersey
LEROY E. HOOD, California Institute of Technology, Pasadena, California
DONALD F. HORNIG, Harvard School of Public Health, Boston, Massachusetts
MARIAN E. KOSHLAND, University of California, Berkeley, California
RICHARD E. LENSKI, University of California, Irvine, California
STEVEN P. PAKES, University of Texas, Dallas, Texas
EMIL A. PFITZER, Hoffmann-LaRoche, Inc., Nutley, New Jersey
THOMAS D. POLLARD, Johns Hopkins Medical School, Baltimore, Maryland
JOSEPH E. RALL, National Institutes of Health, Bethesda, Maryland
RICHARD D. REMINGTON, University of Iowa, Iowa City, Iowa
PAUL G. RISSER, University of New Mexico, Albuquerque, New Mexico
HAROLD M. SCHMECK, JR., Armonk, New York
RICHARD B. SETLOW, Brookhaven National Laboratory, Upton, New York
CARLA J. SHATZ, Stanford University School of Medicine, Stanford, California
TORSTEN N. WIESEL, Rockefeller University, New York, New York

JOHN E. BURRIS, Executive Director

The National Academy of Sciences is a private, nonprofit, self-perpetuating society of distinguished scholars engaged in scientific and engineering research, dedicated to the furtherance of science and technology and to their use for the general welfare. Upon the authority of the charter granted to it by the Congress in 1863, the Academy has a mandate that requires it to advise the federal government on scientific and technical matters. Dr. Frank Press is president of the National Academy of Sciences.

The National Academy of Engineering was established in 1964, under the charter of the National Academy of Sciences, as a parallel organization of outstanding engineers. It is autonomous in its administration and in the selection of its members, sharing with the National Academy of Sciences the responsibility for advising the federal government. The National Academy of Engineering also sponsors engineering programs aimed at meeting national needs, encourages education and research, and recognizes the superior achievements of engineers. Dr. Robert M. White is president of the National Academy of Engineering.

The Institute of Medicine was established in 1970 by the National Academy of Sciences to secure the services of eminent members of appropriate professions in the examination of policy matters pertaining to the health of the public. The Institute acts under the responsibility given to the National Academy of Sciences by its congressional charter to be an adviser to the federal government and, upon its own initiative, to identify issues of medical care, research, and education. Dr. Samuel O. Thier is president of the Institute of Medicine.

The National Research Council was organized by the National Academy of Sciences in 1916 to associate the broad community of science and technology with the Academy's purposes of furthering knowledge and advising the federal government. Functioning in accordance with general policies determined by the Academy, the Council has become the principal operating agency of both the National Academy of Sciences and the National Academy of Engineering in providing services to the government, the public, and the scientific and engineering communities. The Council is administered jointly by both Academies and the Institute of Medicine. Dr. Frank Press and Dr. Robert M. White are chairman and vice chairman, respectively, of the National Research Council.

Preface

Among the recommendations in the Biological Effects of Ionizing Radiation (BEIR) IV report was a statement on the need for further research and analysis on uncertainties in applying the lung cancer risks characterized for underground miners to people in their homes. Specifically, the BEIR IV committee stated:

> Further studies of dosimetric modeling in the indoor environment and in mines are necessary to determine the comparability of risks per WLM [working level month] in domestic environments and underground mines.

Because of the importance to the public of the risks of radon exposure in homes and schools, the U.S. Environmental Protection Agency (EPA) asked the National Research Council (NRC) to initiate a study of the dosimetric considerations affecting the applications of risk estimates, based on studies of miners, to the general population. EPA asked that a panel be assembled to investigate the differences between underground miners and members of the general public in the doses they receive per unit exposure due to inhaled radon progeny.

CHARGE TO THE COMMITTEE

The committee was asked to do the following:

- Review the basis of current dosimetric models for inhaled radon progeny, paying particular attention to the assumptions that are a part of these models and the experimental data that are used to verify them. Compare these models with lung deposition models developed for nonradioactive pollutants.

The committee's study should be directed toward estimating the dose per unit exposure and not exposure levels that may exist in homes or in mines.

- Examine the experimental data base for characterizing aerosols and radon progeny concentrations in underground mines and home environments, taking into account the possible changes in these environments that have occurred in the last 40 years.
- Examine differences in breathing patterns of working miners and individuals in the mining environment, taking into account the level of physical activity, sex, age, and so on. Determine to what extent the relevant physiological parameters are based on experimentally derived values, with particular reference to underground miners and other manual laborers doing similar work.
- Determine which parameters, such as bronchial generations and tissue depth, are most useful for characterizing a dose-risk relationship for radiogenic lung cancers.
- With the foregoing information, estimate the ratio of the dose per working level month to an underground miner to the dose per working level month to people of different ages and sexes in a variety of domestic environments. To the extent possible, characterize the uncertainties in these estimates.
- Define research needs, including judgments on the relative importance of the various parameters needed to calculate the dose from radon progeny. Present the results in an NRC report.

ORGANIZATION OF THE STUDY

The NRC established a committee of seven members with expertise in radon risks, radon dosimetry, aerosol physics, respiratory physiology, carcinogenesis, and lung modeling to carry out the study. General guidance was provided by the Board on Radiation Effects Research of the Commission on Life Sciences.

The committee held five meetings, four in Washington, D.C., and one in Woods Hole, Massachusetts. During an early meeting, the panel invited several outside experts to present the results of their research on lung modeling and lung cancer to ensure that a broad cross-section of the scientific community had provided input.

The study was divided into two parts. The first part was the determination of the differences in atmospheres, breathing rates, and other factors between mines and miners and homes and the people who live in them. The second part included the application of a new lung model to these factors in order to determine a ratio of dose per unit exposure in homes and mines. The report covers these areas and a discussion of other considerations affecting the risk of lung cancer from radon exposure.

ORGANIZATION OF THE REPORT

The report is arranged so that the major elements of the committee's analysis and conclusions are described first, followed by a number of background chapters that include technical details of the data and models used and the reasons for choosing the models used. After a summary of the committee's findings and recommendations for further research and analysis, the main body of the report begins with an introduction to the dosimetry of radon (Chapter 1) and how this differs in mines and homes (Chapter 2). A description of the data used by the committee on exposures of miners and people in their homes follows. Chapter 3, the main element of the report, answers the questions posed to the committee. It includes the major results of the committee's analysis. The main body of the report concludes with a description in Chapter 4 of factors other than those involved in dosimetry, that are important in determining risk.

Five background chapters describe in greater detail many of the elements that went into the first part of the report. Chapter 5 is a general description of radon dosimetry and the lung models that were used. Chapter 6 discusses aerosols in homes and mines, their diffusion and growth, and the means by which they are measured. Oral and nasal breathing and the deposition and clearance of particles in the lung are described in Chapter 7, and Chapter 8 discusses the different types of lung cancers and the cells of origin that are the targets for the dosimetry. Finally, Chapter 9 describes the specific lung model used by the committee and the calculations that led to the results given in the first part of the report.

ACKNOWLEDGMENTS

The panel acknowledges with thanks the scientific input provided by a number of invited participants. These included Dr. Geno Saccomanno, who described his studies on lung pathology; Dr. Naomi Harley, who discussed radon dosimetry; Dr. Elizabeth McDowell, who described the types of lung cells at risk; and Dr. David Swift, who spoke on the fate of inhaled radon and its progeny. All of these scientists gave freely of their data and findings and were of great help in clarifying some of the scientific issues under study.

The lung model used by the panel was developed, in part, by the task group of the International Commission on Radiological Protection (ICRP). The committee would like to thank the ICRP task group and, particularly, M. J. Egan and W. Nixon, who developed the theory for aerosol deposition in the lung upon which the model is based. The committee thanks Doris Taylor for her work on several drafts of this manuscript and for her help in making arrangements for committee meetings and travel.

Finally, thanks are due to the radon research programs sponsored by the U.S. government. Many of the data used by the committee were developed by the radon research program of the U.S. Department of Energy.

Contents

Summary and Recommendations		1
1	Introduction	9
2	Assessment of Exposure to the Decay Products of ^{222}Rn in Mines and Homes	20
3	Extrapolation of Doses and Risk per Unit Exposure from Mines to Homes	31
4	Other Considerations	52
5	Dosimetry and Dosimetric Models for Inhaled Radon and Progeny	60
6	Aerosols in Homes and Mines	90
7	Breathing, Deposition, and Clearance	137
8	Cells of Origin for Lung Cancer	166
9	The Committee's Dosimetric Model for Radon and Thoron Progeny	194
Index		239

Summary and Recommendations

Epidemiological investigations of uranium and other underground miners have provided extensive and consistent data on the quantitative risk of lung cancer associated with exposure to radon progeny in underground mines (Lubin, 1988; National Research Council [NRC], 1988). To extend these data to radon exposure of the general population in the home environment, a series of assumptions with attendant uncertainties must be made (see Figure 1-2). This report compares relations in the home and mining environments between exposure to radon progeny and the dose of alpha radiation to target cells in the respiratory epithelium. A dosimetric model is used to estimate the quantitative uncertainty introduced by differences in exposure-dose relations for miners and the general population who are exposed in their homes (Chapter 3). Additional sources of uncertainty that relate to the likelihood of lung cancer, including age at exposure, sex, cigarette smoking, and effects of environmental contaminants other than radon, are addressed qualitatively in Chapter 4.

The committee's dosimetric model incorporates a wide range of physical and biological parameters (see Chapter 9), some of which plausibly differ for the circumstances of exposure in the mining and home environments. Among the parameters selected by the committee as being different in the two environments are: aerosol size distribution, unattached fraction, and breathing rate and route. For those parameters, the committee reviewed the available evidence to estimate the values most typical in the two environments and to characterize their ranges. The committee also explored the consequences of various underlying model assumptions for the efficiency of nasal deposition, the efficiency of bronchial deposition, the solubility of progeny in mucus, and

TABLE S-1 Summary of K Factors for Bronchial Dose Calculated for Normal People in the General Environment Relative to Healthy Underground Miners

Subject Category	K Factor for the Following Target Cells:	
	Secretory	Basal
Infant, age 1 mo	0.74	0.64
Child, age 1 yr	1.00	0.87
Child, age 5-10 yr	0.83	0.72
Female	0.72	0.62
Male	0.76	0.69

the growth of aerosols in the respiratory tract. Because uncertainty remained after the committee's review concerning the cells of origin of lung cancer, the committee performed the calculations separately for basal and secretory cells in the respiratory epithelium. In addition to adult males, calculations were performed for adult females and for children and infants. Finally, the committee considered the consequences of abnormalities of the airways, such as those that might occur with chronic bronchitis caused by cigarette smoking or exposure to dusts and gases in mines.

The committee's findings on the differences in exposure-dose relations in mines and in homes are expressed as a ratio, termed K in the BEIR IV report (NRC, 1988). This ratio represents the quotient of the dose of alpha energy delivered per unit exposure to an individual in the home to the dose per unit exposure to a male miner in a mine. Thus, if the K factor exceeds unity, the delivered dose per unit exposure is greater in the home; if it is less than unity, the delivered dose per unit exposure is less in the home.

Across the wide range of exposure scenarios considered by the committee, most values of K were less than unity (see Tables 3-4, 3-5, and 3-6). The K factor was also below unity for adult females and children. The K factors for normal people without respiratory illnesses are summarized in Table S-1.

The committee's literature review (see Chapter 2) indicated that homes tended to have higher unattached fractions of radon progeny and room aerosols with smaller aerodynamic diameter than those in mines, although smoking or other activities can lower the unattached fraction and increase the aerosol size. Higher breathing rates were assumed for miners because of the work load imposed by the physical activity of mining. For an adult male exposed in the mining or home environment under typical conditions, the K factors ranged from about 0.8 to 0.6, depending on the assumptions concerning the solubility of radon progeny in mucus.

Similar calculations were performed for adult females and children. For adult females, the K factors tended to be somewhat lower than those for adult males. In several risk assessments, children and infants have been considered to be at increased risk for radon-induced lung cancer in comparison with the risk for adults because of heightened susceptibility (International Commission on Radiological Protection [ICRP], 1987; Puskin and Nelson, 1989). Although the K factors for children and infants were somewhat greater than those for adults, none of the values was above unity.

The committee examined the sensitivities of these findings to changes in the underlying biological assumptions in its model. The K factors remained below unity regardless of whether radon progeny were assumed to be insoluble or partially soluble in the epithelial tissue. The K factor changed little with the assumption that lobar and segmental bronchi ($K = 0.69$) rather than all bronchi ($K = 0.73$) are the target tissues. Similarly, the K values changed little as the assumptions concerning the efficiencies of nasal and airway deposition were changed. The findings were comparable across the range of conditions examined for both secretory and basal cells as well. At the extremes of selection of breathing route, exclusively nasal and exclusively oral, the general pattern of the K values was also unchanged.

Using a dosimetric model, the committee comprehensively compared exposure-dose relations for mining and home environments. The committee's calculations indicate that the dose of alpha energy per unit exposure delivered to target cells in the respiratory tract tends to be lower for the home environment—by about 30% for adults of both sexes and by 20% or less for infants and children. Thus, direct extrapolation of risk estimates from the mining to the home environment may overestimate the numbers of radon-caused lung cancer cases by these percentages.

The limitations of this analysis are addressed in detail in the report. Any dosimetric model, regardless of its sophistication, inevitably simplifies extremely complex physical and biological phenomena. In this report, however, the same model was applied to people in both the mining and home environments. Substantial uncertainty remains concerning the appropriate values for most of the model's parameters. For example, the data on breathing rate and route (oral versus nasal) of the miners and the general population are extremely limited, as are measurements of the unattached radon fractions in the two environments. The distributions of values for most of the model's parameters have not been described in appropriate samples. The committee's recommendations for research to address these uncertainties are provided at the end of this chapter.

Other committees and researchers have also compared exposure-dose relations in mines and homes (see James [1988]). Recent reports of the National Council on Radiation Protection and Measurements (NCRP), the ICRP, and the NRC have addressed this issue. NCRP Report No. 78 (NCRP, 1984) used a dosimetric model to calculate the dose of alpha energy delivered to basal cells. K values calculated from the dose conversion factors provided in that report were 1.40 for adult males, 1.20 for adult females, 2.40 for children, and 1.20

for infants. ICRP Publication No. 50 concluded that K is 0.8 for adult males and females in indoor environments. For children ages 0 to 10 yr, the report's authors suggested that the dosimetric correction might be about 1.5 times larger than that for adults. On the basis of a qualitative analysis, the BEIR IV committee (NRC, 1988) concluded that a K value of 1 could justifiably be assumed. The fact that the findings of the present report diverge from those of earlier reports reflects the further evolution of dosimetric models and the availability to the present committee of additional data on input parameters for the dosimetric model.

This committee also examined other sources of uncertainty in extrapolating risk coefficients from studies of miners to the general population (see Figure 4-2). These sources of uncertainty related not only to aspects of lung dosimetry but to the biology of lung cancer. The committee could not consider these factors directly in the dosimetric model, and their impact on the extrapolation could not be gauged quantitatively. Nevertheless, because the uncertainties associated with these factors are potentially large, the committee synthesized the available evidence on sex, age at exposure and age at risk, exposure rate, cigarette smoking, and agents that cause epithelial injury and promote cell turnover. The BEIR IV report (NRC, 1988) also considered the evidence on sex, age, and cigarette smoking.

The assumption of a greater or lesser risk of lung cancer in children who are exposed to radon has substantial public health implications. ICRP (1987) assumed a threefold greater risk of cancer for exposure during childhood on the basis of dosimetric considerations and the increased lung cancer risk for atomic bomb survivors who were age 20 yr or younger at the time of the blast. The latter evidence is extremely limited, however; on follow-up through 1980, only 10 cases of lung cancer had occurred in persons aged 0 to 10 yr at the time of the bomb (15,564 were at risk) (Yamamoto et al., 1986), and their relative risk of lung cancer through 1985 is less than 1 (Shimizu et al., 1988). Moreover, the relevance of these data, based on low-LET radiation, to the high-LET radiation from radon progeny is uncertain.

The present risk assessment approach of the U.S. Environmental Protection Agency incorporates in part ICRP's threefold greater risk for those aged 0 to 20 yr (Puskin and Nelson, 1989). By contrast, the model in the BEIR IV report (NRC, 1988) reduces the risks as the interval since exposure lengthens, implying a lower risk of lung cancer for exposures during childhood. In its analyses, the BEIR IV committee did not find an effect of age at exposure, but little information was available for miners at young ages. A recent case-control study of lung cancer in radon-exposed Chinese tin miners, 37% of whom were exposed by age 13 yr, offers some relevant information (Lubin et al., 1990). The increase in risk of lung cancer did not vary significantly with age at first exposure. Thus, assumption of either an enhanced or a reduced effect for exposure during childhood is subject to substantial uncertainty.

The committee considered cigarette smoking, which was also reviewed in depth by the BEIR IV committee (NRC, 1988). On the basis of literature review as well as its own analyses, the BEIR IV committee concluded that cigarette smoking and radon progeny interact multiplicatively. That committee noted the uncertainty inherent in extrapolating from a population of males, predominantly smokers, to smokers and nonsmokers of both sexes. Since publication of the BEIR IV report, evidence has not been published that would justify a revised conclusion on the combined effects of smoking and radon exposure. The present committee did address the consequences of bronchitis and hyperplasia of the epithelium with attendant cessation of mucus flow, both of which are effects of cigarette smoking, in the dosimetric model (see Table 3-6).

Underground miners inhale not only radon and progeny but silica and other dusts, blasting fumes, and sometimes engine exhaust as well. Inflammation of the airways with increased cell proliferation may result. This environmental difference should also be considered in extrapolating from miners to the general population in their homes. The increased cell turnover associated with exposures to these other agents in the mining environment may have increased the risk per unit exposure for the miners in comparison with exposure in the absence of these other agents (see Chapter 4).

For the general population, radon exposure occurs throughout the life span at a rate principally dependent on the concentrations in homes. Most of the miners included in the epidemiological studies were exposed underground for only a small proportion of their life spans. The extant risk models do not incorporate terms representing possible changes in risk associated with different exposure rates in the mining and general populations. The committee concluded that it is reasonable to assume that risk is proportional to total exposure (i.e., there is no dose rate effect).

Although the K factors imply a somewhat lower lung cancer risk for exposure in the home environment, the committee's findings do not imply that radon is not carcinogenic. The dosimetric modeling in this report suggests that the BEIR IV committee's risk assessments based on the data from miners may have overestimated to some extent the numbers of radon-associated cases of lung cancer in the general population. However, the degree of overestimation is not large. In applying the data from miners to the general population, it is now likely that assumptions related to other factors (e.g., age at exposure and cigarette smoking) introduce larger uncertainties than the uncertainties related to dosimetric differences between exposures in mines and in homes.

RECOMMENDATIONS FOR FURTHER RESEARCH AND ANALYSIS

To have more certain estimates of the dose per unit exposure and the K factors described here, further data are needed for the input parameters of dosimetric models, as is improved biologic understanding of carcinogenesis due to radon progeny. The committee has reviewed these needs and makes the following recommendations for further studies.

• Assessment of the activity-weighted size distributions of radon progeny in homes, schools, high-rise apartment buildings, and offices is needed. These studies should examine the effects of various indoor aerosol sources on these distributions.

• Additional data on activity-weighted size distributions of particles in underground mines, comparable to those worked in by the miners in the epidemiological cohorts, are also needed.

• Further research is needed on radon carcinogenesis to support the development of biologically based mathematical models of the temporal pattern of tumor formation. The roles of age, tissue injury, and interactions with other environmental toxicants in the carcinogenic response to radon need further investigation.

• Animal experiments are needed to assess modification of the dose-response relation between lung cancer and radon exposure by smoking and other factors.

• Because of the difficulty of determining the dose of alpha energy delivered by radon progeny to the bronchial epithelium, the development of biological markers should be undertaken.

• Data on breathing by humans should be gathered in a variety of settings. Methods have been developed that allow measurements of tidal volume from body surface displacements. Alternatively, once calibrated, heart rate can be used as a surrogate for oxygen consumption. This technology should be adapted to field use, and measurements should be made in many places, including mines and homes. More data are also needed on the range of human activities indoors with associated ventilation rates.

• Additional studies are needed on the anatomic locations in which radon progeny deposit. Little information is available, especially in humans, regarding where these radon progeny are retained.

• Additional information is needed on respiratory deposition of particles less than 0.1 μm in aerodynamic diameter. Sites of interest are the upper airway, trachea, and bronchi.

• More data are needed on the location and morphologic types of lung cancers and on temporal trends in these parameters. Information on the effect of smoking on the anatomic distribution of lung cancers should be obtained, both for underground miners and the general population.

• Additional research should be undertaken to determine whether different histologic types of bronchogenic lung cancer result from different genetic changes in a common bronchial stem cell precursor or from transformation of different types of bronchial cells. For example, further research is needed to determine whether undifferentiated small-cell carcinomas of the lung arise from bronchial Kulchitsky cells of neural crest origin (neurosecretory cells) or from an endodermal precursor or stem cell in the bronchus.

- More information is needed on the hygroscopicity of radon progeny aerosols, on the degree to which they increase in size within the respiratory tract, on the time taken for any growth to occur, and on the effects of hygroscopic growth on deposition within the bronchial airways.
- Estimates of the filtration efficiency of the nasal and oral passageways for unattached radon progeny that have been developed from studies with hollow casts need to be confirmed by in vivo measurements of extrathoracic filtration in sufficient numbers of human subjects. It is also necessary to study the nasal and oral filtration efficiencies in vivo in children.
- It is necessary to extend studies of the deposition efficiency of submicron particles in hollow casts of the human bronchi to much smaller particles in the size range of unattached radon progeny. When coupled with more accurate information on extrathoracic filtration efficiencies, such studies of localized bronchial deposition will resolve some of the uncertainty in evaluating doses from exposure to unattached progeny.
- A better understanding is needed of the respective roles of secretory and basal cells in the etiology of lung cancer, and of the relative sensitivities of the epithelial lobar, segmental, and subsegmental bronchi.
- Estimates of doses received by bronchial basal cells are especially sensitive to uncertainties in the thickness and structure of the epithelium, in the thickness of mucus and its variability, and in the degree to which radon progeny migrate from mucus to be retained in epithelial tissue. Further investigation of these parameters is needed.

REFERENCES

International Commission on Radiological Protection (ICRP). 1987. Lung Cancer Risk from Indoor Exposures to Radon Daughters. ICRP Publ. No. 50. Oxford: Pergamon Press.

James, A. C. 1988. Lung dosimetry. Pp. 259-309 in Radon and Its Decay Products in Indoor Air, W. W. Nazarof and A. V. Nero, Jr., eds. New York: John Wiley & Sons.

Lubin, J. H. 1988. Models for the analysis of radon-exposed populations. Yale J. Biol. Med. 61:195-214.

Lubin J. H., Y. Qiao, P. R. Taylor, S. X. Yao, A. Schatzkin, B. L. Mao, J. Y. Rao, X. Z. Xuan, and J. Y. Li. 1990. Quantitative evaluation of the radon and lung cancer association in a case control study of Chinese tin miners. Cancer Res. 50:174-180.

National Council on Radiation Protection and Measurements (NCRP). 1984. Evaluation of Occupational and Environmental Exposure to Radon and Radon Daughters in the United States. NCRP Report No. 78. Bethesda, Md.: National Council on Radiation Protection and Measurements.

National Research Council (NRC). 1988. Health Risks of Radon and Other Internally Deposited Alpha-Emitters. BEIR IV. Committee on the Biological Effects of Ionizing Radiation. Washington, D.C.: National Academy Press.

Puskin J. S., and C. B. Nelson. 1989. Environmental Protection Agency's perspective on risks from residential radon exposure. J. Air Pollut. Control Assoc. 39:915-920.

Shimizu, Y., H. Kato, and W. J. Schull. 1988. Life Span Study Report II. Part 2. Cancer Mortality in the Years 1980-85 Based on the Recently Revised Doses. DS86 RERF TR 5-88. Hiroshima: Radiation Effects Research Foundation.

Yamamoto, T., K. J. Kopecky, T. Fujikura, S. Tokuoka, T. Monzen, I. Nishimori, E. Nakashima, and H. Kato. 1986. Lung Cancer Incidence Among A-Bomb Survivors in Hiroshima and Nagasaki. RERF TR 12-86. Hiroshima: Radiation Effects Research Foundation.

1

Introduction

Radon (radon-222), an inert gas under usual environmental conditions, is a naturally occurring decay product of radium-226, the fifth daughter of uranium-238 (Figure 1-1). Both uranium-238 and radium-226 are present in most soils and rocks, although the concentrations vary widely (National Council on Radiation Protection and Measurements, 1984a). As radon forms, some atoms leave the soil or rock and enter the surrounding soil or water. Consequently, radon is ubiquitous in indoor and outdoor air. Radon decays with a half-life of 3.82 days into a series of solid, short-lived radioisotopes that are collectively referred to as radon daughters, radon progeny, or radon decay products (Figure 1-1). Two of these progeny, polonium-218 and polonium-214, emit alpha particles. When these radon progeny are inhaled and release alpha particles within the lungs, the cells lining the airways may be damaged and lung cancer may ultimately result.

As information on air quality in indoor environments accumulated, it became apparent that radon and its progeny are invariably present in indoor environments and that concentrations vary widely, even reaching in some dwellings levels found in underground mines. The well-documented excess cases of lung cancer among underground miners exposed to radon raised concern that exposure to the gas might also be a cause of lung cancer in the general population. Although the problem of indoor radon was well known in the scientific community by the late 1970s, it did not receive great public attention in the United States until a widely publicized incident in 1984. During routine monitoring, a worker in a Pennsylvania nuclear power plant was found to be contaminated with radioactivity. This contamination was subsequently

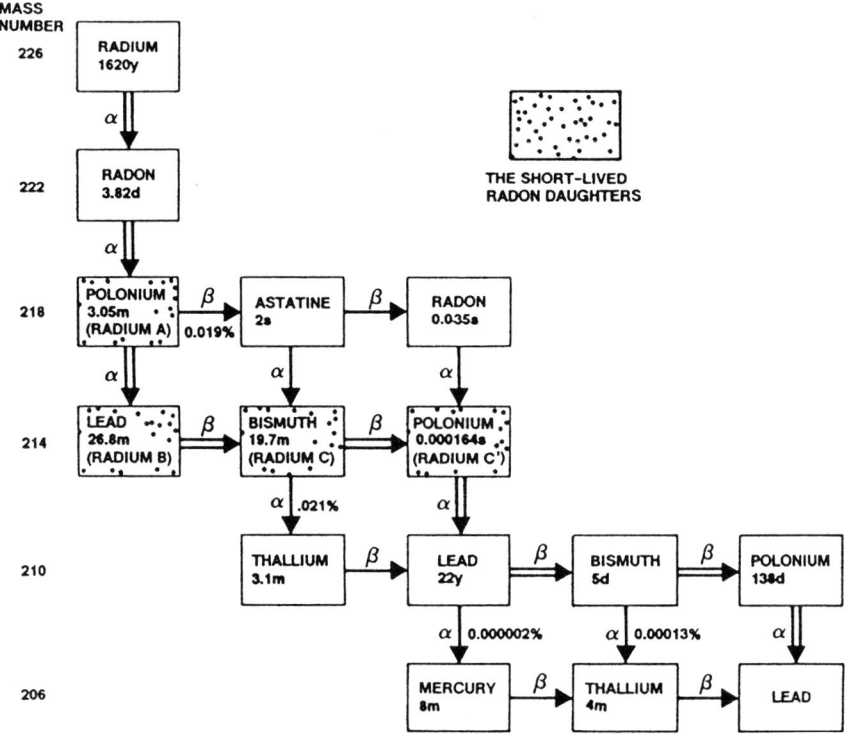

FIGURE 1-1 The radon decay chain.

traced to radon in his home, which was located on a geologic formation known as the Reading Prong. High levels of radon have now been measured in other homes on the Reading Prong and throughout the United States (Logue and Fox, 1985; Cohen, 1986; Nero et al., 1986; U.S. Environmental Protection Agency, 1988).

Recognition that the problem of indoor radon is widespread has prompted action by the EPA and some states. The pamphlet "A Citizen's Guide to Radon," first published by EPA in 1986, sets out the risks of radon and provides guidance on action to be taken at various concentrations (EPA, 1986). Although EPA does not have a statutory basis for regulating concentrations of radon indoors, its action guidelines are the most widely considered national recommendations on acceptable concentrations indoors. In 1988, EPA and the Centers for Disease Control jointly recommended that nearly all homes in the United States be tested for radon and in 1989 the EPA called for testing of schools. In many regions, radon testing is routinely performed at the time of sale of homes. Thus, during the 1980s, the health effects of radon assumed importance for virtually the entire population, and property owners were faced with making a decision about testing and potentially paying for mitigation.

The evidence from epidemiological investigations of miners justifies concern about the hazard posed by indoor radon. Radon has been linked to excess cases of lung cancer in underground miners since the early decades of the twentieth century. As early as the 1500s, Agricola described unusually high mortality from respiratory diseases among underground metal miners in the Erz mountains of eastern Europe. The disease, termed *Bergkrankheit*, probably represented lung cancer and silicosis and tuberculosis, which are common diseases of underground miners. In 1879, Harting and Hesse (1879) reported autopsy findings in miners of Schneeberg in Germany that documented an occupational hazard of lung cancer, although they did not identify the disease as primary cancer of the lung. Early in the twentieth century, further pathological studies showed that the miners developed primary carcinoma of the lung (Arnstein, 1913; Rostoski et al., 1926).

Measurement of radon in the mines in eastern Europe early in this century documented the presence of radon at concentrations that would be considered high by present standards. By the 1930s, excess cases of lung cancer were demonstrated among miners in Joachimsthal, which is on the Czechoslovakian side of the Erz Mountain range, and radon was also found in the air of these mines. Radon was considered to be a likely cause of lung cancer in Joachimsthal miners (Pirchan and Sikl, 1932), but the causal role of radon was not uniformly accepted until the biological basis for carcinogenesis by radon was better understood and further epidemiological evidence documented excess cases of lung cancer in other groups of exposed miners (Seltser, 1965; Hueper, 1966; Lundin et al., 1971). Bale's 1951 memorandum showing that the progeny of radon, rather than radon itself, deliver alpha energy to the respiratory tract was an important advance (Bale, 1980).

Epidemiological evidence on radon and lung cancer, as well as other diseases, is now available from about 20 different groups of underground miners (Samet, 1989). Excess occurrences of lung cancer have been found in uranium miners in the United States, Czechoslovakia, France, and Canada and in other underground miners exposed to radon decay products, including Newfoundland fluorspar miners, Swedish and U.S. metal miners, British and French iron miners, and Chinese and British tin miners (National Research Council [NRC], 1988). Many of these studies include information on the exposure of the miners to radon progeny and provide estimates of the quantitative relationship between exposure and lung cancer risk (Lubin, 1988; Samet, 1989). In view of the contrasting methodologies used in these investigations, the coefficients describing the change in excess relative risk per unit exposure span a remarkably narrow range (Table 1-1).

The risk of indoor radon has been primarily assessed by using risk assessment approaches that extend the findings of the studies of miners to the

TABLE 1-1 Relative Risk Coefficients for Lung Cancer from Epidemiological Studies of Underground Miners[a]

Study	Excess Relative Risk/100 WLM
Colorado Plateau uranium miners	0.5
New Mexico uranium miners	1.1
Ontario uranium miners	1.3
Beaverlodge, Canada, uranium miners	2.6
Port Radium, Canada, uranium miners	0.7
Czech uranium miners	1.9
Malmberget, Sweden, iron miners	1.6
Newfoundland fluorspar miners	3.0
Chinese tin miners	0.9

[a]Taken from Samet (1989).

general population. Epidemiological investigations in the general population can potentially provide direct estimates of the risks of indoor radon. The results of studies of indoor radon have been published (Samet, 1989), and numerous investigations are now in progress throughout the world (U.S. Department of Energy/Commission of the European Communities [DOE, 1989]). However, difficult methodological problems limit the accuracy of these studies (Lubin et al., 1990), and the studies of miners will probably remain the principal basis for estimating the risks of indoor radon for the immediate future.

Extrapolation of the lung cancer risks observed in underground miners to the risks for the general population who are exposed to radon indoors is subject to uncertainties related to the differences between the physical environments of homes and mines, the circumstances and temporal patterns of exposure in the two environments, and the potential biological differences between miners and the general population (Table 1-2 and Figure 1-2).

With regard to the physical differences between homes and mines, the activity-weighted particle size distributions tend to be different, with there being more activity in the ultrafine mode (unattached fraction) in homes. The air of many mines was contaminated by other carcinogens, such as diesel exhaust, and by toxic and irritant dusts and gases, such as silica, uranium, and oxides of nitrogen. In general, mining-related jobs underground would demand breathing rates higher than those demanded by more sedentary indoor activities. A greater minute ventilation of miners would tend to increase the dose to target cells at a particular concentration.

The miners included in the epidemiological studies were generally occupationally exposed to radon progeny for a relatively small proportion of their lives, whereas exposure indoors takes place across the entire life span, albeit at varying rates. The miners were largely adult males with a high proportion of smokers, whereas the general population includes males and females, smokers and nonsmokers, of all ages (Table 1-2). Thus, an assumption must also be

TABLE 1-2 Potentially Important Differences Between Exposure to Radon in the Mining and Home Environments

PHYSICAL FACTORS
    Aerosol characteristics: Greater concentrations in mines; differing size distributions
    Attached/unattached fractions: Greater unattached fractions in homes
    Equilibrium of radon/decay products: Highly variable in homes and mines

ACTIVITY FACTORS
    Amount of ventilation: Probably greater for working miners than for persons indoors
    Pattern of ventilation: Patterns of oral/nasal breathing not characterized

BIOLOGICAL FACTORS
    Age: Miners have been exposed during adulthood; entire spectrum of ages exposed indoors
    Gender: Miners studied have been exclusively male; both sexes exposed indoors
    Exposure pattern: Miners exposed for variable intervals during adulthood; exposure is lifelong for the population
    Cigarette smoking: The majority of the miners studied have been smokers; only a minority of U.S. adults are currently smokers

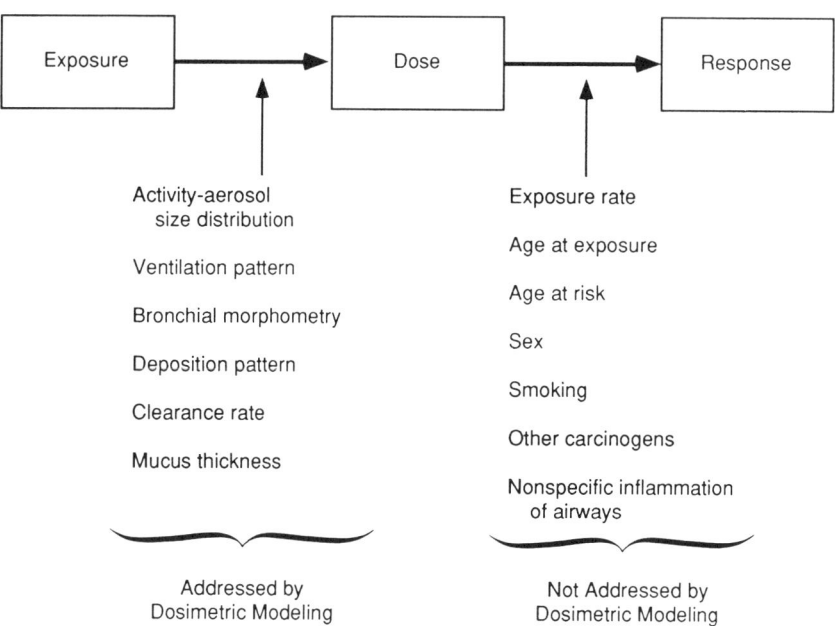

FIGURE 1-2 Factors influencing the relationship between radon exposure and the risk of lung cancer.

made concerning the interaction between radon progeny and cigarette smoking, the principal cause of lung cancer (U.S. Department of Health and Human Services, 1989). Inflammation of the airways of miners by irritant dusts and gases might also have influenced the response to radon progeny.

The need to make assumptions concerning these physical and biological factors that may lead to different risks of radon exposure in mines and in homes and the associated uncertainties have reduced confidence in the projections of the public's risk of radon-related lung cancer; in fact, some scientists and policymakers discard the risk projections based on the studies of miners as too uncertain and call for empiric evidence that radon causes lung cancer in the general population.

Methods are available, however, for characterizing the uncertainties associated with the factors determining the relationship between exposure to radon and the dose of alpha energy delivered to target cells in the respiratory tract. Using models of the respiratory tract, the dose to target cells in the respiratory epithelium, which lines the lung's airways, can be estimated for the circumstances of exposure in the mining and indoor environments. In such comparative analyses reported to date, the relationship between exposure and dose, and hence the potency of radon as a carcinogen, has generally been found to be comparable in these two settings (NCRP, 1984b; James, 1988; NRC, 1988). For some parameters, the input data needed for these analyses are limited, and controversy remains concerning the application of the findings for miners to the entire general population of infants, children, adult females, and adult males. For example, the International Commission on Radiological Protection (ICRP) concluded (ICRP, 1987) that risks were threefold greater for those in the group from 0 to 20 years of age as compared with the risks in older persons; in contrast, the time-dependent risk model developed by the Committee on the Biological Effects of Ionizing Radiation (BEIR IV) (NRC, 1988) suggests that there are reduced effects for exposures received during childhood.

Thus, further consideration of the relation between exposure to radon and the radiation dose and its biological effectiveness as delivered to the respiratory tract is warranted by the continuing scientific controversy concerning the risks of exposure to indoor radon and by the potential policy implications of risk projections of radon-related lung cancer in the general population. Such risk projections serve as the basis for establishing action guidelines for judging the safety of the nation's homes, schools, and offices and for guiding potentially costly mitigation of unacceptable concentrations (EPA, 1986). In response to this immediate and compelling rationale, the present committee was charged with considering the dosimetry of radon and its decay products in the mining and indoor environments. This charge does not fully cover all uncertainties related to extrapolating risks derived from studies of miners to risks for the general population (Figure 1-2 and Table 1-2). To the extent possible, the

INTRODUCTION   15

committee has also considered biological factors relevant to this extrapolation, including age at exposure, duration of exposure, sex, cigarette smoking, and the effects of other contaminants in the air of mines. This chapter provides a general background for the committee's detailed review and conclusions.

CONCENTRATION, EXPOSURE, AND DOSE

For historical reasons, the concentration of radon progeny in mines has generally been expressed as working levels (WL), where 1 WL is any combination of progeny in 1 liter of air that ultimately releases 1.3×10^5 MeV of alpha energy during decay (Holaday et al., 1957). Concentrations of radon in indoor air have most often been expressed as picocuries (pCi) per liter, a unit for the rate of decay; a radon concentration of 1 pCi/liter translates to about 0.005 WL under the usually assumed conditions of equilibrium between radon and its progeny in a home. In international (SI) units, the activity of radon per unit mass of air is expressed as becquerels (Bq) per cubic meter; at radioactive equilibrium between radon and its progeny, 1 WL corresponds to 3.7×10^3 Bq/m^3.

The working level month (WLM) was developed to describe exposure to radon progeny in underground mines (Holaday et al., 1957). Exposure to 1 WL for 170 h equals 1 WLM of exposure. Because most persons spend much more time than 170 h at home each month, a concentration of 1 WL in a residence results in an exposure much greater than 1 WLM on a monthly basis. For example, during 1 month, if the concentration of radon in a home were 1.0 WL, a child spending 75% of the time at home would receive an exposure of 3.2 WLM, whereas an adult spending 60% of the time at home would receive 2.5 WLM. Exposure in SI units is expressed in Joule-hours (Jh) per m^3, and 1 WLM is 3.5×10^{-3} Jh/m^3.

The relationship between exposure to radon progeny, measured as WLM or Jh/m^3, and dose to cells in the respiratory tract, considered as targets for carcinogenesis, is extremely complex and is dependent on both biological and nonbiological factors, including the physical characteristics of the inhaled air, the amount of air inhaled, breathing patterns, and the biological characteristics of the lung (see Chapter 9). Because the dose of alpha energy delivered to target cells in the lungs cannot be measured directly, modeling approaches are used to simulate the sequence of events from inhalation of radon progeny to cellular injury by alpha particles. These complex models generally incorporate biological factors, including airway geometry, mucociliary clearance, particle deposition, ventilation pattern, and location of the target cells, and physical factors, including the aerosol size distribution and the proportion of progeny not attached to particles. In the past, the terms *attached* and *unattached fractions* have been used to refer to radon progeny presumed to be attached and unattached to atmospheric particles. However, these terms are more appropriately replaced by the term *activity-weighted particle size distribution*. Recent evidence suggests that the unattached progeny are in ultrafine clusters in the 0.5- to 3-nm size range and that unattached and attached fractions cannot be distinctly separated (see Chapter 6). Factors for converting exposure to an absorbed radiation dose can be calculated by using dosimetric models of the respiratory tract, but the range of published dose conversion factors is wide (James, 1988). To convert absorbed dose to tissue dose equivalent in rem, or sieverts in the SI system, the absorbed dose in rads or grays is multiplied by a quality factor for alpha radiation.

RISK ASSESSMENT FOR INDOOR RADON

To estimate the lung cancer hazard associated with indoor radon, information on exposure levels in dwellings is used in a risk projection equation, or model, that describes the increment in the occurrence of lung cancer per unit exposure. In the principal models in use at present, the risk coefficients describing the relationship between exposure and lung cancer occurrence are derived from studies of miners (NCRP, 1984b; ICRP, 1987; NRC, 1988). The models project the excess occurrences of lung cancer following exposure across the lifetimes of the exposed persons. Although each of the models incorporates risk coefficients from the studies of miners, the biological assumptions underlying the models differ (Table 1-3), and risk projections from the models may vary substantially (Land, 1989).

The incidence of (and mortality from) lung cancer rises sharply from about

TABLE 1-3 Features of Selected Risk Projection Models for Radon and Lung Cancer

Feature	NCRP	ICRP	BEIR IV
Form of model	Attributable risk	Relative risk	Relative risk
Time-dependent	Yes; risk declines exponentially after exposure	No	Yes; risk declines as time since exposure lengthens
Lag interval	5 years	10 years	5 years
Age at exposure	No effect of age at exposure	Threefold increased risk for exposures before age 20 years	No effect of age at exposure
Age at risk	Risk commences at age 40 years	Constant relative risk with age	Lower risks for ages 55 years and older
Dosimetry adjustment	Increased risk for indoor exposure	Decreased risk for indoor exposure	No adjustment
Risk coefficient	10×10^{-6}/year/WLM	Excess relative risks: 1.9%/WLM at ages 0-20 years and 0.64%/WLM for ages 21 years and above	Excess relative risk of 2.5%/WLM but modified by time since exposure

ages 50 through 80 yr. Risk projection models inherently assume a relationship between the added risk from radon progeny and the background risk of lung cancer. The model of the NCRP (1984b) assumes additivity of the risks and a time-dependent decline in risk following exposure. In contrast, the model of the ICRP (1987) assumes that the background rate is multiplied by the additional risk associated with radon progeny. The model developed by the BEIR IV committee (NRC, 1988) is also multiplicative, but it incorporates a time-dependent decline in risk.

Regardless of the form of a model, its application to the indoor environment requires consideration of the relationship between exposure and dose in the mining environment and in the indoor environment. In this regard, the three models described in Table 1-3 make different assumptions. The BEIR IV model makes no adjustment, whereas the model of the ICRP reduces the risks by 20% for adults in the general population, and the NCRP model increases the risks by 40% for the general population, assuming 0.5 rad/WLM for the mining environment and 0.7 rad/WLM for the indoor environment.

With regard to other factors that potentially influence the risks of radon in homes and in mines, the three models also incorporate different assumptions for some factors. As noted, the ICRP model increases risk for exposures before age 20 yr and the NCRP model assumes that risk commences at age 40 yr. In the BEIR IV model, risk varies with attained age. With regard to smoking, the NCRP model is additive, whereas the other two models are multiplicative.

SUMMARY

The finding that radon and its decay products are invariably present in indoor environments has prompted concern that lung cancer caused by radon is a public health problem. At present, the risks of indoor radon can be estimated best by using risk coefficients derived from epidemiological studies of underground miners in risk models. However, application of the evidence from the mining environment to the indoor environment requires assumptions and introduces substantial uncertainty. Some uncertainties associated with this extrapolation can be estimated by comprehensive consideration of the dosimetry of radon decay products in the indoor and mining environments. However, potentially important biological factors that also introduce uncertainty are not addressed by dosimetric modeling.

REFERENCES

Arnstein, A. 1913. Sozialhygienische Untersuchungen Über die Bergleute in den Schneeberger Kobaltgruben. Wein. Arbeit Geb. Soz. Med. 5:64-83.
Bale, W. F. 1980. Memorandum to the files, March 14, 1951: Hazards associated with radon and thoron. Health Phys. 38:1062-1066.

Cohen, B. L. 1986. A national survey of ^{222}Rn in U.S. homes and correlating factors. Health Phys. 51:175-183.

Harting, F. H., and W. Hesse. 1879. Der lungenkrebs, die Bergkrankheit in den Schneeberger gruben. Vjschr. Gerichtl. Med. Offentl. Gesundheitswesen 31:102-132, 313-337.

Holaday, D. A., D. E. Rushing, R. D. Colcman, ct al. 1957. Control of Radon and Daughters in Uranium Mines and Calculations on Biologic Effects. DHEW Publ. No. (PHS) 57-494. Washington, D.C.: U.S. Government Printing Office.

Hueper, W. C. 1966. Occupational and Environmental Cancers of the Respiratory Tract. New York: Springer-Verlag.

International Commission on Radiological Protection (ICRP). 1987. Lung Cancer Risk from Indoor Exposures to Radon Daughters. ICRP Publ. No. 50. Oxford: Pergamon Press.

James, A. C. 1988. Lung dosimetry. Pp. 259-309 in Radon and Its Decay Products in Indoor Air, W. W. Nazaroff and A. V. Nero, Jr., eds. New York: John Wiley & Sons.

Land, C. E. 1989. The ICRP 50 Model. Pp. 115-126 in Proceedings of the Annual Meeting of the National Council on Radiation Protection and Measurements. Bethesda, Md.: National Council on Radiation Protection and Measurements.

Logue, J., and J. Fox. 1985. Health hazards associated with elevated levels of indoor radon—Pennsylvania. Morbid. Mortal. Weekly Rep. 34:657-658.

Lubin, J. H. 1988. Models for the analysis of radon-exposed populations. Yale J. Biol. Med. 61:195-214.

Lubin, J. H., J. M. Samet, and C. Weinberg. 1990. Design issues in epidemiologic studies of indoor exposure to Rn and risk of lung cancer. Health Phys. 59:807-817.

Lundin, F. E., Jr., J. K. Wagoner, and V. E. Archer. 1971. Radon Daughter Exposure and Respiratory Cancer: Quantitative and Temporal Aspects. NIOSH-NIEHS Monogr. No. 1. Washington, D.C.: National Institute of Occupational Safety and Health-National Institute of Environmental Sciences.

National Council on Radiation Protection and Measurements (NCRP). 1984a. Exposure from the Uranium Series with Emphasis on Radon and Its Daughters. NCRP Report No. 77. Bethesda, Md.: National Council on Radiation Protcction and Measurements.

National Council on Radiation Protection and Measurements (NCRP). 1984b. Evaluation of Occupational and Environmental Exposure to Radon and Radon Daughters in the United States. NCRP Report No. 78. Bethesda, Md.: National Council on Radiation Protection and Measurements.

National Research Council (NRC). 1988. Health Risks of Radon and Other Internally Deposited Alpha-Emitters. BEIR IV. Committee on the Biological Effects of Ionizing Radiation. Washington, D.C.: National Academy Press.

Nero, A. V., M. B. Schwehr, et al. 1986. Distribution of airborne ^{222}Rn concentrations in U.S. homes. Science 234:992-997.

Pirchan, A., and H. Sikl. 1932. Cancer of the lung in the miners of Jachymov (Joachimsthal). Am. J. Cancer 4:681-722.

Rostoski, O., E. Saup, and G. Schmorl. 1926. Die Bergkrankheit der Erzbergleute in Schneeberg in Sacksen ("Schneeberger Lungenkrebs"). Z. Krebsforsch. 23:360-384.

Samet, J. M. 1989. Radon and lung cancer. J. Natl. Cancer Inst. 81:745-757.

Seltser, R. 1965. Lung cancer and uranium mining: A critique. Arch. Environ. Health 10:923-935.

U.S. Department of Energy, Office of Energy Research, Office of Health and Environmental Research. 1989. International Workshop on Residential Radon Epidemiology, Washington, D.C. CONF-8907178. Springfield, Va.: National Technical Information Service.

U.S. Department of Health and Human Services. 1989. Reducing the Health Consequences of Smoking: 25 Years of Progress. A Report of the Surgeon General. Office on Smoking and Health, Center for Chronic Disease Prevention and Health Promotion, Centers for Disease Control, Public Health Service, U.S. Department of Health and Human Services. DHHS Publ. No. (CDC) 89-8411. Washington, D.C.: U.S. Department of Health and Human Services.

U.S. Environmental Protection Agency (EPA). 1986. A Citizen's Guide to Radon. What It Is and What To Do About It. EPA Publ. No. 86-004. Washington, D.C.: U.S. Government Printing Office.

U.S. Environmental Protection Agency (EPA). 1988. Final Report on the 1987 State Survey Results. Washington, D.C.: Office of Radiation Programs, U.S. Environmental Protection Agency.

2
Assessment of Exposure to the Decay Products of ^{222}Rn in Mines and Homes

INTRODUCTION

The purpose of this chapter is to review the available data on the exposures of underground miners and members of the general public to radon progeny. These data are limited by changes in measurement methodologies and the types of measurements that were made. There are relatively few data available to fully characterize the airborne radioactivity in either mines or homes, and there are no data at all that characterize radon progeny in school or workplace atmospheres. Thus, Chapter 6 summarizes in substantial detail those data that are available and highlights the limitations of the existing knowledge base.

BACKGROUND

In order to estimate the dose of alpha energy from inhaled radon progeny received by miners and by the general public in their homes, it is necessary to know both the concentrations of the airborne progeny and their size distributions. The size of the radioactive particles determines their penetration through the upper respiratory system and the pattern of deposition within the tracheobronchial region. As discussed in Chapter 1, the series of short-lived radon decay products can be deposited directly in the respiratory tract or onto the surface of particles that can be deposited in the respiratory tract. The behavior of airborne progeny is described schematically in Figure 2-1.

ASSESSMENT OF EXPOSURE IN MINES AND HOMES

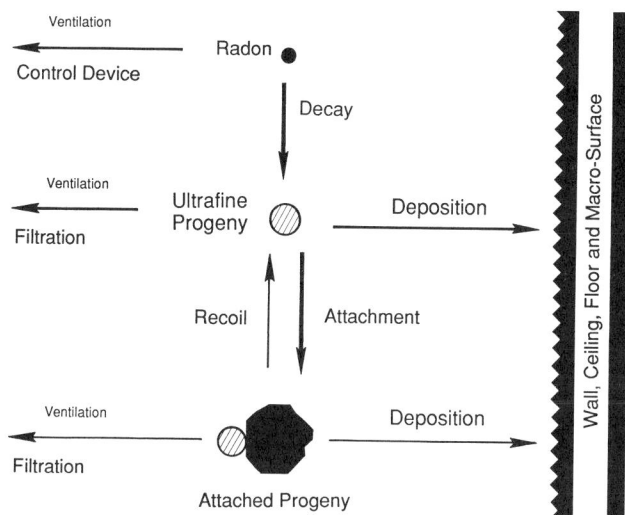

FIGURE 2-1 Schematic representation of the behavior of radon progeny in an enclosed space such as a mine or a room.

Since the discussion of exposure centers on the presence of airborne radioactivity and the aerodynamic behavior of the particles that transport the activity, it is necessary to review the concept of particle size. Although particles can be described in terms of their physical behavior, respiratory deposition and other related problems depend on how the radioactive particles move in the atmosphere. For particles above a few tenths of a micron, the particles are typically described in terms of their aerodynamic diameter. The aerodynamic diameter is the diameter of a unit-density sphere that has the same gravitational settling velocity as the particle. However, for particles below 0.1 μm in size, diffusion is the dominant mechanism for particle deposition. Thus, the diffusion equivalent diameter as defined in Equation 6-1 becomes the appropriate measure of particle size. A detailed discussion of particle size and diffusion coefficient is provided in Chapter 6.

Certain other terms need to be initially defined. The amount of radioactivity is described in units of disintegrations per unit time or activity rather than as the mass of contaminant. The activity (A) is the number of decay events per unit time and is calculated:

$$A = \lambda N \tag{2-1}$$

where λ is the probability of decay of the nucleus of a particular atom in a unit time ($\lambda = 0.693/T_{1/2}$). $T_{1/2}$ is the half-life of the isotope, and N is the total number of the atoms present in the sample. The international (SI) unit of activity is the becquerel (Bq). A becquerel is the quantity of radioactive material in which one atom is transformed per second. However, for radon the unit picocuries (pCi) is commonly used in the United States. A picocurie is 10^{-12} curies (Ci) or 3.7×10^{-2} disintegrations per second. For airborne concentrations (the amount of activity per unit volume), the proper scientific measurement units are becquerels per cubic meter or picocuries per liter. A concentration of 37 Bq/m^3 is equal to 1 pCi/liter.

If all progeny formed by the decay of radon were to remain in air, then equal activities of all of the radon decay products would be present. Such a mixture is said to be in *secular equilibrium*. The quantitative relationship between isotopes in secular equilibrium may be derived in the following manner for the general case for isotopes A, B, and C:

$$A \xrightarrow{\lambda_A} B \xrightarrow{\lambda_B} C \qquad (2\text{-}2)$$

where the half-life of isotope A is very much greater than that of isotope B. The decay constant of isotope A, λ_A, is therefore much smaller than λ_B, the decay constant of isotope B. The product of the decay probability times the number of atoms (N) would be equal for isotopes A and B ($\lambda_A N_A = \lambda_B N_B$) because the larger λ_B compensates for the smaller N_B. For radon decay products, the amount of activity equal to that of Po-218, Pb-214, and Bi-214 in secular equilibrium at a concentration of 100 pCi of each isotope per liter is called 1 working level (WL). This terminology was developed for describing the progeny concentrations in uranium mines.

The WL is a measure of the concentration of the total potential alpha energy of the short-lived progeny. The total exposure from radon progeny can then be expressed as the length of time a person is present in a room or mine with a concentration of radon progeny multiplied by that concentration. Again, for the uranium mining environment, the unit of exposure was called the working level month (WLM), defined as the average WL multiplied by time in units of 170 h. Since people spend more than 170 hrs per month in their homes, the cumulative exposure is given by

$$\text{Cumulative Exposure in WLM} = \sum_{i=1}^{n}(WL)_i \left(\frac{t_i}{170}\right) \qquad (2\text{-}3)$$

where $(WL)_i$ is the average concentration of the radon progeny during the exposure interval i expressed in WL, and t_i is the number of hours of exposure at the ith concentration. The cumulative exposure in a home at a given decay product level could thus be more than four times that for an occupational exposure (8,766 total hours in a year compared with 2,000 working hours in a year).

A better way to express the total activity of all of the radon decay products is as the potential alpha-energy concentration (PAEC) in the air, which is expressed as MeV/m³ or WL. A WL of 100 pCi/liter in equilibrium deposits 1.3×10^5 MeV/liter.

$$\begin{aligned} \text{PAEC (MeV/m}^3) &= 3.62 \times 10^3 A_1 + 1.78 \times 10^4 A_2 + 1.31 \times 10^4 A_3 \\ \text{PAEC(WL)} &= 0.00106 \ I_1 + 0.00513 \ I_2 + 0.00381 \ I_3 \end{aligned} \quad (2\text{-}4)$$

where A_1, A_2, and A_3 (in Bq/m³) and I1, I2, and I3 (in pCi/liter) are the activity concentrations of ^{218}Po, ^{214}Pb, and ^{214}Bi, respectively.

Because of the losses of decay products from the air by deposition on surfaces such as walls, ceilings, and furniture, the decay product activity is less than the radon activity. The term characterizing the airborne concentration of PAEC as a fraction of the radon activity is the equilibrium factor, F. F is defined as the ratio of decay products to radon by

$$F = (100)\text{WL}/10 = 0.106 \ a_1 + 0.513 \ a_2 + 0.381 \ a_3 \quad (2\text{-}5)$$

where a_1, a_2, and a_3 are the relative activities of the three radon progeny relative to that of radon, a_0 (i.e., $a_1 = a_1/a_0$).

Swedjemark (1983) found that for low air exchange rates (<0.3 h⁻¹) in 225 dwellings in Sweden, F was about 0.51 (0.28-0.74). For an average air rate (0.3-0.6 h⁻¹), F is about 0.43 (0.21-0.66), and for high air rate (>0.6 h⁻¹), F was about 0.33 (0.21-0.47). Another study (Porstendörfer, 1987) of the relationship between equilibrium factor and aerosol sources indicated that the mean value of the equilibrium factor F measured in homes without aerosol sources was 0.3 ± 0.1. The equilibrium factor increased up to 0.5 with additional aerosol particles (cigarette and candle smoke) in the room air. Measurements of equilibrium factor determined in Swedish homes (Jonassen and Jensen, 1989) showed that F was 0.51 in a home inhabited by smokers and 0.46 in the homes of eight nonsmokers.

A distinction is often made in the state of the airborne progeny based on apparent attachment to aerosol particles. The unattached fraction (ultrafine mode, 0.5-5 nm) refers to those progeny existing as ions, molecules, or small clusters. The attached fraction (accumulation mode, 0.1-0.4 μm) is regarded as those radionuclides attached to ambient particles. The unattached fraction f_p of the total potential alpha energy of the radon progeny mixture is described as:

$$f_p = \frac{C_{eq}^u}{C_{eq}} \quad (2\text{-}6)$$

where $C_{eq} = 0.105 \ C_1 + 0.516 \ C_2 + 0.379 \ C_3$, and C_j ($j = 1, 2, 3$) is the activity concentration of radon progeny. The superscript u stands for unattached fraction. Typically, most of the unattached activity is that of ^{218}Po. Measurements of the unattached fraction in the domestic environment (Reineking et al., 1985;

Vanmarcke et al., 1987) showed that f_p is between 0.05 and 0.15 without any aerosol sources in the room and can decrease to below 0.05 in the presence of aerosol sources (cigarette smoke, cooking, stove heating). However, Jonassen and Jensen (1989) found that f_p was 0.01 in a house inhabited by smokers and 0.04 in the homes of eight nonsmokers.

As shown schematically in Figure 2-1, the highly diffusive, ultrafine activity (unattached fraction) becomes attached to the particles that are present in the existing aerosol whether it is in the mine or the home. The attachment rate is dependent on both the number of particles suspended in the air as well as their size in a complicated manner (Porstendörfer et al., 1979). The distribution of sizes as measured based on the determination of the amounts of associated radioactivity is referred to as the activity-weighted particle size distribution.

MEASUREMENT METHODS

Because of limitations in the measurement methods, only two physical parameters were used in the earlier lung dosimetry models for estimating alpha doses from inhaled radon decay products: the activity median diameter of the attached radioactive aerosol and the unattached fraction of PAEC (Jacobi and Eisfeld, 1980; James et al., 1980; Harley and Pasternack, 1982). Traditionally defined, the unattached fraction constitutes free molecular daughter atoms or ions possibly clustered with other molecules such as H_2O, as distinct from decay product atoms attached to particles in the preexisting ambient aerosol. The measurement of the unattached fraction of radon progeny has been the subject of extensive research. Diffusion, impaction, and electrostatic deposition methods have been used for the separation of the attached and unattached fractions of radioactive aerosols in ambient and mine atmospheres (Van der Vooren et al., 1982).

Interpretation of unattached fraction measurements has been clouded by limited understanding of the physicochemical behavior of the progeny, particularly ^{218}Po. The studies based on diffusional collection of the unattached activity all used a single, constant diffusion coefficient for unattached ^{218}Po based on the first estimate of 0.054 cm^2/s for the diffusion coefficient of ^{218}Po (Chamberlain and Dyson, 1956). Subsequent investigations have indicated that the unattached ^{218}Po fraction is actually an ultrafine particle or cluster mode, 0.5-3 nm in diameter, rather than free molecular ^{218}Po (Busigin et al., 1981; Reineking and Porstendörfer, 1986). Busigin et al. (1981) were the first to conclude that the diffusion coefficient of unattached ^{218}Po in air could not be adequately described by a single number, given the experimentally observed range of 0.005 to 0.1 cm^2/s. Goldstein and Hopke (1985) have also shown experimentally that the diffusion coefficient for ^{218}Po can be adjusted in the range of 0.03 to 0.08 cm^2/s by controlling the admixed trace gases. Raes (1985) has applied classical nucleation theory to describe the growth of clusters in atmospheres containing ^{218}Po, H_2O, and SO_2. The results of these studies strongly suggest that the so-called unattached fraction is actually an ultrafine particle mode in the size range of 0.5 to 3.0 nm whose nature is dependent upon the gaseous environment surrounding the radon decay products.

Results of recent studies by Chu and Hopke (1988) of the neutralization rate for ^{218}Po$^+$ ions initially present at the end of the polonium nucleus recoil path strongly suggest that the ions are rapidly neutralized in the atmosphere and that electrostatic collection underestimates the unattached fraction. Jonassen (1984) and Jonassen and McLaughlin (1985) found that only about 10% of the unattached fraction, as measured by using a wire screen system, was charged. Thus, the unattached fraction measurements of Blanc et al. (1968) and Chapuis et al. (1970), who used electrostatic collection, found that the unattached fractions were extremely small when they actually measured the fraction of the highly diffusive daughter activity that was still charged. These results are in good agreement with later measurements of Jonassen and McLaughlin (1985). Thus, electrostatic measurements of unattached fractions are not considered further in this assessment.

The detection of these ultrafine clusters by conventional particle detection devices such as condensation nuclei counters is precluded by the lack of sensitivity of these devices below particle diameters of 4 nm (Agarwal and Sem, 1980; Bartz et al., 1985). However, the existence of the ultrafine cluster mode (0.5-3 nm in diameter) in the activity size distribution has been observed in investigations by Reineking and Porstendörfer (1986) and by Tu and Knutson (1988), who used specially developed wire screen diffusion batteries. Wire screens, which are calibrated by using a single, constant value for the ^{218}Po diffusion coefficient and sampling an unattached fraction consisting of an ultrafine cluster mode in the activity size distribution, would thus be unable to separate the unattached and the attached fractions.

In order to measure the amounts of activity in the fractions of various sizes, it is necessary to understand the dynamics of particle deposition in tubes or screens. (The details of the theoretical background to the measurement methods are provided in Chapter 6.) This theoretical framework can then be applied to develop measurement methods. The measurement systems have traditionally attempted to separate the activity into two fractions: a highly diffusive fraction referred to as the unattached fraction and the activity attached to indoor aerosol particles. A review of these measurement methods and their limitations is also presented in Chapter 6.

EXPOSURE TO RADON PROGENY

Mine Atmospheres

The limited data on particle size distribution and unattached fraction measured in uranium mines are presented in Chapter 6. From the study of Bigu and Kirk (1980), who made measurements in active working areas of a mine, under typical working conditions in mines, it appears that the unattached fraction of PAEC is quite small, with values estimated to be below 1%. It is likely that conditions in mines of previous decades were even more dusty as standards were not in place to limit exposures via ventilation. Thus, very low values of the unattached fraction are anticipated to have been typical in the past for active working areas of the mines. For the committee's modeling, an f_p value of 0.005 is assumed for the active working areas of mines. While higher values have been reported, values of about 0.03 are more typical for the unattached fraction of ^{218}Po and not PAEC. In nonworking areas such as haulage drifts (sometimes called haulageways), etc., the values are more on the order of 0.02 to 0.03, and a value of 0.03 is assumed. In lunchrooms and other "clean" areas (workshops, storage areas), the f_p value could be as high as 0.08. It is likely that the miners most highly exposed in the earliest days of underground uranium mining were probably exposed to very low unattached fractions, as the low ventilation that led to high radon concentrations would also produce high aerosol particle concentrations.

The size distribution data of Cooper et al. (1973) show that the activity median diameter (AMD) of the attached mode is on the order of 0.25 μm, with a geometric standard deviation (σ_g) of 2.5. Although transient aerosols with both larger- and smaller-sized modes would be expected immediately following a dynamite blast, the total exposure to aerosols generated by such events should be relatively small compared with the total time underground. In the haulage drifts and other areas away from the areas where active mining is occurring, the distributions are shifted to somewhat smaller sizes, as indicated by Knutson and George (1990). They report an average activity median diameter for unimodal distributions of 151 nm with an average geometric standard deviation of 2.71.

Miners are also exposed to airborne radioactivity from long-lived alpha emitters, including uranium-234, thorium-230, radium-226, and polonium-210 present in the suspended ore dusts in mine atmospheres. Harley and coworkers (1981, 1985) have estimated the contribution of these species to the total lung dose of alpha energy. At the higher concentrations of ^{222}Rn present in mines until 1967, the additional dose from these long-lived radionuclides is estimated to be <10% of that from the short-lived ^{222}Rn decay products. Singh et al. (1985) measured the accumulated long-lived activity in a limited number of autopsied miner's lungs and also estimated that the dose from the long-lived activities is a small fraction of that obtained from the short-lived activities.

Indoor Atmospheres

In general, the data indicate that the unattached fraction of the potential alpha-energy concentration measured in active working areas of uranium mines is much lower than that in typical indoor air. However, the limited number of unattached fraction measurements in unoccupied houses (also summarized in Chapter 6) makes extrapolation to the much larger population of occupied houses subject to substantial uncertainty. The results that have been measured suggest that, in the absence of smoking, unattached fractions are in the range of 0.07 to 0.10. The typical value is taken as 0.08 for this report. In houses with low aerosol particle concentrations, f_p values in excess of 0.16 may be observed. Smoking produces relatively high concentrations of 0.1- to 0.5-μm-diameter particles and can reduce unattached fractions to the range of 0.01 to 0.03. Because smoking is an intermittent source of particles, the higher end of this range is taken and the typical value is assumed to be 0.03.

Recent measurements also suggest that the equilibrium factor in houses without smokers is lower than the commonly used value of 0.5, probably in the range of 0.3 to 0.4. Thus, although there is less airborne activity in indoor air per unit radon activity concentration, more of the activity is in a more diffuse form. However, there are relatively few measurements of the behavior of radon progeny in houses under normal conditions of occupation over a prolonged period of time. The only reported measurements are those from Japan of Kojima and Abe (1988), and the house and life-style examined did not correspond to those in the United States.

To describe better the actual behavior of the airborne activity, multiple wire screens can be used in diffusion battery-type systems to comprehensively determine the activity size distribution in the 0.5- to 500-nm-diameter size range, as described by Reineking and Porstendörfer (1986), Kulju et al. (1986), Tu and Knutson (1988), Strong (1988), and others. These measurements are discussed in Chapter 6. The limited size distributions that have now been measured indicate that there may be unattached activity with an average diffusion coefficient on the order of 0.035 ± 0.010 cm^2/s. However, one or two modes may be present below a diameter of 10 nm, depending on the presence of aerosol sources. Thus, earlier definitions of the unattached fraction as that activity on particles of less than 5 nm in diameter do not adequately describe the actual nature of the aerosol behavior of the progeny. Complete characterization of the activity size distribution can now be accomplished with automated or manual systems. Uncertainties remain in obtaining size distributions from the activity measurements (Ramamurthi and Hopke, 1990), and standardization of measurement systems is currently feasible. Further development and refinement of these systems will lead to incremental improvements in their performances and the reliability of the information extracted from them. Only a limited number of such systems are available, and there are few data on both unattached fractions and complete activity-weighted size distributions in indoor air. The available measurements have mostly been made under artificial conditions, not during normal occupancy. Therefore, current estimates of unattached fractions cannot be assumed, with a high degree of confidence, to represent typical values in indoor air.

From the available size distribution data, the AMD of the attached mode in houses without cigarette smokers is estimated to be 0.15 μm, with a σ_g of 2.0. With cigarette smokers present, the average aerodynamic diameter increases to 0.25 μm, with a σ_g of 2.5. The presence of active aerosol sources such as a vacuum cleaner motor or burning gas burners yields an activity mode at 20 nm, with a σ_g of 1.5, containing approximately 10 to 15% of the PAEC.

A summary of the characteristics of the radon progeny aerosols assumed by the committee is given in Chapter 3. Listed there are the different scenarios that are assumed, the f_p values for these scenarios, and the AMD's of the aerosols in the rooms and in the human respiratory tract.

REFERENCES

Agarwal, J. K., and G. J. Sem. 1980. Continuous flow, single-particle counting condensation nuclei counter. J. Aerosol Sci. 11:343-358.

Bartz, H., H. Fissan, C. Helsper, Y. Kousaka, K. Okuyama, N. Fukushima, P. B. Keady, S. Kerrigan, S. A. Fruin, P. H. McMurry, D. Y. H. Pui, and M. R. Stolzenburg. 1985. Response characteristics for four different condensation nucleus counters to particles in the 3-50 nm diameter range. J. Aerosol Sci. 5:443-456.

Bigu, J., and B. Kirk. 1980. Determination of the unattached radon daughter fractions in some uranium mines. Presented at the Workshop on Attachment of Radon Daughters, Measurement Techniques and Related Topics, October 30, 1980, University of Toronto. (Report available from CANMET, P.O. Box 100, Elliot Lake, Ontario, Canada.)

Blanc, D., J. Fontan, A. Chapuis, F. Billard, G. Madelaine, and J. Pradel. 1968. Dosage du radon et de ses descendants dans une mine d'uranium. Repartition granulometrique des aerosols radioactifs. Pp. 229-238 in Symposium on Instruments and Techniques for the Assessment of Airborne Radioactivity in Nuclear Operations. Vienna: International Atomic Energy Agency.

Busigin, A., A. W. Van der Vooren, J. C. Babcock, and C. R. Phillips. 1981. The nature of unattached [218]Po (RaA) particles. Health Phys. 40:333-343.

Chamberlain, A. C., and E. D. Dyson. 1956. The dose to the trachea and bronchi from the decay products of radon and thoron. Br. J. Radiol. 29:317-325.

Chapuis, A., A. Lopez, J. Fontan, F. Billard, and G. J. Madelaine. 1970. Spectre granulometrique des aerosols radioactifs dans mine d'uranium. J. Aerosol Sci. 1:243-253.

Chu, K. D., and P. K. Hopke. 1988. Neutralization kinetics for polonium-218. Environ. Sci. Technol. 22:711-717.

Cooper, J. A., P. O. Jackson, J. C. Langford, M. R. Petersen, and B. O. Stuart. 1973. Characteristics of attached radon-222 daughters under both laboratory and field conditions with particular emphasis upon underground mine environments. Report to the U.S. Bureau of Mines under contract H0220029. Richland, Wash.: Battelle Pacific Northwest Laboratories.

Goldstein, S. D., and P. K. Hopke. 1985. Environmental neutralization of polonium-218. Environ. Sci. Technol. 19:146-150.
Harley, N. H., and I. M. Fissenne. 1985. Alpha dose from long-lived emitters in underground uranium mines. Pp. 518-522 in Occupational Radiation Safety in Mining, Vol. 2, H. Stocker, ed. Toronto: Canadian Nuclear Association.
Harley, N. H., and B. S. Pasternack. 1982. Environmental radon daughter alpha dose factors in a five-lobed human lung. Health Phys. 42:789-799.
Harley, N. H., D. E. Bohning, and I. M. Fissenne. 1981. The dose to basal cells in bronchial epithelium from long-lived alpha emitters in uranium mines. In Radiation Hazards in Mining: Control, Measurement, and Medical Aspects, M. Gomez, ed. New York: American Institute of Mining, Metallurgical, and Petroleum Engineers, Inc.
Jacobi, W., and K. Eisfeld. 1980. Internal dosimetry of radon-222, radon-220 and their short-lived daughters. GSF Report S-626. Munich-Neurherberg, Germany: Gesellschaft für Strahlen- und Umweltforschung.
James, A. C., J. R. Greenhalgh, and A. Birchall. 1980. A dosimetric model for tissues of the human respiratory tract at risk from inhaled radon and thoron daughters. Pp. 1045-1048 in Radiological Protection—Advances in Theory and Practice, Vol. 2, Proceedings of the 5th Congress IRPA, Jerusalem, March 1980. Oxford: Pergamon Press.
Jonassen, N. 1984. Electrical properties of radon daughters. Presented at the International Conference on Occupational Radiation Safety in Mining, Toronto, Canada.
Jonassen, N., and B. Jensen. 1989. Radon daughters in indoor air. Final Report to Vattenfall Technical University of Denmark. Lyng, Denmark.
Jonassen, N., and J. P. McLaughlin. 1985. The reduction of indoor air concentrations of radon daughters without the use of ventilation. Sci. Total Environ. 45:485-492.
Knutson, E. O., and A. C. George. 1990. Reanalysis of data on the particle size distribution of radon progeny in uranium mines. Proceedings of the 29th Hanford Life Sciences Symposium, Indoor Radon and Lung Cancer: Reality or Myth?, October 16-19, 1990, Richland, Washington.
Kojima, H., and S. Abe. 1988. Measurement of the total and unattached radon daughters in a house. Radiat. Prot. Dosim. 24:241-244.
Kulju, L. M., et al. 1986. The detection and measurement of the activity size distribution of ultrafine particles. Paper No. 86-40.6. Pittsburgh, Pa.: Air Pollution Control Association.
Porstendörfer, J., G. Röbig, and A. Ahmed. 1979. Experimental determination of the attachment coefficients of atoms and ions on monodisperse particles. J. Aerosol Sci. 10:21-28.
Porstendörfer, J. 1987. Free-fractions, attachment rates, and plate-out rates of radon daughters in houses. Pp. 285-300 in Radon and Its Decay Products: Occurrence, Properties and Health Effects, P. K. Hopke, ed. Symposium Series 331. Washington, D.C.: American Chemical Society.
Raes, F. 1985. Description of properties of unattached ^{218}Po and ^{212}Pb particles by means of the classical theory of cluster formation. Health Phys. 49:1171-1187.
Ramamurthi, M., and P. K. Hopke. 1990. Simulation studies of reconstruction algorithms for the determination of optimum operating parameters and resolution of graded screen array systems (non-conventional diffusion batteries). Aerosol Sci. Technol. 12:700-710.
Reineking, A., and J. Porstendörfer. 1986. High-volume screen diffusion batteries and α-spectroscopy for measurement of the radon daughter activity size distributions in the environment. J. Aerosol Sci. 17:873-879.

Reineking, A., et al. 1985. Measurements of the unattached fractions of radon daughters in houses. Sci. Total Environ. 45:261-270.

Singh, N. P., et al. 1985. Concentrations of ^{210}Pb in uranium miners lungs and its states of equilibria with ^{238}U, ^{234}U, and ^{230}Th. In Occupational Radiation Safety in Mining, Vol. 2, H. Stocker, ed. Toronto: Canadian Nuclear Association.

Strong, J. C. 1988. The size of attached and unattached radon daughters in room air. J. Aerosol Sci. 19:1327-1330.

Swedjemark, G. A. 1983. The equilibrium factor. F. Health Phys. 45:453-462.

Tu, K. W., and E. O. Knutson. 1988. Indoor radon progeny particle size distribution measurements made with two different methods Radiat. Prot. Dosim. 24:251-255.

Van der Vooren, A. W., A. Busigin, and C. R. Phillips. 1982. An evaluation of unattached radon (and thoron) daughter measurement techniques. Health Phys. 42:801-808.

Vanmarcke, H., A. Janssens, F. Raes, A. Poffijn, P. Perkvens, and R. Van Dingenen. 1987. The behavior of radon daughters in the domestic environment. Pp. 301-323 in Radon and Its Decay Products: Occurrence, Properties and Health Effects, P. K. Hopke, ed. Symposium Series 331 Washington, D.C.: American Chemical Society.

3

Extrapolation of Doses and Risk per Unit Exposure from Mines to Homes

INTRODUCTION

It is believed that initiating or promotion of the genetic changes in target cells that lead to radiation-induced cancer is caused by specific interactions of ionizing radiation events with DNA in the cell nucleus (NRC, 1990). At the low doses of concern in the assessment of exposure to radon progeny in the home, it is presumed that the likelihood of these carcinogenic events occurring is directly proportional to the radiation dose received by the cell nucleus. More fundamental discussion of the dosimetry and radiobiology of alpha particles can be found in the BEIR IV report (NRC, 1988).

The purpose of this chapter is to translate exposures to radon progeny under the different environmental conditions encountered in mines and homes into the biologically effective dose delivered in each environment. The committee's approach to this task was to apply a dosimetric model based on an up-to-date interpretation of the relevant physiological and biological factors (see Chapters 7 and 8) with current knowledge of the variability of radon progeny aerosol conditions in mines and homes (reviewed in Chapter 2). Details of the dosimetric model together with a summary of the experimental data and underlying assumptions on which the model is based are given in Chapter 9.

This chapter is concerned with evaluating dose conversion coefficient that links exposure to dose. More specifically, the aim is to compare values of the exposure-dose conversion coefficient that apply to subjects exposed in the home with those applicable to underground miners, for the purpose of extrapolating to the home environment the exposure-risk relationships found in epidemiological studies of underground miners. Using the terminology of the BEIR IV report (NRC, 1988), if exposure is expressed in the commonly used unit working level month (WLM), the risk per unit exposure in the home, $(\text{Risk})_h/(\text{WLM})_h$, can be related to that in mines, $(\text{Risk})_m/(\text{WLM})_m$, by a dimensionless factor, K.

$$K = \frac{(\text{Risk})_h/(\text{WLM})_h}{(\text{Risk})_m/(\text{WLM})_m}, \quad (3\text{-}1)$$

On the premise that risk is related primarily to the doses received by the appropriate cellular targets, the value of the risk extrapolation factor K is given by the ratio of the doses that result from unit exposure in each environment.

$$K = \frac{(\text{Dose})_h/(\text{WLM})_h}{(\text{Dose})_m/(\text{WLM})_m}, \quad (3\text{-}2)$$

The dose conversion coefficients (Dose/WLM) for miners and for adults and children exposed in the home and the corresponding values of the factor K are evaluated below. The factor K is referred to as the dosimetric risk extrapolation factor or, for convenience, as the K factor. The variability of dose per WLM exposure and the K factor over a range of representative exposure scenarios in mines and homes are examined in this chapter. The sensitivities of these predicted values to parameters of dosimetric modeling that remain uncertain are also determined.

REPRESENTATIVE EXPOSURE CONDITIONS

Radon Progeny Aerosols

Underground mining involves a variety of work activities, from which three representative and characteristically different exposure scenarios are identified in Chapter 6. These scenarios are rock-breaking, ore-winning, and other dusty activities (referred to below as mining); transport and maintenance work in haulageways; and time spent in less active and dusty areas (referred to as lunchrooms). The respective unattached fractions of potential alpha energy (f_p) and the activity median thermodynamic diameter (AMTD) of the attached radon progeny aerosols that are assumed to characterize each of these categories of exposure are given in Table 3-1.

Chapter 6 also identifies several characteristically different exposure scenarios in the home. These are distinguished as time spent in living rooms and in bedrooms. The normal living room is considered to be free of strong sources of aerosol particles. The radon progeny aerosol is characteristically different in a room in which cigarettes are smoked (referred to as a smoker), one in which an electric motor is being used (e.g., during vacuuming), or a

TABLE 3-1 Summary of Radon Progeny Aerosol Characteristics Assumed to Represent Exposure Conditions in Mines and Homes

Exposure Scenario	f_p	AMD of Room Aerosol (μm)	AMD of Aerosol in Respiratory Tract (μm)
Mine			
Mining	0.005	0.25	0.5
Haulage drifts[a]	0.03	0.25	0.5
Lunch room	0.08	0.25	0.5
Living Room			
Normal	0.08	0.15	0.3
Smoker—average	0.03	0.25	0.5
Smoker—during smoking	0.01	0.25	0.5
Cooking/vacuuming	0.05	0.02/0.15[b] (15%/80%)	0.02/0.3 (15%/80%)
Bedroom			
Normal	0.08	0.15	0.3
High	0.16	0.15	0.3

[a] As noted in Chapter 2, a reconsideration of early aerosol size measurements has indicated that the aerosol AMTD in haulage drifts and other areas away from active mining is shifted to 0.15 μm. The effect on calculated K factors of this revision of the committee's assumed value of 0.25 μm is noted at the end of the chapter.

[b] The radon progeny aerosol produced by cooking/vacuuming has three size modes; 5% of potential alpha energy is unattached, 15% has an AMD of 0.02 μm, and 80% has an AMD of 0.15 μm. The 0.02-μm AMD mode is hydrophobic and does not increase in size within the respiratory tract.

kitchen and connecting rooms during cooking. The radon progeny aerosol in a normal bedroom is considered to be the same as that in the normal living room. However, in well-insulated homes with low exchange between indoor and outdoor air, the particle loading of bedroom air may be substantially less, giving rise to a high value of the unattached fraction. The corresponding values of the unattached fraction and radon progeny aerosol size that are assumed to represent these various exposure scenarios in the home are given in Table 3-1.

In both mine and home atmospheres, unattached radon progeny are considered to have a characteristic diffusion coefficient of 0.035 cm²/s (see Chapter 6). This corresponds to a particle of diameter 0.0011 μm. The size of the unattached progeny is assumed to remain constant in the respiratory tract. However, the condensation nuclei to which radon progeny attach are unstable in saturated air. They are assumed to grow rapidly on inhalation, such that the AMTD of the radon progeny aerosol within the respiratory tract is double that in ambient air (Sinclair et al., 1974). To evaluate the contributions to respiratory tract deposition made by impaction and sedimentation (see Chapter 9), the density of these hygroscopically enlarged particles is taken to be unity.

TABLE 3-2 Levels of Physical Exertion and Average Minute Volumes Assumed for Underground Miners and for Adults in the Home

Exposure Scenario	Level of Exertion	Average V_E (liters/min) Man	Woman
Underground mine			
Mining	25% heavy work/75% light work	31	—
Haulageway	100% light work	25	—
Lunchroom	50% light work/50% rest	17	—
Home—Living room			
Normal and smoker	50% light work/50% rest	17	14
Cooking/vacuuming	75% light work/25% rest	21	17
Home—Bedroom			
Normal and high	100% sleep	7.5	5.3

BREATHING RATES AND CALCULATION OF DOSE PER UNIT EXPOSURE

The breathing rates of miners at work underground and adults and children under various circumstances in the home are discussed in Chapter 7. To evaluate dose per unit exposure in these situations, the committee assumed the levels of exertion and corresponding minute volumes (V_E) given in Table 3-2 for adults and in Table 3-3 for children and infants. The dose per unit exposure for each subject and for each exposure scenario (defined in Table 3-1) was obtained by combining individual values of the primary exposure-dose conversion coefficient that were calculated using the dosimetric model described in Chapter 9. Examples of these primary exposure-dose conversion coefficients are tabulated in Chapter 9. An individual conversion coefficient relates to a specific radon progeny aerosol size and level of exertion (i.e., sleep, rest, light exercise, or heavy exercise). The dependence of primary exposure-dose conversion coefficients on radon progeny aerosol size and level of exertion for male and female adults, children, and infants is illustrated and discussed in Chapter 9.

In order to derive exposure-dose conversion coefficients for a population of underground miners, the committee calculated the individual time-weighted doses for mining and haulageway work. The corresponding average dose-weighted minute volume corresponds to a minute volume of 28 liters/min. To represent overall exposure in the home, the committee used a time-weighted average of individual values of the exposure-dose conversion coefficient calculated for the normal living room and normal bedroom. The corresponding time-averaged dose-weighted minute volumes are 12.3 liters/min for an adult male, 9.7 liters/min for an adult female, 8.8 liters/min for a 10-yr-old child,

TABLE 3-3 Levels of Physical Exertion and Average Minute Volumes Assumed for Children and Infants in the Home

Exposure Location and Subject	Level of Exertion	Average V_E (liters/min)
Living room		
Child, age 10 yr	50% light work/50% rest	12.4
Child, age 5 yr	50% light work/50% rest	7.4
Infant, age 1 yr	50% light work/50% rest	4.8
Infant, age 1 mo	30% light work/70% sleep	1.5
Bedroom		
Child, age 10 yr	100% sleep	5.2
Child, age 5 yr	100% sleep	4.0
Infant, age 1 yr	100% sleep	2.6
Infant, age 1 mo	100% sleep	1.3

5.7 liters/min for a 5-yr-old child, 3.7 liters/min for a 1-yr-old infant, and 1.4 liters/min for a 1-mo-old infant.

COMPARATIVE DOSES FROM RADON PROGENY IN MINES AND HOMES

Target Cells in Bronchial Epithelium

In uranium miners, and also in the general population, the majority of lung cancers arise from the epithelium of bronchial airways (Chapter 8). In this tissue both secretory cells and basal cells are considered to be targets for lung cancer development. The bronchial airways include all airways in which the epithelium is supported by a thick wall of connective tissue and cartilage. These airways are generally larger than 2 mm in caliber in an adult male (see Chapter 9 for the corresponding dimensions in other people). The bronchi are represented in the dosimetric model by airway generations one through eight, in which the trachea is termed generation zero. The committee used the model to calculate doses received by target cells in each generation and then averaged the results to express an exposure-dose conversion coefficient for bronchial epithelium as a whole.

The exposure-dose conversion coefficients calculated for a miner and for an adult male exposed to radon progeny at home are shown in Figure 3-1. Values are shown separately for secretory and basal cells, which are treated as discrete targets. Figure 3-1 also shows the effect of plausible, but different, assumptions about the clearance behavior of radon progeny after deposition on an airway surface (see Chapter 9 for a discussion of the clearance model). The alternative assumptions considered here are (1) radon progeny are effectively insoluble, i.e., they are retained in mucus and cleared progressively toward the throat, and (2) that the progeny are partially soluble, i.e., 30% of the activity deposited in each airway generation is assumed to be taken up rapidly by epithelial tissue.

⊠ Secretory Cells / Insoluble Rn-progeny: K=0.81
☐ Basal Cells / Insoluble Rn progeny: K=0.77
☰ Secretory Cells / Partially Soluble Rn-progeny: K=0.72
■ Basal Cells / Partially Soluble Rn-progeny: K=0.60

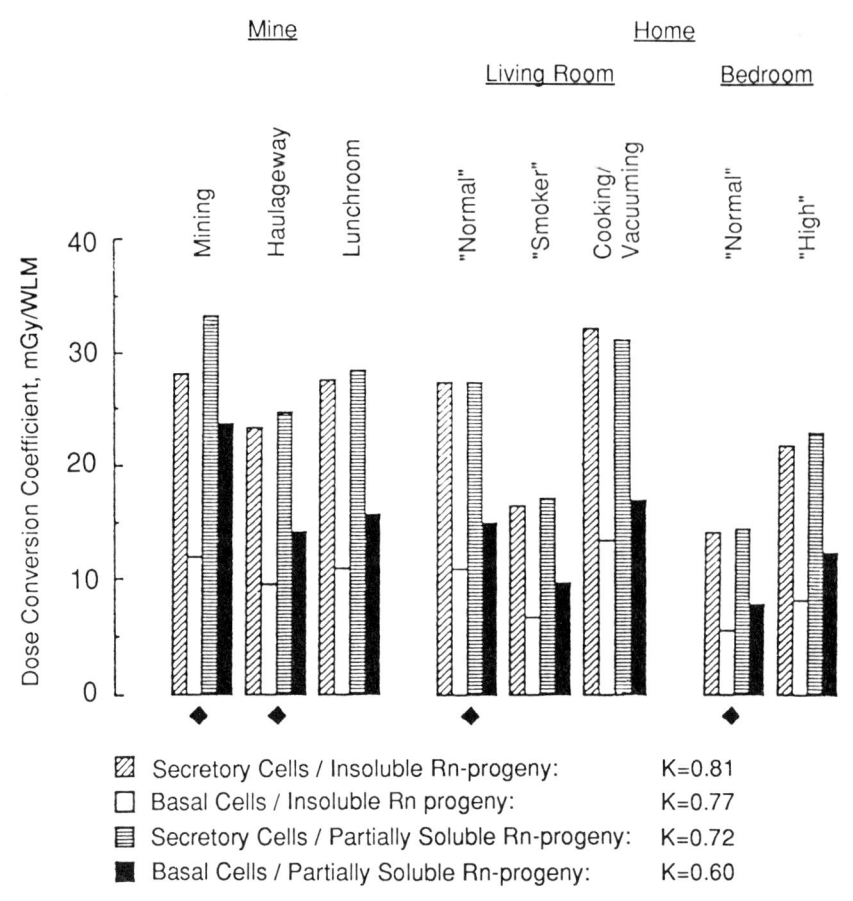

FIGURE 3-1 Comparison of dose conversion coefficients for a miner and for a man exposed to radon progeny at home. The histograms represent various conditions of exposure in each environment, which are described in the text. Four values of the dose conversion coefficient are shown for each situation. From left to right, these represent the doses calculated for secretory and basal cell nuclei in bronchial epithelium, if radon progeny are assumed to be insoluble, and for the same target cell nuclei if part of the radon progeny activity is assumed to be taken up by the epithelium. The diamond-shaped symbol indicates conditions that are considered to represent a "typical" exposure in each environment: for the mine, an equal mixture of active "mining" and work in a "haulageway" and for the home, equal exposure in the "normal" living room and bedroom. Additional conditions of exposure are also shown, to indicate the degree of variability of the dose conversion coefficient in each environment. The numerical factor K represents the ratio of the "typical" dose conversion coefficient for the man exposed at home to that for the miner.

It is seen from the data in Figure 3-1 that the calculated exposure-dose conversion varies substantially with exposure conditions and with the choice of target cells (secretory or basal cells), but varies to a lesser extent with the assumed clearance behavior. In absolute terms, the conversion coefficient varies from a maximum of 33 milligrays (mGy)/WLM (to secretory cells) during active mining, with similar values for exposure in the home to the radon progeny aerosol produced by cooking/vacuuming, to a minimum of 5.6 mGy/WLM (to basal cells) for an adult male sleeping in a normal bedroom. However, for the purpose of deriving a dosimetric risk extrapolation factor (the K factor), the choice of target cell population and the assumptions about clearance behavior are less influential. Figure 3-1 shows that the corresponding values of the K factor, which represent the ratios of dose conversion coefficients from exposure in the home to that in a mine, vary from 0.81 to 0.60 for secretory cell and basal cell targets, respectively.

As noted earlier, the K factor is derived by averaging the exposure-dose conversion coefficients calculated for the normal living room and bedroom and comparing the result with the average value calculated for mining and haulageway work. Exposure-dose conversion coefficients calculated for additional exposure scenarios (defined in Table 3-1) are also shown in Figure 3-1 (see also Figures 3-2 to 3-8). These are included to illustrate the variability of exposure-dose conversion coefficients, and, potentially, the K factor as well, with particular conditions in the home or mine.

LOCALIZED VERSUS REGIONAL DOSES

The extent of sensitive tissue within the lung and the factors that influence this are not well understood (see also Chapter 8). It is therefore prudent to examine the effect on risk extrapolation of choosing alternative tissues as the reference target. Figure 3-2 shows values of the exposure-dose conversion coefficient and K factor that are obtained when the reference tissue is taken to be the epithelium in just the lobar and segmental bronchi (generations 2-5 in the model), the epithelium in the bronchioles (generations 9-15 in the model), or epithelial cells in the alveoli. These values are compared in Figure 3-2 with the dose conversion coefficients derived for the bronchial airways as a whole by averaging doses received by secretory and basal cell targets. It is seen from the data in Figure 3-2 that exposure-dose conversion coefficients calculated for the lobar and segmental bronchi are uniformly higher than those calculated for the bronchi as a whole (by about 30%). Values calculated for the bronchioles are lower than those for the bronchi, on average by about 50%, and the coefficients calculated for alveolar epithelial cells are only about 2% of those for the bronchi. However, this marked variation of the exposure-dose conversion coefficient between different target tissues has a relatively small impact on the extrapolation of risk from the mine to the home. The K factor is

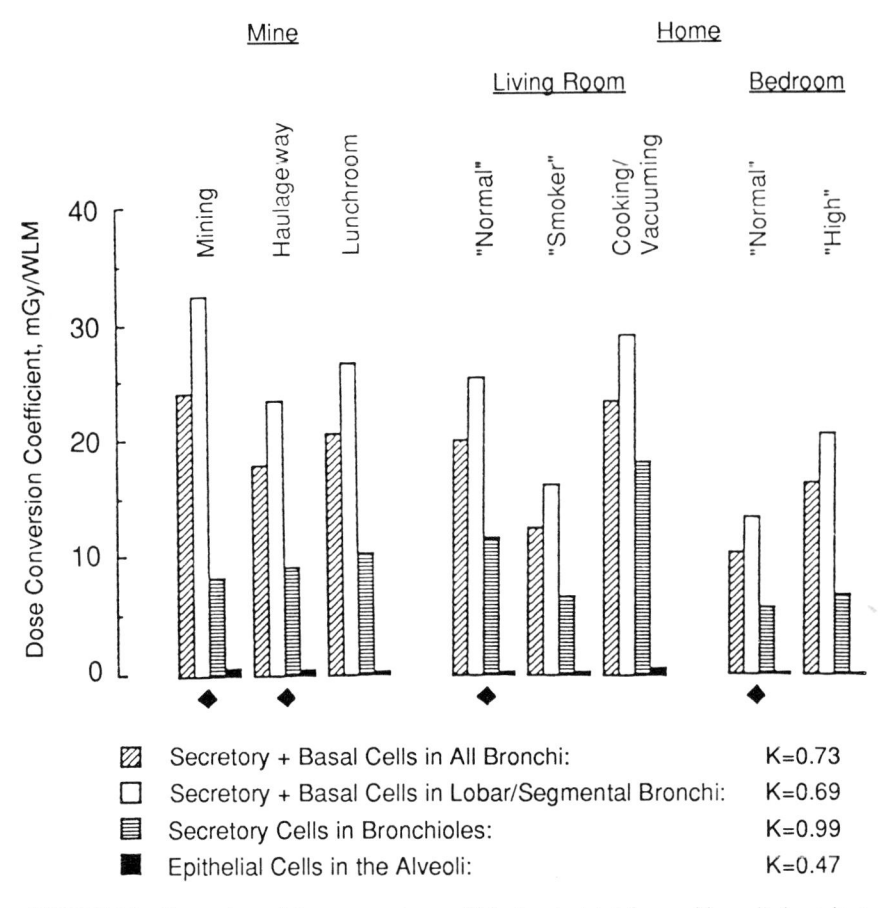

FIGURE 3-2 Comparison of dose conversion coefficients calculated for sensitive cells in various tissues of the lung. Values are shown by histograms for four cell populations. They represent the average doses received by: secretory and basal cell nuclei throughout the bronchial epithelium, the subpopulation of these cell nuclei in the epithelium of just the lobar and segmental bronchi, secretory cell nuclei in the bronchioles, and the alveolar epithelium. The various exposure conditions in the mine and the home are the same as those described for Figure 3-1.

found to be 0.73 for the bronchi as a whole, 0.69 for just lobar and segmental bronchi, 0.99 for the bronchioles, and 0.47 for alveolar epithelial cells (Figure 3-2). In the dosimetric comparisons that follow, the reference target tissue is assumed to be the bronchial epithelium as a whole.

INFLUENCE OF MODELING UNCERTAINTIES

As discussed in Chapter 9 neither the efficiencies of the nose and mouth in filtering unattached radon progeny from inhaled air nor the theoretical calculation

EXTRAPOLATION OF DOSES AND RISK PER UNIT EXPOSURE

[Figure 3-3: Bar chart showing Dose Conversion Coefficient (mGy/WLM) vs. various exposure scenarios]

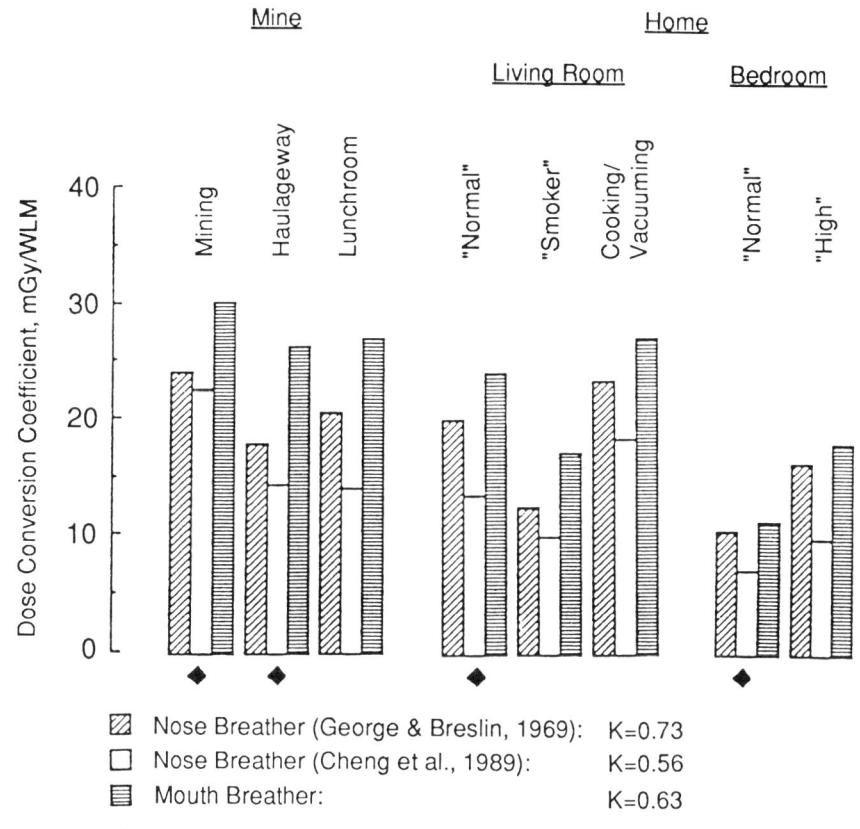

Mine: Mining, Haulageway, Lunchroom
Home — Living Room: "Normal", "Smoker", Cooking/Vacuuming
Home — Bedroom: "Normal", "High"

Legend:
- ▨ Nose Breather (George & Breslin, 1969): K=0.73
- ☐ Nose Breather (Cheng et al., 1989): K=0.56
- ☰ Mouth Breather: K=0.63

FIGURE 3-3 Influence of the assumed filtration efficiencies of the nasal and oral airways on the calculated bronchial dose conversion coefficient.

of radon progeny deposition in the bronchi (where airflow is complex) is well established. The impact of these uncertainties on calculated exposure-dose coefficients and the K factor are discussed here. A further source of uncertainty is the degree to which the attached radon progeny aerosols in various exposure situations are hygroscopic and increase in size within the respiratory tract.

The influence of radon progeny filtration by the nasal and oral passages is shown in Figure 3-3. Values of the exposure-dose coefficient are shown in Figure 3-3 for secretory cell targets in the bronchi. The values calculated by using a model of nasal deposition consistent with the human experimental data obtained by George and Breslin (1969) are compared in Figure 3-3 with values based on the recent data from nasal casts that were reported by Cheng et al. (1989). The respective data and the empirical nasal deposition models that are derived from them are discussed in Chapter 9. The lower exposure-dose conversion coefficients implied by the data of Cheng et al.'s result from corresponding reductions in the dose contributed by unattached progeny. According to Cheng et al., only about 15% of the inhaled unattached progeny are able to pass through the nasal passages without depositing there, whereas George and Breslin's data, and also other data from nasal casts (Strong and Swift, 1990), indicate that the nasal penetration efficiency is between 30 and 40%. Values of the exposure-dose conversion coefficient shown in Figure 3-3 for a mouth breather were calculated on the assumptions that the oral filtration efficiency is half that estimated for the nose from George and Breslin's (1969) data and that a typical mouth breather inhales partly through the nose (see Chapter 9 for a discussion of these assumptions). The resulting K factors are 0.73 according to the committee's preferred estimate of nasal deposition efficiency (George and Breslin, 1969), 0.56 according to the data of Cheng et al. (1989) from nasal casts, and 0.63 for a mouth breather.*

The finding that the K factor for mouth breathers is lower than the value of 0.73 estimated by the committee for nose breathers is at first surprising. However, breathing partly through the mouth has a more complex effect for underground miners than it does for a subject exposed to radon in the home. The mine aerosol is assumed to grow rapidly to reach an AMD of 0.5 μm in nasal or oral passages. At the higher rates of airflow that occur when all inspired air passes through the nose, deposition of the attached radon progeny within the nose is significant. Larger particles, which would otherwise tend to deposit in the bronchi, are lost from the aerosol. Filtration of larger aerosol particles is lower on two counts if the inhaled air is split between the nose and mouth: the airflow rate through the nose is reduced (decreasing the inertial deposition efficiency of the nose), and the inertial deposition efficiency of the oral passageway is comparatively low. However, if the incidence of mouth breathing among underground miners is substantially greater than that among subjects at home, K factors derived purely for nose breathers may be biased toward artificially high values.

The effect on calculated exposure-dose conversion coefficients by using bronchial deposition efficiencies based on empirical observations made from experiments with bronchial casts (Cohen et al., 1990), rather than a purely theoretical analysis, is shown in Figure 3-4 (see Chapter 9 for a discussion of the data and theoretical deposition models). It is seen from the data in Figure 3-4 that the assumption of enhanced bronchial deposition efficiencies reported by Cohen et al. gives uniformly higher conversion coefficients. However, the

---

*As noted in Chapter 9, new experimental studies of the penetration of unattached radon progeny through nasal casts, carried out after the committee completed its work, are found to support the data of Cheng et al. (1989) and no longer to support the lower values of nasal deposition that are obtained from George and Breslin's (1969) study. After further experimental verification, it may become preferable to evaluate K factors based on the higher nasal deposition efficiency reported by Cheng et al. The effect of higher nasal deposition is evaluated in a note to Table 3-4.

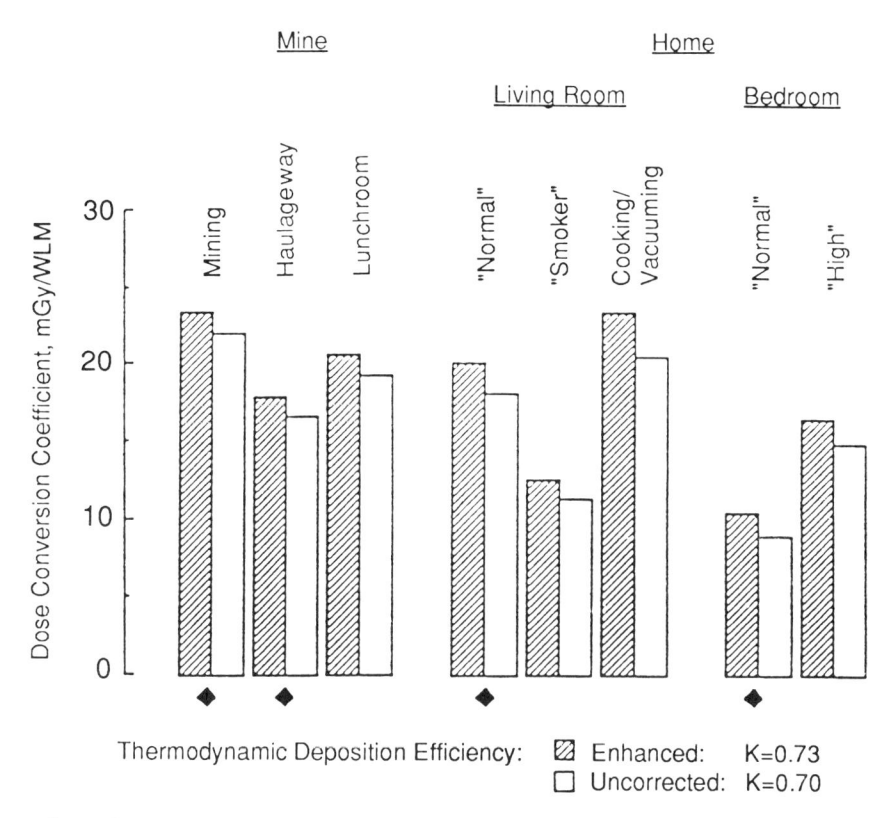

FIGURE 3-4 Influence of the enhanced bronchial deposition efficiency observed experimentally for submicron aerosol particles by Cohen et al. (1990) on the calculated dose conversion coefficient.

effect on the K factor is marginal. The empirically enhanced value of the K factor is 0.73, and the uncorrected theoretical value is similar at 0.70.

The sensitivities of calculated exposure-dose conversion coefficients and the K factor to the assumption made by the committee that the attached radon progeny aerosol grows in the respiratory tract to double its size in ambient air are examined in Figure 3-5. In this case, the values shown in Figure 3-5 are averages of dose conversion coefficients calculated for secretory and basal cell targets. It is seen that this assumed aerosol growth tends to increase conversion coefficients for exposures in mines but to decrease those calculated for exposures in the home. These effects arise because an aerosol of 0.5 μm in AMD (the size that a mine aerosol is assumed to attain by growth in the respiratory tract) is deposited more efficiently in the bronchi than is an aerosol of 0.25-μm AMD (the ambient size of the mine aerosol), whereas the reverse holds for growth of the smaller ambient aerosol generally found in the home.

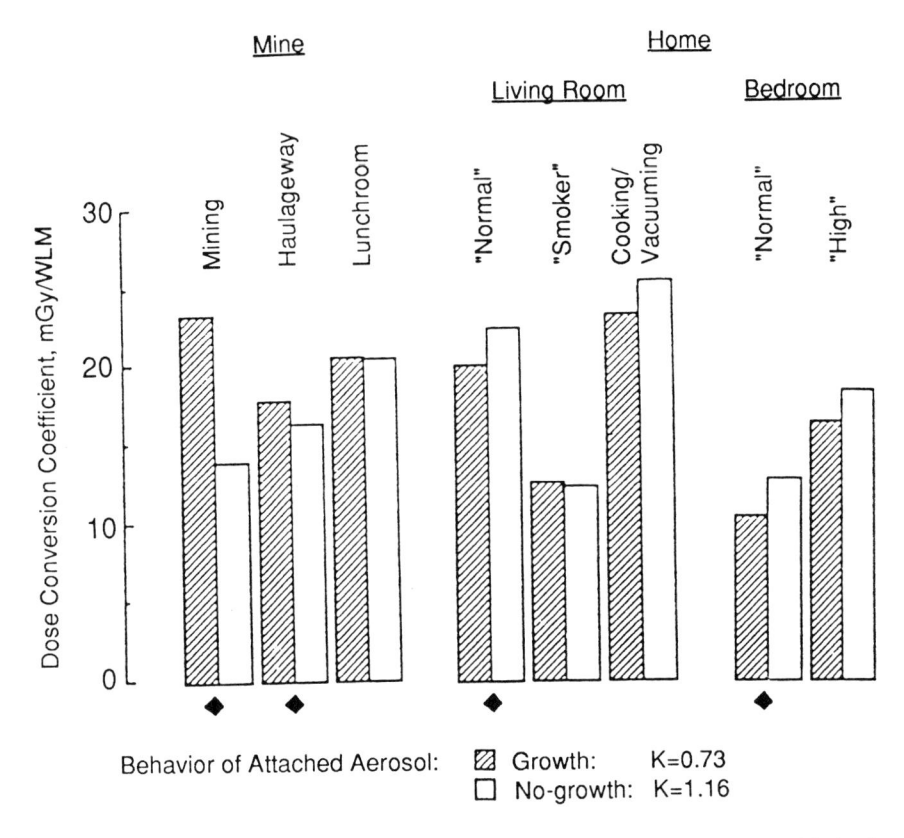

FIGURE 3-5 Influence of the assumed growth of radon progeny aerosols in the humid air of the respiratory tract on the calculated bronchial dose conversion coefficient.

If the attached radon progeny aerosols in both mines and homes were, in fact, stable in the respiratory tract, the K factor would increase to 1.16, which is significantly higher than the committee's estimate of 0.73.

DOSES TO ADULT FEMALES

Bronchial exposure-dose conversion coefficients calculated for a female exposed to radon progeny in the home (a female homemaker) are compared in Figure 3-6 with the values presented in Figure 3-1 for an underground miner. Variations of the dose conversion coefficient with exposure conditions and with the type of target cell considered (secretory or basal cell) are similar to those calculated for a male (compare Figure 3-6 with Figure 3-1). However, the K factors are somewhat lower for a female. The K factor is 0.72 for secretory

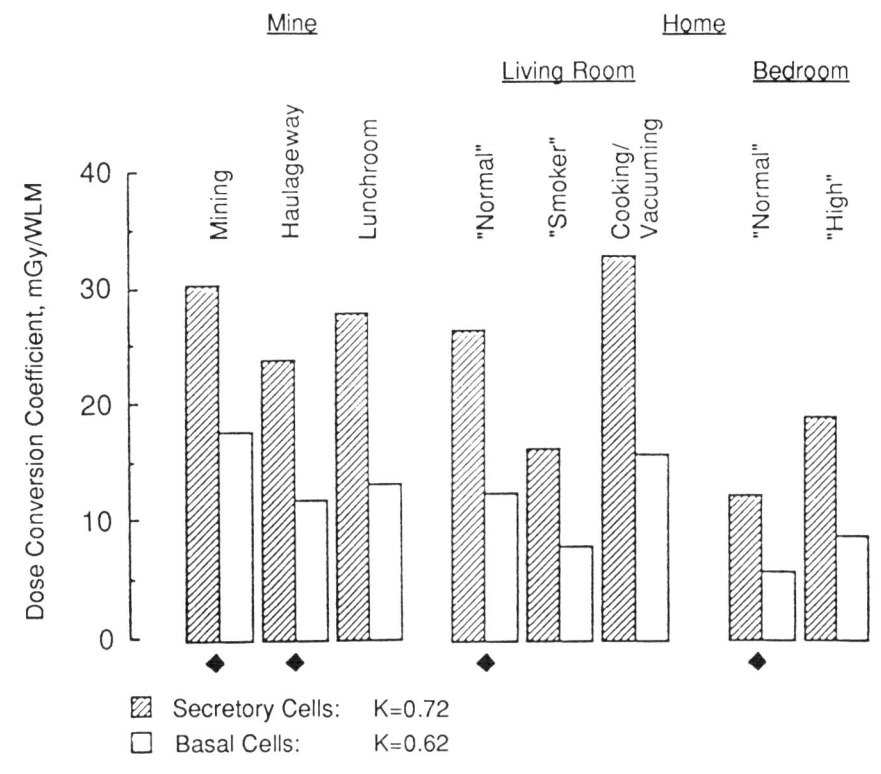

FIGURE 3-6 Comparison of the bronchial dose conversion coefficients calculated for exposure of a female homemaker to radon progeny with those for a miner.

cell targets and 0.62 for basal cells in a female compared with 0.76 and 0.69, respectively, for a male.

DOSES TO CHILDREN AND INFANTS

Exposure-dose conversion coefficients calculated for children and infants exposed to radon progeny in the home are compared in Figure 3-7 with those for adult males and females. The values shown in Figure 3-7 apply to secretory cell targets in the bronchial epithelium of each subject. In this example, the conversion coefficients calculated for subjects breathing entirely through the mouth are compared with those calculated for those breathing entirely through the nose. Although it is unlikely that any subject breathes entirely through the mouth, this behavior is examined here to indicate an upper bound for the dose conversion coefficient in each case. The ratio of dose conversion coefficients calculated for 100% oral breathing relative to 100% nasal breathing is 1.44 ± 0.02 (standard deviation [SD] between subjects) for exposure in the living room

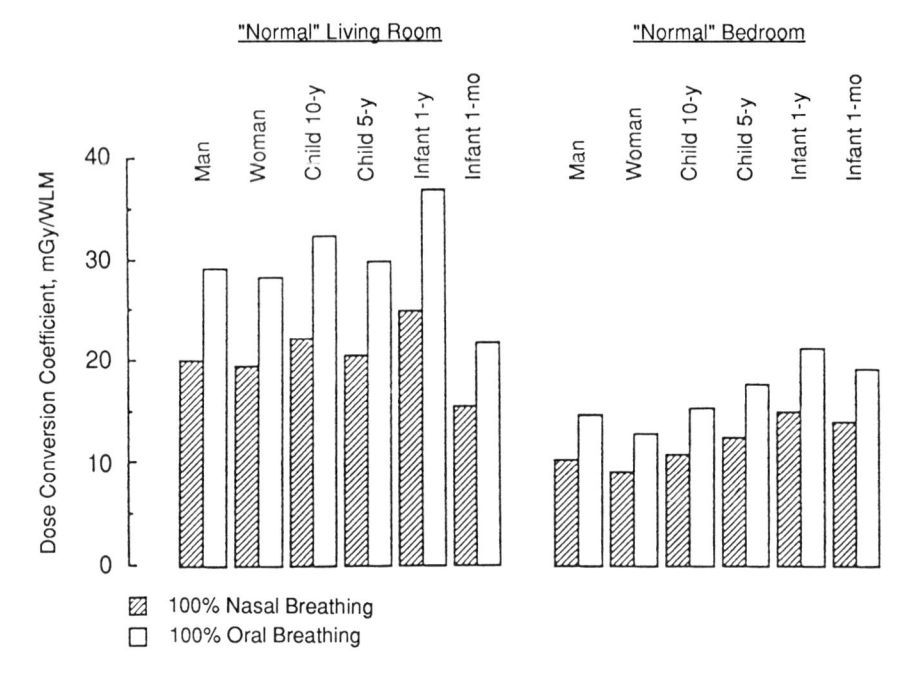

44   COMPARATIVE DOSIMETRY OF RADON IN MINES AND HOMES

FIGURE 3-7 Influence of subject age and gender on the bronchial dose conversion coefficient calculated for exposure to radon progeny under "normal" conditions at home in the living room and bedroom. Values are compared for subjects breathing entirely through the nose or mouth.

and 1.40 ± 0.01 for exposure in the bedroom. During activities not demanding high work levels, the majority of people breathe through the nose unless they are talking (or crying in the case of an infant) or suffering from nasal congestion (see Chapter 7).

It is seen from Figure 3-7 that the exposure-dose conversion coefficient is generally higher for children than for adults, but only slightly so. The corresponding values of the K factor (relative to exposures of underground miners) for each subject are summarized in the concluding section of this chapter. The K factors are given in Tables 3-4 and 3-5, for normal subjects and for nose and mouth breathers, respectively.

VARIABILITY OF THE DOSE CONVERSION COEFFICIENT IN HOMES

The dose conversion coefficients shown in Figure 3-7 apply to exposure conditions in living rooms and bedrooms that are assumed to represent normal (or typical) situations (these were defined in Table 3-1). The effect of more extreme aerosol conditions is examined in Figure 3-8. In Figure 3-8 the dose

TABLE 3-4 Summary of K Factors Calculated for Normal Healthy Subjects[a]

Subject	Target Region	Target Cell	Radon Progeny Solubility	K Factor ± SD[b]
Man	Bronchi	Secretory	Insoluble	0.81 ± 0.25
			Part-soluble	0.72 ± 0.22
		Basal	Insoluble	0.77 ± 0.25
			Part-soluble	0.60 ± 0.21
Man	Lobar/segmental Bronchi	Secretory	Insoluble	0.78 ± 0.24
			Part-soluble	0.66 ± 0.20
		Basal	Insoluble	0.75 ± 0.24
			Part-soluble	0.59 ± 0.19
Man	Bronchioles	Secretory	Mean	0.99 ± 0.48
Man	Bronchi	Secretory + basal	Mean	0.73 ± 0.23
Man	Bronchi	Secretory	Mean	0.76 ± 0.23
	"	Basal	"	0.69 ± 0.22
Woman	"	Secretory	"	0.72 ± 0.26
	"	Basal	"	0.62 ± 0.24
Child, 10 yr	"	Secretory	"	0.83 ± 0.28
	"	Basal	"	0.72 ± 0.26
Child, 5 yr	"	Secretory	"	0.83 ± 0.23
	"	Basal	"	0.72 ± 0.22
Infant, 1 yr	"	Secretory	"	1.00 ± 0.29
	"	Basal	"	0.87 ± 0.28
Infant, 1 mo	"	Secretory	"	0.74 ± 0.21
	"	Basal	"	0.64 ± 0.20

[a]Use of the nasal deposition efficiency reported by Cheng et al. (1989), and also on a revised estimate of 0.15 μm for the characteristic aerosol size in the underground haulage drifts (see Table 3-1), would yield smaller K factors. The mean values of the K factor calculated for bronchial target cells become: Adult male, 0.58 vis à vis 0.73; adult female, 0.55 vis à vis 0.67; child age 10 yr and age 5 yr, 0.64 vis à vis 0.78; infant age 1 yr, 0.78 vis à vis 0.94; infant age 1 mo, 0.56 vis à vis 0.69.
[b]±SD refers to the standard deviation of the calculated K factor for the five home environments and three mine environments shown in Figure 3.1.

conversion coefficients calculated for each subject exposed to radon progeny in a living room in the presence of an active cigarette smoker (where the unattached fraction of potential alpha energy, f_p, is assumed to be only 1%, and the AMD of the attached aerosol is 0.25 μm) are compared with the values shown in Figure 3-7 for nose breathers exposed to the normal living room atmosphere (where f_p is 8% and the aerosol AMD is 0.15 μm). In this case, when cigarettes are being smoked in a room, the exposure-dose conversion coefficient is, on average only 47% of the normal value (±1% standard deviation between subjects). Figure 3-8 also shows the effect of the higher unattached fraction (16%) expected in a bedroom when the rate of exchange with outdoor air is low. The exposure-dose conversion coefficient is then calculated to be 51% higher (±5% standard deviation) than the values applicable to the normal bedroom atmosphere.

TABLE 3-5 Summary of the Effect on Calculated K Factors of Nasal or Oral Breathing Habit

Subject	Target Cell	Breathing Habit of Reference Miner Nasal	Oral
K Factor for Nasal Breathers in the Home			
Man	Secretory	0.76	0.56
	Basal	0.66	0.51
Woman	Secretory	0.72	0.53
	Basal	0.62	0.48
Child, 10 yr	Secretory	0.83	0.61
	Basal	0.72	0.55
Child, 5 yr	Secretory	0.83	0.61
	Basal	0.72	0.55
Infant, 1 yr	Secretory	1.00	0.73
	Basal	0.87	0.66
Infant, 1 mo	Secretory	0.74	0.54
	Basal	0.64	0.49
K Factor for Oral Breathers in the Home			
Man	Secretory	0.88	0.64
	Basal	0.81	0.59
Woman	Secretory	0.84	0.61
	Basal	0.76	0.55
Child, 10 yr	Secretory	0.97	0.71
	Basal	0.88	0.64
Child, 5 yr	Secretory	0.95	0.70
	Basal	0.87	0.63
Infant, 1 yr	Secretory	1.16	0.85
	Basal	1.06	0.77
Infant, 1 mo	Secretory	0.82	0.60
	Basal	0.75	0.55

EFFECTS OF AIRWAY DISEASE ON DOSE

The results of modeling the effects of airway diseases on the exposure-dose conversion coefficients calculated for underground miners are shown in Figure 3-9. Three distinct disease conditions are considered:

Bronchitis, in which the epithelium is assumed to be normal, but the overlying mucus is assumed to have an abnormal thickness of 30 μm and the mucous clearance rates in each airway generation are assumed to be one-half the normal values. In this case, secretory cells are assumed to be the principal targets.

Hyperplasia of bronchial epithelium, in which the bronchial epithelium is assumed to be twice the normal thickness (at 100 μm) and is devoid of secretory cells. Basal cell nuclei, which are assumed to be the sensitive targets, are located in an abnormally thick layer (30-μm thick) at a depth of 70 μm. It is assumed that the epithelial surface is devoid of cilia and is lined only by a

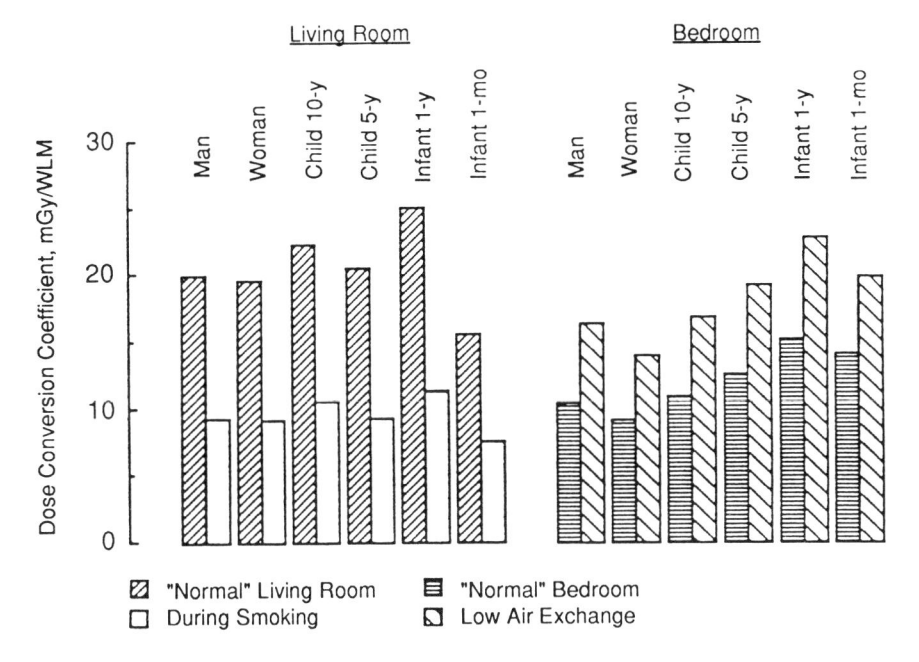

FIGURE 3-8 Influence of aerosol conditions in the living room and bedroom on the dose conversion coefficients calculated for various subjects exposed to radon progeny.

thin layer of fluid (3-μm thick). In this case, the committee assumed that there is no movement of deposited radon progeny toward the throat.

Local regeneration of bronchial epithelium in response to epithelial injury, in which case the epithelium is assumed to be one-half the normal thickness, such that secretory cell nuclei occur in a 15-μm-thick layer at a 5-μm depth and basal cell nuclei occur in an 8-μm-thick layer at a 17-μm depth. The epithelium is again assumed to be devoid of cilia and to be covered only by a 3-μm layer of fluid for which clearance is ineffective. The principal targets are assumed to be the nuclei of secretory cells.

Comparison of the conversion coefficients calculated for miners under these assumed conditions of disease with values calculated for healthy subjects yields the somewhat complex effects shown in Figure 3-9. The dose to secretory cell nuclei in the bronchial epithelium of a bronchitic miner is estimated to be between 40 and 50% of that for a healthy subject. The ratio of doses depends to this rather limited extent on the conditions of exposure and the assumed solubility of radon progeny. In the case of epithelial hyperplasia, however, the deep-lying basal cell targets receive no dose if the deposited radon progeny remain in the thin layer of fluid that is assumed to cover the epithelium. Alternatively, if the progeny are assumed to be partially soluble and 30% of

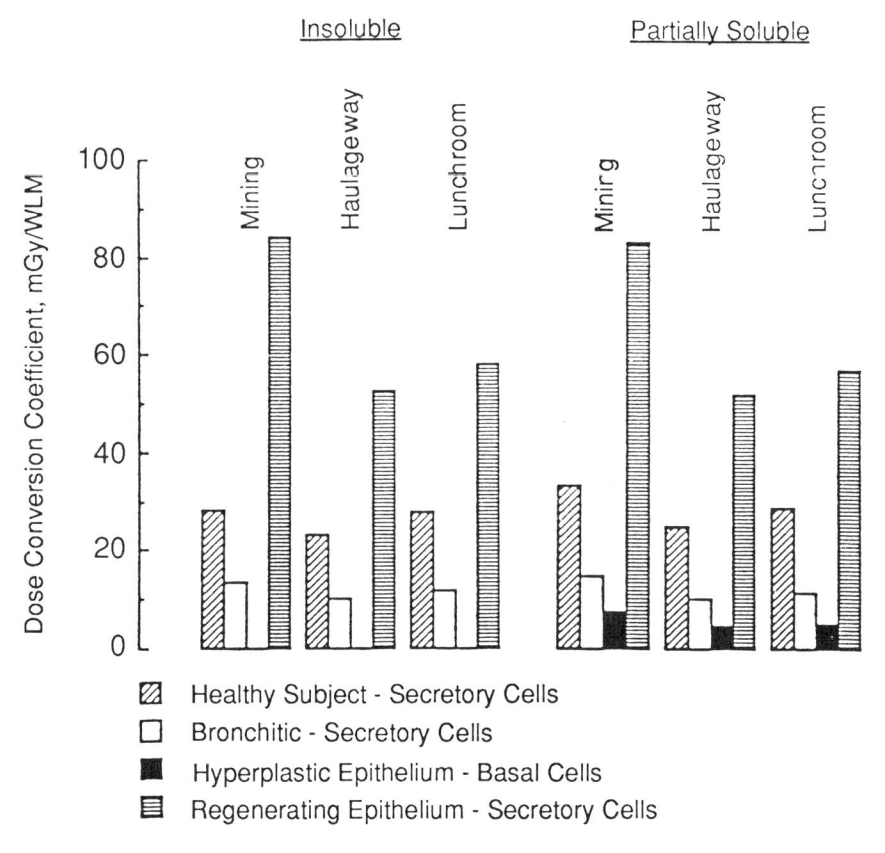

FIGURE 3-9 Modeled effects of bronchial airway disease in miners on the calculated dose conversion coefficient for exposure to radon progeny. Values calculated for subjects with bronchitis (secretory cells are the assumed targets), or for those with hyperplasia of bronchial epithelium (where the targets are assumed to be basal cells) or regenerating epithelium (secretory cell targets), are compared with values calculated for miners with healthy bronchial epithelium. Dose conversion coefficients calculated on the assumption that radon progeny are insoluble (i.e., that they remain in mucus) are shown separately from the corresponding values calculated if part of the progeny activity is assumed to be taken up by the bronchial epithelium.

the deposited activity is taken up by the epithelial tissue, the dose received by the basal cell nuclei at the site of hyperplasia is calculated to be approximately 20% of the average dose received by secretory cell nuclei in a healthy subject. In the case of local areas of regenerating epithelium, it is estimated that the target secretory cell nuclei receive between two- and threefold higher doses than the average dose for a healthy subject. The actual ratio of doses within this range again depends on the exposure conditions and the assumed solubility characteristics of the deposited radon progeny.

Similar dosimetric effects are calculated to arise for subjects with these

TABLE 3-6 Summary of K Factors Calculated for Bronchial Target Cells in Adult Males with Airway Disease Compared to Healthy and Diseased Miners

Subject/Disease	Target Cell	Radon-Progeny Solubility	K Factor ± SD[a]
Diseased subject in the home (cf. healthy miner)			
Bronchitis	Secretory	Insoluble	0.34 ± 0.10
		Part-soluble	0.28 ± 0.08
Hyperplasia	Basal	Insoluble	0
		Part-soluble	0.18 ± 0.06
Epithelial injury	Secretory	Insoluble	1.58 ± 0.38
(regeneration)		Part-soluble	1.38 ± 0.35
Healthy subject in the home (cf. diseased miner)			
Bronchitis	Secretory	Insoluble	1.75 ± 0.56
		Part-soluble	1.70 ± 0.53[b]
Hyperplasia	Basal	Insoluble	
		Part-soluble	1.84 ± 0.63
Epithelial injury	Secretory	Insoluble	0.30 ± 0.11
(regeneration)		Part-soluble	0.31 ± 0.11

[a]±SD refers to the standard deviation of the calculated K Factor for the five home environments and three mine environments shown in Figure 3-1.
[b]No dose to diseased miner.

airway diseases when they are exposed in the home. The overall effects of airway disease on the estimated K factor, for dosimetric extrapolation of risk, are summarized in Table 3-6.

COMPARATIVE DOSES FROM RADON AND THORON PROGENY

The dose conversion coefficients calculated for exposure to thoron progeny in a mine or a home are compared in Figure 3-10 with the values applicable for an adult male exposed to radon progeny. For thoron progeny, the unattached fraction of potential alpha energy, f_p, is assumed to be 0.1% during active mining, 1% in an underground haulageway, and normally, 2% indoors in the home. The AMD of the attached thoron progeny aerosol is assumed to be 0.25 μm in all cases in ambient air and 0.5 μm within the respiratory tract.

It is found that the dose from exposure to thoron progeny is determined principally by the behavior of lead-212, which has a relatively long radioactive half-life (10.6 h). The assumed solubility characteristics and clearance behavior are then found to have a substantially greater effect on calculated doses than they do on calculated doses for radon progeny. Dose conversion coefficients calculated separately on the alternative assumptions that thoron progeny are insoluble (i.e., remain in mucus) or are partially soluble and partially taken up

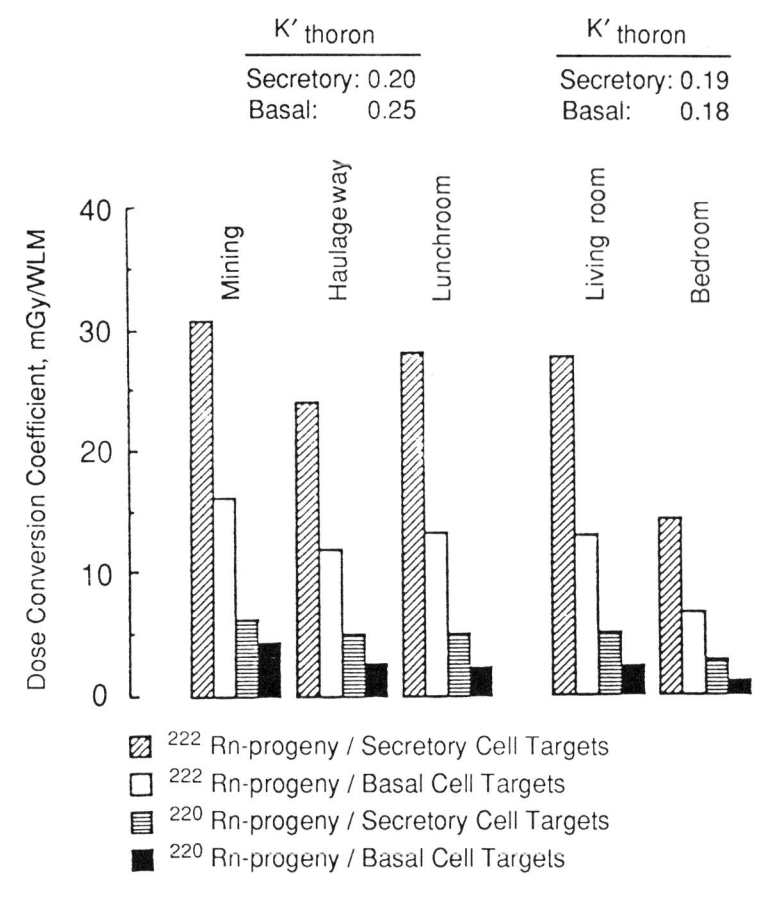

50 COMPARATIVE DOSIMETRY OF RADON IN MINES AND HOMES

K'_{thoron}
Secretory: 0.20
Basal: 0.25

K'_{thoron}
Secretory: 0.19
Basal: 0.18

[Bar chart showing Dose Conversion Coefficient, mGy/WLM for Mining, Haulageway, Lunchroom, Living room, Bedroom]

▨ ^{222} Rn-progeny / Secretory Cell Targets
☐ ^{222} Rn-progeny / Basal Cell Targets
☰ ^{220} Rn-progeny / Secretory Cell Targets
■ ^{220} Rn-progeny / Basal Cell Targets

FIGURE 3-10 Comparison of bronchial dose conversion coefficients calculated for exposure of a man to radon progeny in a mine or at home with those calculated for exposure to thoron (^{220}Rn) progeny.

by epithelial tissue are given in Chapter 9. The exposure-dose conversion coefficients shown in Figure 3-10 were obtained by averaging the values calculated for both types of assumed clearance behavior.

It is seen from the data in Figure 3-10 that, for unit exposure to thoron progeny in a mine or in the home, secretory cell nuclei are expected to receive only about 20% of the dose that they receive from the same exposure to radon progeny. Basal cell nuclei are estimated to receive a somewhat higher dose from unit exposure to thoron progeny in a mine than they do in the home (25% of the dose from radon progeny in a mine compared with 18% in the home). The dosimetric risk extrapolation factors K'_{thoron} for exposure to thoron progeny in the home relative to radon-222 in mines, Figure 3-10 were obtained by substituting these relative doses in the following equation:

$$K'_{\text{thoron}} = K_{\text{radon}} \times \frac{[(\text{Dose})_{\text{home}}/(\text{WLM})_{\text{home}}]\text{thoron}}{(\text{Dose})_{\text{home}}/(\text{WLM})_{\text{home}}]_{\text{radon}}}. \qquad (3\text{-}3)$$

SUMMARY OF DOSIMETRIC RISK EXTRAPOLATION FACTORS

Dosimetric risk factors (K factors) for extrapolating the observed risk of radon progeny exposure in mines to domestic settings that were derived in this chapter for various reference target tissues and different subjects are summarized in Table 3-4. K factors derived with respect to miners and subjects in the home who breath habitually through both the mouth and nose are summarized in Table 3-5. Finally, the K factors derived with respect to subjects with airway disease are summarized in Table 3-6.

REFERENCES

Cheng, Y. S., D. L. Swift, Y. F. Su, and H. C. Yeh. 1989. Deposition of radon progeny in human head airways. Pp. 29-30 in Inhalation Toxicology Research Institute Annual Report 1988-89. LMF-126, Albuquerque, N.M.: Lovelace Biomedical and Environmental Research Institute.

Cohen, B. S., R. G. Sussman, and M. Lippmann. 1990. Ultrafine particle deposition in a human tracheobronchial cast. Aerosol Sci. Technol. 12:1082-1091.

George, A. C., and A. J. Breslin. 1969. Deposition of radon daughters in humans exposed to uranium mine atmospheres. Health Phys. 17:115-124.

National Research Council (NRC). 1988. Health Risks of Radon and Other Internally Deposited Alpha-Emitters. BEIR IV. Committee on the Biological Effects of Ionizing Radiation. Washington, D.C.: National Academy Press.

National Research Council (NRC). 1990. Committee on the Biological Effects of Ionizing Radiations. Health Effects of Exposure to Low Levels of Ionizing Radiation. BEIR V. Washington, D.C.: National Academy of Press.

Sinclair, D. R., R. J. Countess, and G. S. Hoopes. 1974. The effect of relative humidity on the size of atmospheric aerosol particles. Atmos. Environ. 8:1111-1117.

Strong, J. C., and D. L. Swift. 1990. Deposition of 'unattached' radon daughters in models of the human nasal and oral airways. Paper presented at the 29th Hanford Life Sciences Symposium, Indoor Radon and Lung Cancer: Reality or Myth? October 16-19, 1990, Richland, Wash.

4

Other Considerations

BIOLOGICAL FACTORS

Using the lung cancer risk due to radon progeny that is observed among miners to estimate this risk among members of the general population requires the consideration of two different kinds of variables:

1. Physical and biological factors that affect the dose to the bronchial mucosa where lung cancers occur. A comparison of the effect of these factors on the bronchial dose in miners and in the general population is the primary subject of this report.

2. The tumorigenic responsiveness of miners' lung tissues to given alpha particle doses from radon decay products as compared to tumorigenic responsiveness in the more heterogeneous general population.

The second of these components is discussed in the BEIR IV report (NRC, 1988) and this chapter as it is a major source of uncertainty in extrapolating risks due to radon and its decay products.

Factors that need to be considered here include age, gender, and concurrent exposure to domestic sources of radon and airborne pollutants such as cigarette smoke. This chapter is not meant to be comprehensive in scope or in its discussion of any of these issues. Rather, the intent is to provide a perspective on the potential importance and the extent of the evidence on some of these factors.

GENDER

Because the data on lung cancer in underground miners involves observations in males only (Samet, 1989), the question arises as to whether females are likely to exhibit a significantly different susceptibility to lung cancer induction by radon. The effect of gender on lung cancer risk has not been comprehensively studied, but there is no strong evidence that females have lung cancer susceptibilities different from those of males (NRC, 1988). In the United States, females have had lower lung cancer incidence and mortality than males, the lower rates reflecting differences in smoking by gender across the twentieth century (U.S. Department of Health and Human Services, 1989). Susceptibility to the induction of lung cancer by ionizing radiation in atom bomb survivors in Japan appears to be about the same for males and females in terms of absolute risk, although the relative risk is higher in females because they have a lower incidence of lung cancer (NRC, 1990; Kopecky et al., 1988). Comparable responsiveness to radon progeny in males and females was assumed in the BEIR IV (NRC, 1988) and ICRP 50 reports (International Commission on Radiation Protection [ICRP], 1987).

AGE, GROWTH, AND INJURY IN RELATION TO CELL PROLIFERATION

The issue of greater susceptibility to lung cancer induction by radon progeny in infants and children requires consideration; growing individuals have relatively high rates of cell proliferation and growth has been implicated as a factor in carcinogenesis. Cell proliferation can also be an expression of the regenerative repair of tissues that have sustained injury from chemical toxicants or wounding. Tissue damage can also promote tumorigenesis, as discussed below. Increased cell proliferation is a characteristic manifestation of the action of chemical and physical agents that are promoters of tumorigenicity (Slaga et al., 1974) and may play a role in the conversion of normal cells to neoplastic transformants and in the progression of such transformants to increasing degrees of malignancy. Heightened cell proliferation may also reflect a state of tissue injury that may facilitate the outgrowth of transformed cells into tumors, but the proliferation per se may not be the proximate cause of the enhanced clonal expansion of transformed cells.

The importance of tissue factors on tumor formation has been shown in experiments that demonstrate the presence of a large number of transformed cells in the carcinogen-treated rat tracheal epithelium under conditions that produce few tumors (Terzaghi, 1979); also, neoplastic transformants have been shown to be nontumorigenic when used, together with normal cells, to repopulate the denuded tracheal epithelium (Terzaghi-Howe, 1987). It is possible that tissue damage may be relevant to the extrapolation of lung cancer responses to radon in miners and the general population since miners and cigarette smokers are exposed to agents which cause injury of the bronchial mucosa. Such damage is probably infrequent among nonsmokers in the general population. As discussed below, there is evidence at the cellular level that while cell proliferation has a marked effect on the neoplastic transformation caused by low-LET radiation, it is less important for high-LET radiation such as the alpha particles emitted by radon progeny in the lung. However, smoking appears to be an important risk factor even if proliferation is not (see below).

AGE

While there is little evidence of any age effect for tumorigenesis due to radon progeny, the relevant experimental and epidemiological data supporting the absence of an age effect are not extensive. Children under the age of 13 who worked in a Chinese tin mine, where they were exposed to radon and arsenic, did not show a higher incidence of lung cancer than those who began work as adults (Lubin et al., 1990), but the interpretation of this finding is limited by the number of subjects in the study. The BEIR IV committee (NRC, 1988) found there was no improvement in fit of the data when a term for age-at-exposure was added to their regression model for respiratory cancer among underground miners. By contrast, bone tumor induction in humans by radium-224 shows only a marginal increase in susceptibility at early ages (Mays et al., 1978). The induction of liver cancer in humans by alpha-emitting (high-LET) thorotrast does not show an age dependence (Kaick et al., 1989). The experimental studies of the lung tumor response of dogs exposed to the high-LET alpha radiation of plutonium (^{239}PuO$_2$) at an early age (3 months) was not significantly different from the response of older dogs (18 months), basing the comparison on the cumulative radiation dose (Guilmette, 1990).

CELL PROLIFERATION

There is in vitro evidence that cell proliferation affects the tumorigenic response to low-LET but not high-LET radiations. Cells that are held in a confluent state in tissue culture, with very little cell proliferation, rapidly repair tumorigenic damage caused by low-LET radiation (Borek and Sachs, 1968); in contrast, alpha-particle irradiation of confluent cells in culture shows no evidence of such repair (Robertson et al., 1983). Reversal of the tumorigenic effect of low-LET radiation, ranging from 0.1 to 10 keV/μm, in the rat skin with dose fractionation has a half-life of several hours and eventually reaches about 90% completion; the time pattern is about the same as that in tissue culture and may correspond to the repair of chromosomal breaks (Burns and Albert, 1986; Burns et al., 1979). In contrast, skin tumorigenesis in the rat shows no recovery with fractionated irradiations when high-LET radiation is used (Burns and Albert, 1987). Thus it can be inferred that, with low-LET radiation, for which rapid and almost complete repair of tumorigenic injury occurs, the rate of cell transformation depends on the proliferative rate since the cells that become neoplastically transformed are those that were close to or in the S period of the cell cycle at the time of irradiation. A similar dependence has been shown with chemical carcinogens in the liver (Kaufman and Cordiero-Stone, 1990). Cells of the bronchial mucosa, where most lung cancers arise, have one of the lowest proliferative rates in the adult body (Kaufman, 1980); thus a very small fraction of the cells are candidates for neoplastic transformation by low-LET radiation, unless the proliferation rate is increased by growth. On the other hand, damage due to high-LET alpha radiation, such as that from radon, is less repairable so that the effect of cell proliferation would be less. These arguments suggest that there may be little or no age dependence on the susceptibility of the lung to tumor induction by radon.

DOSE RATE, FRACTIONATION, AND DURATION OF EXPOSURE

Fractionation of low-LET radiation exposure reduces the carcinogenic action because time is available for repair. This reduction of carcinogenic action with fractionation of the radiation does not occur with high-LET radiation, as that from radon progeny, because repair does not appear to take place.

The Armitage-Doll multistage model (Doll, 1971), currently used for carcinogen risk assessment, assumes the same mechanism of action of specific carcinogens as for the determinants of background cancer. Thus, as the dose rate diminishes, the time of carcinogen-induced tumor formation is assumed to gradually approach that of background tumor formation with a resultant linear, nonthreshold relationship between dose rate and tumor incidence (Crump and Howe, 1984). However, this assumption is not based on strong evidence, and the temporal and tumor incidence patterns have not been characterized at very low dose rates. The BEIR IV committee's examination of the miners experience indicates that the tumorigenic effect of radon exposure fades with time after the exposure is discontinued (NRC, 1988). Nevertheless, uncertainty remains concerning the effects of dose rate and duration of exposure at very low levels of exposure.

TISSUE DAMAGE

Tissue wounding by mechanical injury is clearly a strong promoting agent for tumorigenesis by carcinogens of many types. In the case of lung cancer induction in rodents by the intratracheal instillation of benzo[a]pyrene, cancers do not arise in the trachea unless it is traumatized by the endotracheal injection needle, which causes wounding with increased cell proliferation (Keenan et al., 1989). The induction of bronchial tumors in the hamster lung by instillation of radioactive polonium is markedly enhanced by subsequent intratracheal injections of saline, which also cause waves of cell proliferation (Little et al., 1978). Urethane administered to adult mice does not induce liver tumors unless there is a stimulation of cell replication in this proliferatively static organ by partial hepatectomy (Chernozemski and Warwick, 1970). Wounding of the mouse skin after initiation by a small dose of chemical carcinogen is an effective cancer-promoting factor (Argyris, 1980).

Tissue damage induced by chemicals can also be a tumor promoter. After exposure to neutrons, the mouse liver responds to the proliferative stimulus of a damaging dose of carbon tetrachloride with enhanced tumor induction (Cole and Nowell, 1964). Tumor induction in the nasal mucosa of the rat by the inhalation of formaldehyde does not occur at doses that do not increase the normally very low level of cell proliferation in that tissue (Monticello and Morgan, 1990). Lung tumor induction by plutonium-238 appears to be much enhanced in regions of the lung where the radioactive particles cluster and produce local tissue damage (Sanders et al., 1988).

Tissue wounding or damage with its associated regeneration appears to be an exaggerated form of tissue growth, because like growth, regenerative proliferation is associated with retention of the multiplying cells, and thus, clonal expansion of neoplastically transformed cells is favored. Tissue damage also frequently has an inflammatory reaction that may constitute an enhancing factor for tumorigenesis. Inhalation exposure of dusts, usually considered inert such as titanium and activated charcoal, produce hyperplasia or irritation and have been implicated in lung cancer (Heinrich, 1990; Lee et al., 1986).

CIGARETTE SMOKING

Cigarette smoking, the cause of most cases of lung cancer in the general population, has been shown to interact with exposure to radon progeny in a synergistic fashion. The BEIR IV report (NRC, 1988) reviewed potential mechanisms underlying this interaction as well as the relevant epidemiological evidence from studies of miners. Cigarette smoking has numerous effects on the lung that could contribute to the synergism between smoking and radon progeny. Chronic inhalation of cigarette smoke, even without other irritants, causes tissue damage in the respiratory mucosa, changing the relatively nonproliferative secretory and ciliary pattern to squamous metaplasia, with an increased rate of cell proliferation (Wehner, 1983). Patches of squamous metaplasia may not have a moving protective sheath of mucus to clear deposited radon progeny. Cells in such areas might not only be proliferating faster than the epithelial cells in the normal mucosa, but the nuclei might also be closer to the deposited progeny; an increased dose of alpha-particle energy could result. Increased central deposition of inhaled particles in smokers might also increase doses to the central airways where lung cancers arise.

In extrapolating data from the studies of miners to the general population, consideration must be given separately to the risks of lung cancer for smokers and for never smokers. The risk estimates from the studies of miners largely reflect the occurrence of lung cancer in smokers; thus, in extrapolating risks from miners to the general population, an assumption needs to be made concerning the modification of the effect of radon by smoking. The BEIR IV committee assumed a multiplicative interaction between these two factors in producing lung cancer. The committee based this decision on the weight of the evidence from the literature and on its own analysis of data from two studies of uranium miners. Data from the study of atomic bomb survivors in Japan may not be relevant to the interaction between radon and smoking.

REFERENCES

Argyris, T. S. 1980. Tumor promotion by abrasion induced epidermal hyperplasia in the skin of mice. J. Invest. Dermatol. 75:360-362.
Borek, C., and L. Sachs. 1968. The number of cell generations required to fix the transformed state in X-ray induced transformation. Proc. Natl. Acad. Sci. 53:83-86.
Burns, F. J., and R. E. Albert. 1986. Pp. 199-214 in Rodent Carcinogenesis in Rat Skin in Radiation Carcinogenesis, A. E. Upton, R. E. Albert, F. J. Burns, and R.E. Shore, eds. New York: Elsevier.
Burns, F. J., and R. E. Albert. 1987. Dose-response for radiation-induced cancer in rat skin. Pp. 51-70 in Radiation Carcinogenesis and DNA Alterations, F. J. Burns, A. C. Upton, and G. Silini, eds. NATO ASI Series, Series A: Life Sciences, Vol. 124.
Burns, F. J., R. E. Albert, M. Vanderlaan, and P. Strickland. 1979. The dose response curve for tumor induction with single and split doses of 10 Mev protons. Radiat. Res. 62:598-599.
Chernozemski, I. M., and G. P. Warwick. 1970. Liver regeneration and induction of hepatomas in B6AF1 mice by urethane. Cancer Res. 30:2685-2690.
Cole, L. J., and P. C. Nowell. 1964. Accelerated induction of hepatomas in fast neutron-irradiated mice injected with carbon tetrachloride. Ann. N.Y. Acad. Sci. 114:259-267.
Crump, K., and R. B. Howe. 1984. The multistage model with a time dependent dose pattern: Applications to carcinogen risk assessment. Risk Anal. 4:163-176.
Doll, R. 1971. The age distribution of cancer: Implications for models of carcinogenesis. J. R. Stat. Soc. 134:133-155.
Guilmette, R. 1990. Personal communication, Lovelace Inhalation Toxicology Research Institute.
Heinrich, U. 1990. Tumorigenic effects of carbon black in the lung. Presented at an U.S. EPA Workshop on Diesel Emissions, Chapel Hill, N.C., July 18-19.
International Commission on Radiation Protection (ICRP). 1987. Lung Cancer Risk from Indoor Exposures to Radon Daughters. ICRP Publ. No. 50. Oxford: Pergamon Press.
Kaick, G. van, H. Wesch, H. Luhrs, D. Liebermann, A. Kaul, and H. Muth. 1989. The German Thorotrast Study—report on 20 years follow-up. Pp. 98-104 in Brit. Inst. Radiol. Report 21.

Kaufman, D. G., and M. Cordiero-Stone. 1990. Variations in suceptibility to initiation of carcinogenesis during the cell cycle. Pp. 1-27 in Transformation of Human Diploid Fibroblasts: Molecular and Genetic Mechanisms, G. E. Milo, ed. Boca Raton, Fla.: CRC Press.

Kauffman, S. L. 1980. Cell proliferation in the mammalian lung. Int. Rev. Exp. Pathol. 22:131-191.

Keenan, K. P., U. Saffiotti, S. F. Stinson, C. W. Riggs, and E. M. McDowell. 1989. Multifactorial hamster respiratory carcinogenesis with interdependent effects of cannula-induced mucosal wounding, saline, ferric oxide, benzo(a)pyrene and N-methyl-N-nitrosourea. Cancer Res. 49:1528-1540.

Kopecky, K. J., E. Nakashima, T. Yamamoto, and H. Kato. 1988. Pp. 1-31 in Lung Cancer, Radiation, and Smoking Among A-Bomb Survivors, Hiroshima and Nagasaki. Radiation Effects Research Foundation.

Lee, K. P., N. W. Henry, H. J. Trochimowitz, and C. F. Reinhardt. 1986. Pulmonary response to impaired lung clearance in rats following excessive TiO_2 dust deposition. Environ. Res. 41:144-167.

Little, J. B., R. B. McGandy, and A. R. Kennedy. 1978. Interactions between polonium-210 alpha radiation, benzo(a)pyrene, and 0.9% NaCl solution instillations in the induction of experimental lung cancer. Cancer Res. 38:1929-1935.

Lubin, J. H., Y. Qiao, P. R. Taylor, S-X Yao, A. Schatzkin, B.-L. Mao, J.-Y. Rao, X.-Z. Xuan, and J.-Y. Li. 1990. Quantitative evaluation of the radon and lung cancer association in a case control study of Chinese tin miners. Cancer Res. 50:174-180.

Mays, C. W., H. Spiess, and A. Gerspach. 1978. Skeletal effects following Ra224 injections into humans. Health Phys. 35:83-90.

Monticello, T. M., and K. T. Morgan. 1990. Correlation of cell proliferation and inflammation with nasal tumors in F344 rats following chronic formaldehyde exposure. P. 138 in American Association of Cancer Research Proceedings No. 826.

National Research Council (NRC). 1988. Health Risks of Radon and Other Internally Deposited Alpha-Emitters. BEIR IV. Washington, D.C.: National Academy Press.

National Research Council (NRC). 1990. Health Effects of Exposure to Low Levels of Ionizing Radiation. BEIR V. Washington, D.C.: National Academy Press.

Peraino, C., E. F. Staffeldt, and V. A. Ludeman. 1981. Early appearance of histochemically altered hepatocyte foci and liver tumors in female rats treated with carcinogens one day after birth. Carcinogenesis 2:463-475.

Robertson, B., A. Koehler, J. George, and J. B. Little. 1983. Oveogenic transformation of mouse Balb 3T3 cells by plutonium-238 alpha particles. Radiat. Res. 96:261-274.

Samet, J. S. 1989. Radon and lung cancer. J. Natl. Cancer Inst. 81:745-757.

Sanders, C. L., K. E. McDonald, and K. E. Lauhala. 1988. Promotion of pulmonary carcinogenesis by plutonium article aggregation following inhalation of $^{239}PuO_2$. Radiat. Res. 116:393-405.

Slaga, T. J., J. D. Scribner, S. Thompson, and A. Viaje. 1974. Epidermal cell proliferation and promoting ability of phorbol esters. J. Natl. Cancer Inst. 52:1611.

Terzaghi, M. 1979. Dynamics of neoplastic development in carcinogen exposed tracheal mucosa. Cancer Res. 39:4003-4010.

Terzaghi-Howe, M. 1987. Inhibition of carcinogen altered rat tracheal epithelial cell proliferation by normal epithelial cells in-vivo. Carcinogenesis 8:145-150.

U.S. Department of Health and Human Services. 1989. Reducing the Health Consequences of Smoking, 25 Years of Progress. A Report of the Surgeon General. Rockville, Md.: U.S. Department of Health and Human Services, Public Health Service, Centers for Disease Control, Center for Chronic Disease Prevention and Health Promotion, Office on Smoking and Health.

Wehner, A. P. 1983. Cigarette smoke-induced alterations in the respiratory tract of man and experimental animals. Pp. 1-42 in Comparative Respiratory Tract Carcinogenesis. Vol. II: Experimental Respiratory Tract Carcinogenesis, H. Reznik-Schuller, ed. Boca Raton, Fla.: CRC Press.

5

Dosimetry and Dosimetric Models for Inhaled Radon and Progeny

Dosimetry is defined as the measurement of the amount, type, rate, and distribution of radiation emitted from a source of ionizing radiation and the calculation of both spatial and temporal patterns of energy deposition in any material of interest as a result of ionizing radiation. Instruments and techniques exist to measure fields of penetrating radiation such as X rays or gamma rays that are external to the body, and provide means for directly quantifying the amount of energy deposited per unit mass of material (air, tissue, water). These dose measurements can then be related to a person present in the radiation field and the radiation dose that he or she would receive. In the case of internally deposited radionuclides, however, direct measurement of the energy absorbed from the ionizing radiation emitted by the decaying radionuclide is rarely, if ever, possible. Therefore, one must rely on dosimetric models to obtain estimates of the spatial and temporal patterns of energy deposition in tissues and organs of the body. In the simplest case, when the radionuclide is uniformly distributed throughout the volume of a tissue of homogeneous composition and when the size of the tissue is large compared with the range of the particulate emissions of the radionuclide, then the dose rate within the tissue is also uniform and calculation of absorbed dose can proceed without complication. However, if nonuniformities in the spatial and temporal distributions of radionuclide are coupled with heterogeneous tissue composition, then calculation of absorbed radiation dose becomes complex and uncertain. Such is the case with the dosimetry of inhaled radon and radon progeny in the respiratory tract.

The objective of this chapter is to provide both background and a historical perspective of the development of dosimetric models for radon and radon progeny and to provide a basis for more detailed descriptions of the scientific issues that relate to the different component parts of the present radon progeny dosimetric model. The physical dosimetry and microdosimetry of alpha particles in tissue are discussed in Appendix I of the BEIR IV report (NRC, 1988).

SOME PHYSICAL CHARACTERISTICS OF RADON AND RADON PROGENY

Three isotopes of radon occur naturally in the environment as a result of being part of the so-called uranium (^{238}U), thorium (^{232}Th), and actinium (^{235}U) decay series. These are ^{222}Rn, ^{220}Rn, and ^{219}Rn, respectively. Their importance as environmental sources of radioactivity stems from the fact that radon is a noble gas that can be carried in air or water streams far from its source of creation in soil and rock and can reach the indoor environment where people work and live. Radon's importance as an environmental source of radiation depends principally on the local concentrations of the parent radionuclide, the physical characteristics of the rocks and soil, and on its half-life. The last two factors determine the time available to migrate into air and water. The decay characteristics of the isotopes of the three decay series are summarized in Figures 5-1A through C. Of least importance is ^{219}Rn, which, because of its 3.96 s half-life, has limited capacity for migration in the environment. Moreover, its progenitor, ^{235}U, has a relatively low concentration in the environment. ^{220}Rn, with a 55.6 s half-life, is more readily available to the environment, and in geographic regions that have shallow deposits of thorium-rich soils, it can be a major contributor to the potential radiation dose to people. However, the majority of the concern for risks from radon exposure is due to the environmental presence of the isotope ^{222}Rn, which has the longest half-life of the radon isotopes, 3.824 days, and is ubiquitous because of the pervasive presence of its precursor radionuclides, ^{226}Ra and ^{238}U, in the earth's crust.

The apparent complexity of the ^{222}Rn decay scheme in Figure 1-1 can be simplified by neglecting the decays ^{218}Po→^{218}As and ^{214}Bi→^{210}Tl, both of which occur with very low probabilities ($\leq 0.02\%$), and by truncating the decay scheme at ^{210}Pb, which has a 22-yr half-life and is of little consequence as a respiratory hazard. Only three alpha-emitting radionuclides of consequence, then, remain, ^{222}Rn, ^{218}Po, and ^{214}Po. Table 5-1 provides additional physical data on ^{222}Rn and its progeny, including the historical designations for the decay products. Note that because the very short half-life for ^{214}Po (164 s), it is in virtual equilibrium with its parent ^{214}Bi under any conditions. Thus, decay chain calculations need only be done for the chain ^{218}Po → ^{214}Pb → ^{214}Bi.

MORPHOMETRIC MODELS OF THE RESPIRATORY TRACT

A morphometric model of the respiratory tract is a fundamental component

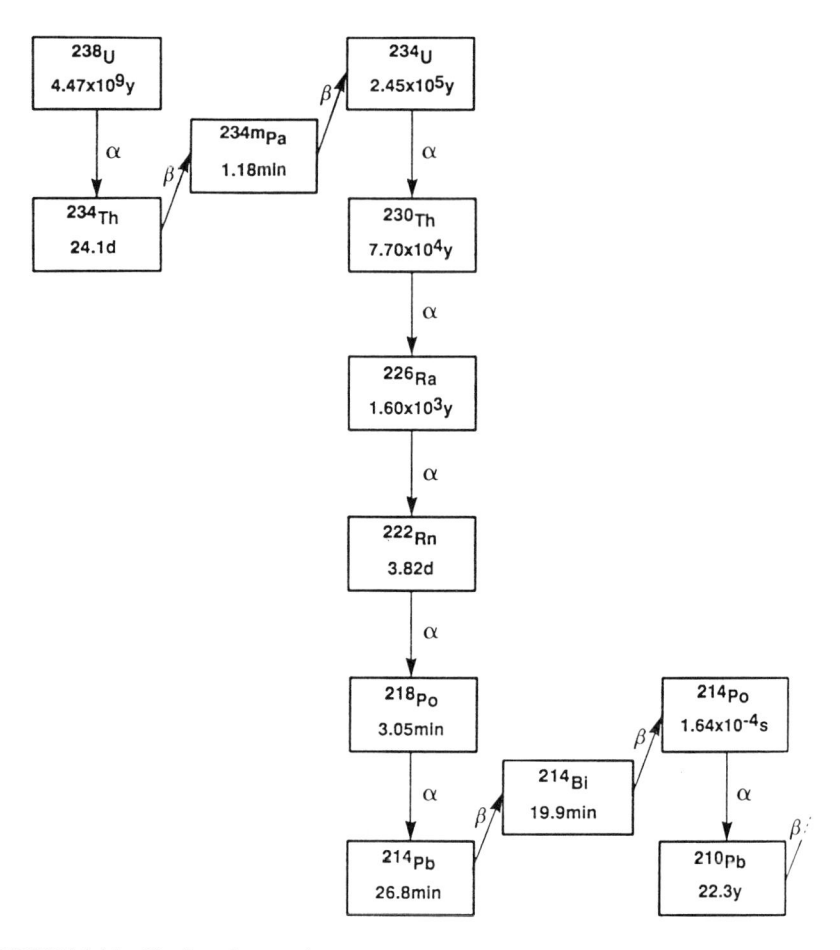

FIGURE 5-1A Uranium decay series.

of any dosimetric model intended for use in estimating alpha-radiation doses from radon and radon progeny, because of the heterogeneous distribution of deposited activity that occurs upon inhalation and the diverse structures, airway sizes, and geometries that are found at the different levels of the respiratory tract. Without an accurate anatomic description of the respiratory tract, one is limited in the ability to apply theoretical or empirical principles of airflow patterns and aerosol deposition at the level of resolution that is presumed to be necessary for properly modeling the radiation dose distribution from inhalation of radon and radon progeny, i.e., at the millimeter and submillimeter levels (see Chapter 9). Although morphometric models of the respiratory tract below the level of the larynx have existed for some time, there is currently no

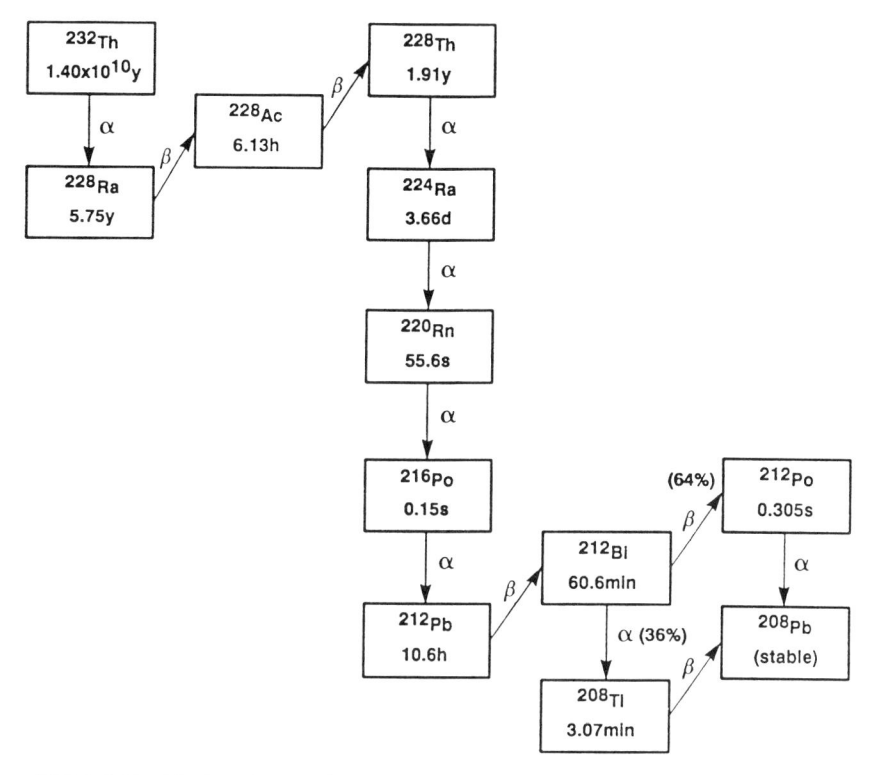

FIGURE 5-1B Thorium decay series.

single morphometric model considered to be fully adequate for describing the morphometry of the entire respiratory tract, which includes the nasal airways, oral cavity, nasopharynx and oropharynx, larynx, trachea, bronchi, bronchioles, and alveoli. Additionally, the roles of subject size and age on the dimensions of the various regions of the respiratory tract have only recently received attention by researchers.

HEAD AIRWAY MORPHOMETRY

For several reasons, morphometric models have not yet been developed for either the nasal or the oral airways. First nasal airways in humans, although much simpler geometrically than those in most other animal species, are complex anatomic structures that do not lend themselves to description in simple geometric terms, e.g., as tubes with circular or elliptical cross sections. Second, significant interindividual variabilities in the size and shape of nasal airways, particularly as a function of the size of the individual, have been recognized.

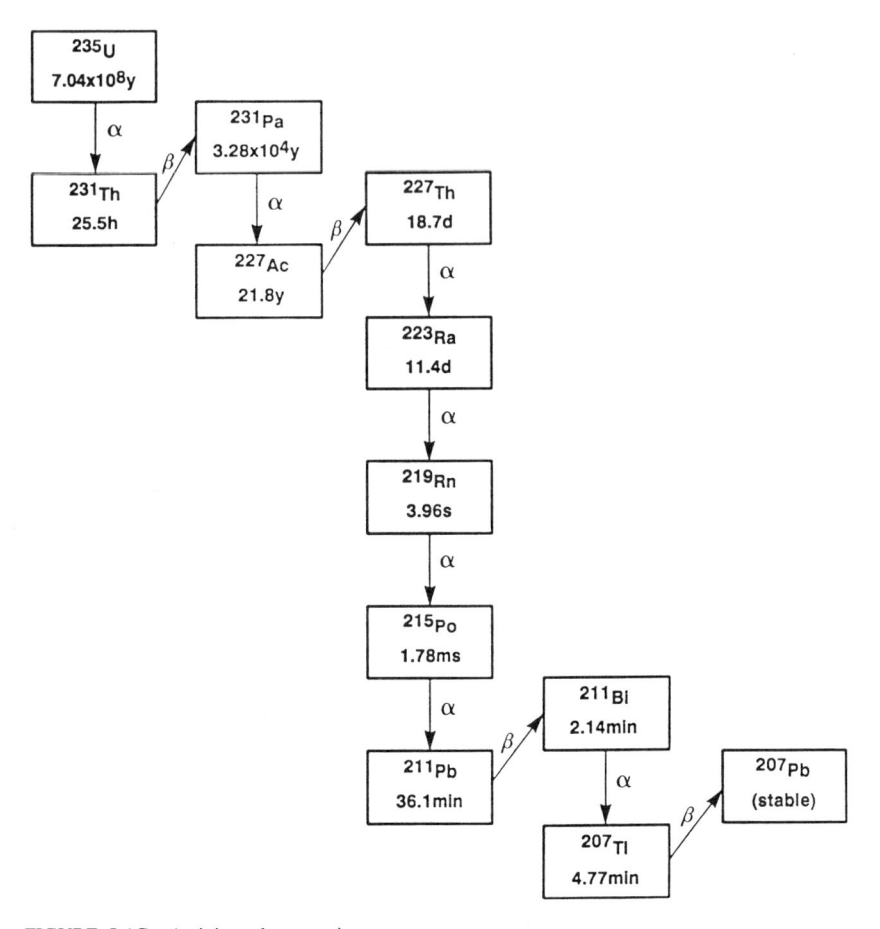

FIGURE 5-1C Actinium decay series.

Third, although the nasal airways have long been recognized as an important deposition site for large-sized aerosol particles (≥ 10 μm), their importance in filtering out smaller particles (<0.1 μm) has only recently been appreciated (Cheng et al., 1988).

The nasal airways consist of several distinct anatomic regions that differ in terms of their size, shape, physiological function, and types of epithelial cells lining the airways (Swift and Proctor, 1976). Air enters through elliptically shaped openings, the anterior nares, and passes into a conically shaped vestibular region that extends approximately 10 to 15 mm in the posterior direction. The cross-sectional area decreases posteriorly to a minimum value at the location of the nasal valve or ostium internum, which has been determined by Haight and Cole (1983) to be the location of maximum airway resistance. Proctor

TABLE 5-1 Some Physical Properties of ^{222}Rn and Its Short-Lived Decay Products

Element	Historical Symbol	Principal Radiation(s)	Decay Energies (MeV)	Half-Life	No. of Atoms Per μCi	Per Bq
^{226}Ra	Ra	α	4.8	1,620 yr	2.7×10^{15}	7.4×10^{10}
^{222}Rn	Rn	α	5.5	3.82 day	1.8×10^{10}	4.8×10^{5}
^{218}Po	RaA	α	6.0	3.05 min	9.77×10^{6}	2.6×10^{2}
^{214}Pb	RaB	β, γ	1.0 max	26.8 min	8.58×10^{7}	2.3×10^{3}
^{214}Bi	RaC	β, γ	3.3 max	19.7 min	6.31×10^{7}	1.7×10^{3}
^{214}Po	RaC	α	7.7	164 μs	8.8	2.4×10^{-4}

and Swift (1971) estimated that the average cross-sectional area at this point is probably between 20 and 40 mm² per side. Posterior to the nasal valve are the nasal fossae, which contain the inferior, medial, and superior turbinates attached to the lateral walls. The left and right nasal airway cavities are separated by a thin cartilaginous nasal septum, whose thickness and shape can vary significantly among individuals, but it is typically 3- to 7-mm thick in the anterior cartilaginous part and 2- to 3-mm thick in the posterior bony part (International Commission on Radiological Protection [ICRP], 1974). Because of the intrusion of the turbinates, the air passages of the nasal fossae have ribbon-like cross sections that are complexly folded within the turbinate structures. Because of this complexity, there is a high surface area-to-volume ratio in this region of the airways (ICRP, 1974). At the posterior end of the turbinate region, about 50 mm in length, the nasal septum terminates, and the two nasal passages combine (nasal choanae) and join the pharynx, which then bends 70° downward to the remainder of the respiratory tract. The length of the nasal passage from nostril to oropharynx is about 130 mm (Swift, 1981).

Limited morphometric information is available regarding the sizes of the different parts of the nasal airways. Additionally, most of the published measurements on nasal airway dimensions were obtained from cadaver studies. Postmortem measurements result in overestimation of the cross-sectional areas and airway volumes of the region containing the turbinates, because of the marked shrinkage of the nasal mucosa after death (Guilmette et al., 1989).

For calculational purposes, Landahl (1950) assumed the following schematic representation of the nasal airways: Each of the external nares had a cross-sectional area of 75 mm², with hairs of 100 μm in diameter occupying one-half of the projected area. The second region was the region containing the nasal valve, 2 cm behind the opening of the nares, and was assumed to have a rectangular cross section 1.2-cm high and 0.25-cm wide and to bend at an angle of 30°. The third region was also assumed to have a cross-sectional profile of a rectangular tube 3-cm tall, 0.2-cm wide, and 1-cm long, with a bend of 20°. The fourth region was subdivided into two elements. The first was the narrow and tortuous upper passage of 1 mm in width; the second was the more direct lower passage of 2 mm in width. The total height for both elements was 40 mm, with a length of 50 mm, so that the total effective surface area for both nasal passages was 80 cm². The appropriateness of the various simplifying assumptions used in this nasal airway model has not been evaluated to date.

Scott et al. (1978) also formulated a geometrical model for calculating aerosol particle deposition within the nasal airways. In this model, a symmetrical, bilateral model was divided into five contiguous parts that represented the anterior nares, the nasal valve, the expansion region just posterior to the nasal valve, the turbinate area, and the posterior bend. There are many similarities between this model and that of Landahl (1950). The major differences between them are the increased geometrical complexity of the turbinate areas, which is accomplished by using coupled rectangular shapes that simulate the meatal folds, the graded cross-sectional area in the region of the nasal valve, and the presence of the posterior nasopharyngeal bend. Again, as with the Landahl model, the adequacy of this particular model in representing nasal deposition has not been evaluated experimentally. However, the model, together with the accompanying theory of particle deposition in the nasal airways, did show reasonable agreement with the data available from studies of aerosol deposition in human nasal airways (Scott et al., 1978).

Except for the dosimetry model of Bailey (1984), in which the nasal airways were assumed to be two sets of parallel disks, representing the anterior and posterior portions of the airway, no other morphometric models of the nasal airways have been presented. To date, no morphometric models of the oral airways have been used either experimentally or for theoretical calculations. The reasons for this lack are not clear. It is clear, however, that oral breathing can be important, and even predominant, in cases of acute or chronic nasal airway obstruction and in cases of significant physical exertion, in which increased ventilation can only be accomplished by decreasing respiratory resistance, i.e., by compensatory oral breathing (see Chapter 7). Intuitively, it would appear to be difficult to construct a single morphometric model of the oral cavity that would be adequate for deposition and dosimetry purposes, as the cavity dimensions are likely to depend significantly on the size of the individual, as well as the position of the tongue and jaw. These latter positions depend on the ventilation rate requirements and the fractionation of breathing between the nasal and oral routes.

TRACHEOBRONCHIAL AIRWAY MORPHOMETRY

Anatomically, the tracheobronchial airways consist of a series of bifurcating tubes with circular or near-circular cross sections and ever-decreasing diameters that extend from the trachea at the proximal end to the terminal bronchioles in the distal region of the lung. The trachea divides into the left and right major bronchi, which in turn divide by dichotomous branching into smaller bronchi and bronchioles. In humans, there are about 18 to 20 branchings before the level of the respiratory bronchioles is reached. The bronchi have a ciliated epithelium covered by a layer of mucus, which is produced by the goblet cells and mucous glands. The bronchi also have irregularly shaped cartilage plates situated on the outside of the bronchial walls that, along with the smooth muscle layers, provide structural support to maintain airway patency during ventilation. The smallest conducting airways, the bronchioles, occur at the ends of the bronchi and have neither cartilage nor mucous glands. They do, however, have a ciliated epithelium and mucus-producing cells, although the ciliated regions and the mucus blanket are no longer continuous, as they tend to be in the larger bronchi.

Morphometric models of the conducting airways have been developed based on measurements made by using airway corrosion casts or in vivo bronchoscopic or radiographic measurements or at autopsy. These measurements have included airway diameters, segment lengths, and branching angles. From these, luminal surface areas and volumes can be calculated. The earliest models of the respiratory airways were made assuming symmetry of size and length for airways of a given generation (Findeisen, 1935; Landahl, 1950; Davies, 1961). These models have been summarized by Raabe (1982).

More recent morphometric models have been based on sets of data obtained primarily from measurements of conducting airway dimensions by using corrosion cast techniques (Weibel, 1963; Horsfield et al., 1971; Yeh and Schum, 1980; Phalen et al., 1985). Weibel (1963) used a plastic airway cast prepared by Liebow and measured the lengths, diameters, and branching angles completely for the first 5 airway generations and incompletely through 10 airway generations. Sampling frequencies decreased gradually from 91 to 95% for generations 6 to 18 to 21% for generation 10. The results confirmed earlier views that the conducting airway system has a characteristic irregular dichotomous branching.

For a given generation, there was a dispersion of both airway diameters and airway lengths, with a greater variability being seen in the lengths of the airway segments. The above measurements were then combined with data collected from the more peripheral regions of the lung by quantitative morphometry of histologically prepared sections. These measurements were used to construct two morphometric models of the human lung: the first (Weibel lung model A), which emphasized the regular features of the airways and their patterns, and the second (Weibel lung model B), which attempted to take into account the irregularities encountered in the measurements (Weibel, 1963). The Weibel morphometry models, particularly Weibel lung model A, have been incorporated into deposition and dosimetric models by many investigators.

Horsfield et al. (1971) described a morphometric model based on data obtained from measurement of a resin cast of a normal human bronchial tree (Horsfield and Cumming, 1968; Parker et al., 1971). Their measurements, which included airway diameters, segment lengths, and branching patterns, indicated that asymmetric branching was needed to produce a more realistic model of the airway structure. Thus, they developed two models that included asymmetric branching, unequal numbers of airways in different lung lobes, and variations in the airway segment lengths (Horsfield et al., 1971). Of note was their determination that the mean diameter of the smaller of a pair of daughter bronchi was similar to that of a larger bronchial branch occurring four generations distal to the original. Their model has been implemented mathematically by Yeates and Aspin (1978).

Yeh and Schum (1980) prepared a flexible silicone rubber airway cast of the lungs of a 60-yr-old male. Their measurements included diameters and lengths of each airway segment together with the respective branching angles and angle of each airway with respect to gravity. Measurements were made of all airways to 2 or 3 mm in diameter, with randomized sampling of 20% of the smaller-diameter airways being made to the level of the terminal bronchioles. They then developed a typical path lung model in which n number of identical pathways (down to the level of the terminal bronchioles) was used to represent n number of actual or estimated pathways of a complicated tree structure such as those present in each lobe of a lung or the lung as a whole. The concept is similar to that of Weibel lung model A, except that it does not impose a symmetry requirement on the bronchial tree structure.

Phalen et al. (1985) developed a morphometric model of the tracheobronchial airways based on measurements obtained from 20 humans ranging in age from 11 days to 21 yr. Complete measurements of diameters, lengths, branching angles, and angles with respect to gravity were made for the first three generations; successive generations were sampled, and pathways to the level of the terminal bronchioles were marked and measured at random. Data were then analyzed with respect to the height of the subject, and linear regressions were performed on the data sets grouped by airway generation.

Although several studies have measured morphometric parameters of human conducting airways, no single study has made a complete set of measurements for the length of the tracheobronchial tree. This is understandable, considering that analysis of all airways for 15 generations of a symmetrical dichotomous tree would require more than 65,000 measurements. Thus, each investigator must make assumptions as to the structure of the tracheobronchial airways and how representative their sampling is with respect to the real morphology of the conducting airways. Additionally, very few human lung specimens have been analyzed, and their ages and sizes have differed significantly. Thus, a standard human conducting airway model is not available, nor has variability been adequately described by age, size, gender, or other anthropomorphic characteristic. However, James (1988) compared the generation-specific surface areas and volumes of airways from Weibel lung model A, the Yeh and Schum model, and the Phalen (University of California, Irvine [UCI]) model. The data for the Weibel and Yeh and Schum models were derived from Yu and Diu (1982), in which the respective models were scaled to an initial lung volume of 3,000 cm^3. Figure 5-2 illustrates the relationships between the generation-specific volumes for the three models. Although there is relatively good agreement between the model values for the first four generations, there is a significant divergence of the volumes of the Yeh and Schum model for generations 5 to 12 from those of the other two models, with a maximum variance of a factor of 3 in generation 8.

It is not known whether the present models accurately reflect the morphology of the conducting airways of the human lung, nor is it known to what degree the variabilities in the morphometric models can affect dosimetry calculations. Since most of the models currently being used for deposition and dosime-

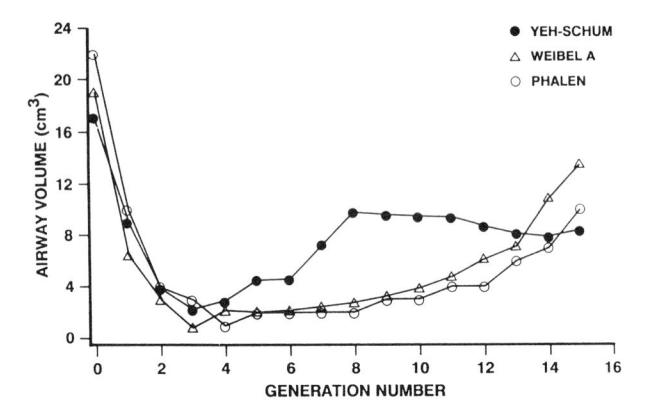

FIGURE 5-2 Calculated airway volumes of tracheobronial airways using different models.

try calculations rely on simplifying assumptions about the generation-specific structure and size of the airways, it would be useful to evaluate the effect of introducing asymmetry assumptions into the morphometric models. These assumptions, which can now be characterized by using statistical distributions for the generation-specific variables of airway size, length, and branching angle, would undoubtedly add complexity to the models. However, with current computer technologies, it would not be difficult to perform the necessary calculations. Weibel, in lung model B, attempted to account for the irregular dichotomy of the lung by applying a binomial distribution to the occurrence of airways with 2-mm diameters with respect to generation number. The distance from the root of the trachea at which these 2-mm bronchi occurred was then described by a normal distribution with a mean of 24.5 cm. Asymmetries for the smaller airways (1.0- and 0.5-mm diameters) were then estimated based on the assumption of normal distributions analogous to that used for the 2-mm-diameter airways. Thus, a model accounting for airway diameter and length asymmetries was developed. To date, however, this model has not been applied to deposition or dosimetry calculations, although there is no reason why this could not be done.

Soong et al. (1979) also used Weibel lung model A as a basis for constructing a tracheobronchial airway morphometry model that included variability in airway size. Initially, they estimated coefficients of variation for 24 generations of airways based on data obtained from available morphometric studies, and then applied the coefficients to the mean diameters and lengths for the different generations of airways taken from Weibel (1963). They later determined that the distribution of airway diameters and lengths could be characterized by either log-normal or gamma probability distributions. Yu et al. (1979) then applied the above statistical morphometric model to human data on total and regional deposition of inhaled aerosols and stated that the model agreed well with the experimental data, suggesting that intersubject variability in deposition was due primarily to differences in airway dimensions.

Koblinger and Hofmann (1985) also developed a statistical morphometric model for calculating aerosol deposition. Their model, which was based on the extensive data set developed by Raabe et al. (1976), included the asymmetries measured in human airways in terms of airway diameters, lengths, and branching angles. They then used Monte Carlo methods to select pathways randomly, but the pathway selections were weighted by the measured statistical distributions for each of the morphometric parameters.

Nikiforov and Schlesinger (1985) made morphometric measurements of the upper airways of right lungs from eight young adult males ages 17 to 45 yr to estimate the range of interindividual variability that could be expected from a human population. Their mean morphometric values agreed reasonably well with those of the Weibel (1963) and Yeh and Schum (1980) models, particularly considering that their measurements involved only right lungs, whereas both lungs were used by Weibel (1963) and Yeh and Schum (1980). The greatest intersubject and intrasubject variability was found in branching angles, and the least variability was found in airway diameter, with airway length variability being moderate.

Significant effort has been expended over the last 30 yr in determining quantitatively the structure of the conducting airways in humans. It is clear that the structure can be described as asymmetric dichotomy. More recent studies have attempted to quantitate the bounds of variability that characterize the morphometric parameters in the lung airways. Some attempts have been made to incorporate statistical variability into morphometric parameters into deposition and dosimetric models. It is not clear, however, what level of accuracy in the morphometric model is needed for the specific case of constructing an adequate dosimetric model for exposure of the conducting airways to radon and radon progeny. In this context, other important facets of the dosimetric model, e.g., thickness of epithelium and mucus, deposition and clearance parameters, and identification of cells at risk, must also be considered.

RESPIRATORY AIRWAY MORPHOMETRY

The respiratory or gas-exchange region of the respiratory tract is generally considered to include the airway structures that occur beyond the terminal bronchioles. Each terminal bronchiole supplies air to a pulmonary lobule or acinus, which consists of about three generations of bronchioles that are partially alveolated (respiratory bronchioles), followed by three more generations of ducts that are fully alveolated (alveolar ducts) and that lead to the final generation of alveolar sacs (Weibel, 1963). The vast majority of the airway volume and surface area occurs in this region of the lung. Morphometric analyses have indicated that there are about 3×10^8 alveoli in a human lung (Weibel, 1963). The total surface area of the respiratory airways is in the range of 70 to 100 m^2 (Weibel, 1963; Thurlbeck, 1967; Gehr et al., 1978). More recent morphometric studies done by electron microscopy have focused on the size and structure of the pulmonary acinus (Haefeli-Bleuer and Weibel, 1988; Ciurea and Gil, 1989). Based on their studies, Haefeli-Bleuer and Weibel have constructed an idealized model of the acinar airways that features a regularized dichotomy between transitional (respiratory) bronchiole and alveolar ducts. Their model also includes airway diameters and lengths, as well as total cross-sectional areas and volumes of the ducts per generation, and the total alveolar surface area.

Past radon dosimetric models have not emphasized radiation doses to the respiratory region of the lung, probably because of the distribution of lung cancers that have been observed in uranium and other metal miners, in whom most of the tumors have been said to originate in the upper bronchial airways (see Chapter 8). Although this may have been appropriate for this population, it is not clear whether differences in exposure aerosol characteristics, age and size of respiratory tracts, and health status may affect the distribution of alpha-radiation dose. It is clear from dose-response studies in several animal species that the cells in the parenchymal lung are capable of being irradiated as a result of exposure to aerosols containing alpha-emitting radionuclides and that such irradiation can lead to lung cancer (Sanders et al., 1977; Lambert et al., 1982; Lundgren et al., 1983; National Council on Radiation Protection and Measurements [NCRP], 1984; Dagle et al., 1989). Therefore, it may be prudent to include some form of morphometric model for the respiratory airway portion of the respiratory tract, if for no other purpose than to facilitate comparison of results from experimental animal studies to humans.

DEPOSITION MODELS OF THE RESPIRATORY TRACT

One of the important factors controlling the distribution of alpha-radiation dose to the different portions of the human respiratory tract is the deposition pattern of radon progeny-containing aerosols. Much research has been done to determine experimentally the deposition fractions for different sizes and types of aerosols in different experimental animal models, including humans. This large body of information has been summarized in numerous articles and chapters (e.g., Raabe et al., 1977; Brain and Valberg, 1979; Lippmann et al., 1980; Clarke and Pavia, 1984) and is considered in detail from the point of view of the special particle characteristics associated with radon progeny-containing aerosols later in this volume (see Chapter 7).

Deposition of different sizes of aerosols in the nasal and oral airways of humans has been measured experimentally both in vivo (Landahl and Black, 1947; Pattle, 1961; Lippmann, 1970; Hounam et al., 1971; Fry and Black, 1973;

Heyder and Rudolf, 1977; Bowes and Swift, 1989) and in physical replica casts (Cheng et al., 1988; Gradon and Yu, 1989; Phalen et al., 1989). Most of the early data obtained in humans consisted of deposition of aerosol particles whose sizes ranged from 0.5 to 10 μm in diameter (Landahl and Black, 1947; Pattle, 1961; Lippmann, 1970; Hounam et al., 1971; Fry and Black, 1973; Heyder and Rudolf, 1977; Bowes and Swift, 1989). Their results showed that the deposition fractions could be modeled by assuming that impaction mechanisms were dominant. Deposition fractions increased with increasing particle diameter and with increasing flow rate. Scott et al. (1978) developed a five-stage theoretical model of the nasal airways and calculated deposition fractions based on considerations of impaction and secondary flow generation posterior to the area of the nasal valve. Their calculations were in general agreement with published deposition data in humans. Their model considered the particle size range of 0.2 to 10 μm. Yu et al. (1981) used all of the available data on nasal and oral airway deposition of particles ranging from 0.3 to 23 μm and developed empirical fits of deposition fraction with the impaction parameter $\rho d^2 Q$, where ρ is the particle density, d is the particle diameter, and Q is the flow rate. They also used the variabilities in the data sets that were probably due to intersubject and intrasubject variabilities to calculate variances for the calculated mean deposition fractions, assuming normal statistics. Although these data and models are relevant to many types of inhalation exposures, the particle size range for the studies described above does not include the sizes of most radon progeny-containing aerosols, which are significantly smaller.

Most of the data relative to the deposition of ultrafine aerosols in the nasal or oral airways have thus far been obtained by using physical replicas of the nasal airways (Cheng et al., 1988; Gradon and Yu, 1989). These data showed that deposition within the nasal airways increased with decreasing particle diameter and with decreasing flow rate, indicating that diffusion was the dominant deposition mechanism. In particular, both Cheng et al. (1988) and Gradon and Yu (1989) have stated that turbulent diffusion was the dominant mechanism, and each group has developed empirical descriptions of their research results. Because of the small sizes of most radon progeny-containing aerosols, these new results concerning ultrafine aerosol deposition need to be considered in any radon dosimetric model.

The experimental basis for modeling the deposition of aerosols in the conducting airways arises from analysis of deposition data from a large number of studies in which people have inhaled test aerosols under controlled experimental conditions. The tracheobronchial region can be simply defined from an anatomic viewpoint, i.e., airways beginning at the trachea and extending to the level of the terminal bronchioles. However, it is not straightforward to define what fraction of an aerosol has deposited in these conducting airways as a result of an inhalation exposure, mainly because of deposition of aerosols in the other parts of the respiratory tract. Using radioactively labeled aerosols, several investigators have functionally defined tracheobronchial deposition as being equal to the amount of total aerosol deposited in the lung that has been cleared from the thorax in 24 h after a brief inhalation exposure (Albert et al., 1967; Lippmann et al., 1971; Camner and Philipson, 1978). Although it is reasonable to assume that essentially all of the aerosol that is cleared in this 24-h period is due to particles that were deposited on the ciliated airway surfaces, it cannot be easily demonstrated that all of the aerosol particles that had originally deposited on the conducting airways had been cleared by that time. It is, therefore, likely that tracheobronchial deposition has been underestimated by the 24-h clearance criterion; however, it is unlikely that the underestimate is of significant magnitude.

Considerable effort has been expended in developing deposition models for the conducting airways. Despite the morphological complexity of the airways and the known variability in airway size and geometry among different people, many investigators have used a simplified morphometry that consists of a series of bifurcating cylindrical tubes with varying degrees of symmetry (Findeisen, 1935; Landahl, 1950; Beekmans, 1965; Task Group on Lung Dynamics, 1966; Gerrity et al., 1979; Yeh and Schum, 1980; Egan and Nixon, 1985). Because of the preponderance of human deposition data available for particle sizes greater than 0.5 μm, most of the theoretical models have focused on the range of particle sizes in which inertial impaction and gravitational settling are the predominant mechanisms of deposition (D), i.e., for diameters $\geq$0.5 μm. Many of these model predictions, compared to the available human deposition data, have been summarized in review articles (Mercer, 1975; Chan and Lippmann, 1980; Raabe, 1982; Stuart, 1984; Morrow and Yu, 1985). In general, for nasal breathing and for a particle size range of greater than 0.5 μm, less than 10% deposition of aerosol is found in the conducting airways, within wide limits of assumed airway sizes and breathing patterns. Oral breathing tends to result in increased deposition of large aerosol particles in the conducting airways by impaction; this is a result of increased penetration into the lung because of the absence of nasal filtration.

To supplement the data obtained in vivo in controlled inhalation studies in humans, several investigators have constructed physical models of the conducting airways (e.g., Martin and Jacobi, 1972; Ferron, 1977; Schlesinger et al., 1977; Chan and Lippmann, 1980; Martonen, 1983; Gurman et al., 1984). The models were constructed by either corrosion techniques by using human lung specimens or idealized bifurcating tube models built according to published morphometry models, usually Weibel A lung model (Weibel, 1963). The use of such models under conditions of controlled aerosol exposure provides opportunities for measuring regional deposition patterns within the models and comparing the qualitative and quantitative deposition patterns with the predictions of existing theoretical deposition models. These types of data cannot be obtained for the conducting airways in vivo by noninvasive methods.

Conducting airway models have been used experimentally both to determine deposition efficiencies for aerosols with different particle sizes as well as to evaluate fluid dynamic parameters such as flow patterns or velocity profiles within the model tubes. Olson et al. (1973) used replica casts to determine velocity profiles, dynamic and static pressures, and flow instabilities from the oral cavity to the segmental bronchi. Lee and Wang (1977) used a two-level bifurcation model based on Weibel morphometry to measure airflow characteristics and deposition velocities; they showed the importance of considering the effects of consecutive serial bifurcations on the prediction of local particle deposition patterns. Such localized deposition patterns were measured in replica hollow casts by Schlesinger et al. (1982) using a five-generation model. Their results with large respirable aerosols and results of other investigations (Ferron, 1977; Johnson et al., 1977; Chan and Lippmann, 1980; Martonen, 1983; Gurman et al., 1984) have been used to validate several theoretical deposition models that feature deposition by inertial impaction and gravitational sedimentation.

Although these previous studies, which were done with large aerosol particles, have been important in validating theoretical predictions of deposition, they are probably of lesser importance when considering the particle size distributions typically observed for radon progeny-containing aerosols, which tend to be significantly smaller than 1.0 μm. Very few experimental studies of the deposition of particles of $\leq$0.2 μm in diameter in the human respiratory tract have been done (Dautrebande and Walkenhorst 1961; George and Breslin, 1969; Tu and Knutson, 1984), and only Tu and Knutson (1984) used monodisperse aerosols. The total respiratory tract deposition measurements of Tu and Knutson (1984) for hydrophobic particles were in general agreement with the mathematical predictions of Yeh and Schum (1980) and Yu et al. (1981), but were significantly less than those predicted by the Task Group on Lung Dynamics (1966). With respect to the deposition of ultrafine aerosols in the conducting airways, only cast studies provide data that can be compared with theoretical models (Chamberlain and Dyson, 1956; Martin and Jacobi, 1972; Cohen, 1987). In general, their results indicate that deposition efficiency for diffusion-controlled deposition is higher than that predicted by using assumptions of laminar flow within the airways (Ingham, 1976). This increased deposition is important to consider with respect to the deposition of radon progeny aerosols within the tracheobronchial region of the respiratory tract.

Nonuniform enhanced deposition of aerosol particles within the regions of conducting airway bifurcations has been reported in studies in which cast replica models of the upper airways have been used (Martin and Jacobi, 1972; Gurman, 1983; Martonen and Lowe, 1983; Schlesinger et al., 1983; Cohen et al., 1988). Because of the short range of the alpha particles associated with radon progeny, such increased deposition, expressed in terms of particles per unit surface area, could result in radiation doses to these anatomical structures that exceed those calculated on the basis of an assumed uniform distribution of deposited particles. For particles depositing primarily by impaction mechanisms, increased deposition bias toward bifurcations has been shown for increasing particle sizes and increasing flow rate (Schlesinger and Lippmann, 1972; Martonen and Lowe, 1983) for inhalation and for cyclic inspiratory airflow patterns (Gurman, 1983). For particle sizes ranging from 2 to 11 μm, the enhancement in deposition relative to that in the adjoining straight-tube sections of the airway models ranged from 20 to 360%. Using smaller particle sizes, Cohen et al. (1988) found an average of about 20% increase in deposition at bifurcations compared with airway lengths for 0.15- and 0.2-μm particles at low flow rates (<20 liters/min). However, the degree of enhancement varied significantly between the different generations of the airway cast model that they used. Additionally, at higher flow rates, no enhanced deposition was found. For 0.04-μm particles, there did not appear to be enhanced deposition at any of the flow rates measured (Cohen et al., 1988). These data agree with modeling predictions of Gradon and Orlicki (1990), who concluded that deposition enhancement in bifurcation regions was insignificant for particles with diameters of $\leq$0.1 μm, in which diffusional mechanisms dominate. As diffusional mechanisms increasingly predominate with decreasing particle size, the effect of localized flow patterns becomes less important, leading to more and more uniform localized depositions (Gradon and Orlicki, 1990). Thus, although the data are very limited, it appears that enhanced deposition at airway bifurcations may not be a significant factor to be considered for the expected particle sizes to which radon progeny would be attached.

The experimental basis for the development of deposition models for the alveolar or nonciliated regions of the respiratory tract arises from the human deposition data of Lippmann and Albert (1969), Lippmann (1977), Foord et al. (1978), Camner and Philipson (1978), Emmett and Aitken (1982), Stahlhofen et al. (1980, 1983), and Heyder et al. (1986). These studies have focused on the inhalation of particles with sizes ($\geq$2 μm) that deposit primarily by impaction and sedimentation mechanisms. However, the more recent studies of Heyder et al. (1986) and Schiller et al. (1988) have extended the particle size range down to 0.005 μm.

Many theoretical models of the deposition of aerosols in the alveolar regions of the lung have been reported (e.g., Findeisen, 1935; Landahl, 1950; Altshuler, 1959; Beekmans, 1965; Task Group on Lung Dynamics, 1966; Taulbee and Yu, 1975; Gerrity et al., 1979; Yeh and Schum, 1980; Rudolf et al., 1986; Egan and Nixon, 1985, 1988). The models differ in terms of their selection of morphometric model, gas dynamics model, particle transport model, and physiological model (Heyder and Rudolf, 1984). Although the number of theoretical models is large, basically they are all derived from three primary deposition models (Heyder and Rudolf, 1984), which differ in their basic mathematical formalisms. The Findeisen (1935) formalism depicts the human respiratory tract as a series of discrete morphometric compartments through which air containing aerosol particles flows. The model is, therefore, spatially discrete; it is also temporally discrete in that it assumes constant airflow for the inspiratory and expiratory phases of the breathing cycle and includes a respiratory pause after inspiration. There have been over 20 secondary deposition models derived by using Findeisen's formalism, including that of the Task Group on Lung Dynamics (1966). The second formalism was developed by Altshuler (1959), who regarded the respiratory tract as a continuous filter bed. As such, this formalism is spatially continuous. However, it is temporally discrete in that it assumes constant airflow conditions for the inspiratory and expiratory parts of the breathing cycle. This formalism has only been used by Altshuler et al. (1967). The third formalism was developed by Taulbee and Yu (1975), who described the respiratory tract as a single channel with a variable cross-sectional area as a function of the depth of penetration, z, into the model. This morphometric "trumpet" model, which was originally proposed by Scherer et al. (1972), features a terminal volume component due to the alveoli that expand and contract uniformly with breathing. Scherer et al. (1972) also used an integral expression for the volumetric flow rate at the entrance of the trumpet model, so that the model is therefore both spatially and temporally continuous. Several investigations have used this model formalism in their deposition models (Yu, 1978; Yu and Diu, 1982; Egan and Nixon, 1985).

Comparison of the results of calculations using different deposition models with experimental data have been published (Yu and Diu, 1982; Heyder and Rudolf, 1984; Ferron et al., 1985). These comparisons have shown that the calculation of total deposition within the respiratory tract is relatively insensitive to the assumptions of the models and do not differ significantly from the original calculations of Findeisen (1935). The major differences between model predictions and experimental data are seen for regional depositions in the tracheobronchial and alveolar regions, where the models tend to overestimate the deposition in the tracheobronchial region and underestimate deposition in the alveolar region when the results are compared with the experimental results of Stahlhofen et al. (1980, 1983). It is difficult, however, to attempt to differentiate between model predictions strictly by comparison with the results from human experimental deposition studies. Those data arise from studies with substantially different experimental designs and with varying degrees of control imposed on the respiratory breathing patterns of the experimental subjects. Additionally, as pointed out earlier, it has not been possible to measure fractional deposition directly within the alveolar and conducting airway compartments of the lung. Such deposition values have been functionally defined based on observed clearance rates of particles from the thorax region.

CLEARANCE MODELS OF THE RESPIRATORY TRACT

Models of clearance of inhaled particles from the respiratory tract differ significantly from those previously described for the deposition of particles. Clearance models are mostly empirical and reflect the lack of knowledge concerning mechanisms of clearance compared with knowledge of deposition models, which are based on a fluid dynamic understanding of the behavior of micrometer and submicrometer particles in moving fluids. As such, the models are based on the results of limited human studies.

Clearance of particles deposited in the nasopharyngeal region depends on the site of deposition. Studies of mucous and particle clearance have shown that the nasal airways can be divided into two anatomically distinct regions: (1) the anterior region, which includes the nostrils, vestibule, and nasal valve, and (2) the posterior nasal airways, which include the nasal turbinates and extend to the pharyngeal area. Because of the lack of mucociliary clearance owing to the absence of ciliated and secretory cells in the anterior region of the nose, clearance occurs via extrinsic means in this region and occurs generally toward the nostrils (Proctor et al. 1977). Particles deposited in this region may remain there for many hours.

Measurements of mucous clearance velocities and particle clearance for the ciliated region of the nasal airways have shown that clearance rates are large, but with substantial intersubject variability (Ewert, 1965; Proctor and Wagner, 1965; Bang et al., 1967; Quinlan et al., 1969; Andersen et al., 1971; Fry and Black, 1973; Proctor, 1973; Yergin et al., 1978). These results have been obtained by a combination of methods, including radiographic imaging of radiopaque particles, direct optical visualization of the streaming of dye particles, radiometric imaging of instilled radionuclide-containing particles, and subjective perception of a substance by taste (saccharine test). Proctor et al. (1977) have summarized the results of several studies and have determined that the average mucous clearance velocity for the ciliated portion of the nasal airways in normal adults without respiratory disease is 5.3 mm/min; however, the range of values extends from 0.5 to 24 mm/min. For a nasal passage of 10 cm in length, the average clearance time would be 19 min. The degree of variability in clearance rates should not be underestimated. Distributions of clearance rates are not normally distributed; i.e., there appears to be a bias toward slower clearance rates, giving the distribution a log-normal appearance, although statistical analysis of measured populations of normal people has not been done. In the study of Andersen et al. (1971), in which over 200 measurements were made in 58 subjects, 57% could be described as having constant unobstructed mucous flow rates, 25% as having heterogeneous flow rates in different regions of the nasal airways, and 18% as having uniformly slow clearances. However, they also pointed out that repeated measurements on the same individual often resulted in a shift of the classification. The reasons for the large variabilities in clearance rates in different individuals are not clear, although variability in mucous rheology and in ciliary beat frequency have been suggested (Bang et al., 1967).

Measurement of mucous clearance rates in the conducting airways below the larynx has been limited to measurements made in the trachea and major bronchi, but primarily the former. The techniques used to make these measurements have been similar to those used to measure nasal mucous clearance, i.e., roentgenographic localization of insufflated powder (Gamsu et al., 1973) or radiopaque Teflon disks emplaced bronchoscopically (Friedman et al., 1977; Goodman et al., 1978); direct observation of the movement of Teflon disks through a bronchoscope (Santa Cruz et al., 1974; Wood et al., 1975); or measurement of the clearance of radioactively labeled microscopic particles by using gamma imaging devices, the particles being emplaced either by bronchoscopy (Chopra et al., 1979) or by inhalation (Thomson and Short, 1969; Yeates et al., 1975; Foster et al., 1976; van Hengstum et al., 1989). The measured average values of tracheal mucous clearance velocities varied between studies that used normal nonsmoking adults: 10.1 ± 3.5 (young adult) and 5.8 ± 2.6 mm/min (Goodman et al., 1978), 15.5 ± 0.69 mm/min (Chopra et al., 1979), 21.5 ± 5.5 mm/min (Santa Cruz et al., 1974), 20.1 ± 1.4 mm/min (Wood et al., 1975), and 5.5 ± 0.4 mm/min (Foster et al., 1980). Yeates et al. (1982) described their measured tracheal mucous velocities in terms of a log-normal distribution with a geometric mean of 4.0 mm/min with a coefficient of variation of 48%. It is not clear whether the differences in the measured velocities among the various studies were due to differences in the subjects or in the different measurement methods used. In addition to the values for normal nonsmoking adults given above, other measurements have been made on smokers and on subjects with lung disease.

Foster et al. (1980) measured clearance velocities for ferric oxide particles deposited in the large bronchi by gamma camera imaging and found average velocities to be 2.4 ± 0.5 mm/min, compared with tracheal velocities of 5.5 mm/min. However, no studies have measured the mucous clearance rates for the different generations of the bronchial and bronchiolar airways. Since clearance of particles from the different levels of the tracheobronchial tree has been considered to be an important variable in assessing the radiation dose to the bronchial epithelium from radon progeny-containing aerosols, investigators have resorted to theoretical models to calculate the generation-by-generation mucous clearance rates for the conducting airways. Clearance velocities calculated by Harley and Pasternack (1972), Lee et al. (1979), Haque and Geary (1981), Yu et al. (1983), and Cuddihy and Yeh (1988) are summarized in Table 5-2. In general, the values differ within a range of a factor of three. The basis for the differences seen in these models is the selection of the appropriate tracheal mucous velocity, since with all of the model structures there is a mathematical scaling of generation-specific velocities to that of the trachea, which is the only experimentally measured value. The second basis for the differences lies in the assumptions that were made regarding the dynamics of mucous and lung water production and absorption. For example, Lee et al. (1979) assumed a

TABLE 5-2 Mucous Clearance Velocities Derived from Theoretical Models

Mucous Clearance Velocity (mm/min) from the Following Models

Weibel Airway Generation	Harley and Pasternack	Lee et al.	Haque and Geary	Yu et al.	Cuddihy and Yeh
0	15.0	5.5	10	13	5.5
1	8.0	4.1	4.9	8.98	3.5
2	2.5	3.0	4.9	3.45	2.0
3	2.5	2.2	4.9	1.52	1.1
4	0.9	1.4	0.4	0.94	0.9
5	0.9	0.88	0.4	0.53	0.9
6	0.9	0.55	0.4	0.28	0.7
7	0.9	0.34	0.09	0.15	0.6
8	0.25	0.21	0.09	0.14	0.4
9	0.25	0.13	0.09	0.14	0.3
10	0.01	0.074		0.15	0.2
11	0.01	0.044		0.14	0.1
12	0.01	0.025		0.13	0.05
13	0.01	0.015		0.12	0.02
14	0.01	0.0082		0.11	0.007
15	0.01	0.0046		0.04	0.001

SOURCES: Harley and Pasternack (1972), Lee et al. (1979), Haque and Geary (1981), Yu et al. (1983), and Cuddihy and Yeh (1988).

uniformly thick mucous blanket that undergoes neither secretion nor absorption throughout the length of the conducting airways; thus, the generation-specific mucous velocities scale as the ratio of the total airway perimeters, i.e., $v_\alpha = v_0 d_0 / 2^\alpha d_\alpha$, where v_α is the velocity in airway generation α, d_α is the diameter of an airway of generation α, and v_0 and d_0 are the velocity of clearance and diameter of the trachea, respectively. On the other hand, Altshuler et al. (1964) assumed that mucus is produced throughout the length of the conducting airways and at different generation-specific rates. Their methodology was used by Harley and Pasternack (1972), albeit with a different morphometric model. Currently, there is no firm basis on which to select one theoretical clearance model because of the lack of biological understanding of the mechanisms of secretion and absorption of either lung water or mucus. Nor are the sites of production and absorption well known. It is interesting to note, however, that the elimination of clearance in the Harley and Pasternack (1972) model resulted in a 50% increase in alpha-radiation dose to the larger bronchi compared with that obtained by assuming normal clearance.

Clearance of particles from the nonciliated or parenchymal regions of the lung have generally been done by quantitating the long-term retention of inhaled gamma-emitting radioactive particles within the thorax region by using appropriate gamma-ray detectors. As described above, the clearance of particles during the first 24 h after a brief exposure have been attributed to clearance from the ciliated tracheobronchial airways. Likewise, decreases in radioactivity beyond the first 24 h has been assigned to clearance of particles from the nonciliated regions of the lung. The use of insoluble particles in which the radioactive label has been firmly bound has facilitated the measurement of long-term clearance, as the confounding influence of clearance via translocation of the radiolabel to blood becomes minimized. Studies have been reported in which humans have inhaled ^{51}Cr-labeled Teflon particles (Philipson et al., 1985), ^{198}Au-Fe$_2$O$_3$ (Stahlhofen et al., 1986), or ^{85}Sr- or ^{88}Y-labeled fused aluminosilicate particles (Bailey et al., 1985). The half-lives for retention have varied somewhat, but all have been long. Philipson et al. (1985) measured a mean half-life of 1,200 days for 68% of the deposited particles; Bailey et al. (1985) estimated a mechanical clearance half-life of about 700 days beyond 200 days after inhalation. This latter value was corroborated by the modeling of Cuddihy and Yeh (1988). The Task Group on Lung Dynamics (1966) has assigned a clearance half-life of 500 days to "class Y" insoluble particles. Thus, it is clear that retention of particles that deposit in the respiratory region of the lung is very long, i.e., hundreds of days. When compared with the effective radiological half-life of radon progeny, ≈30 min, then retention of the progeny in the pulmonary region can be considered to be infinite.

DOSIMETRY MODELS OF THE RESPIRATORY TRACT

Radon and radon progeny dosimetric models have evolved since the first efforts were made to develop themin the 1940s. During this radon period, a large number of models have been published. The impetus for the continued development of increasingly sophisticated models stems from the fact that the actual alpha-radiation doses to the presumed target tissue, the epithelial cells of the bronchial airways, cannot actually be measured, thus requiring dosimetric modeling, and because better data relating to the key components of the dosimetric models have become available, i.e., the physical characteristics of mine and home atmospheres, improved morphometric measurements of the respiratory tract, better deposition modeling of aerosols in the different regions of the respiratory tract, more detailed modeling of clearance phenomena, particularly in the tracheobronchial airways, and evolving views concerning the critical cells at risk to the development of alpha-radiation-induced lung cancer.

Several documents have summarized the radon dosimetric models that have developed over the years. NCRP (1984) lists 48 references through 1981 and remarks on the evolution in thinking and increasing sophistication of the models as well as the dose conversion factors in rad per working level month (WLM), and derived limits for exposure to both radon gas and radon progeny. The investigators noted a trend of decreasing values of the dose conversion factor toward 1 rad/WLM as the unattached fraction decreased and the activity becomes attached to intermediate-sized aerosols (tenths of a micrometer).

James (1988) has updated the listing of radon dosimetry models to include the newer works of Hofmann (1982), Harley and Pasternack (1982), and the NEA Experts Report (1983), as well as the model of James (1988). In examining the dose conversion factors, James (1988) pointed out the importance of determining separately the dose conversion factors for attached and unattached radon progeny. This facilitates comparison of the different dosimetric models on a more equivalent basis. Thus, for the attached fraction, the dose conversion factors have converged somewhat since 1980, to a range of 0.2 to 1.3 rad/WLM; the variations were attributed mainly to the assumed aerosol particle size and the depth of the target cells. For unattached radon progeny, the dose conversion factors for most estimates were mostly in the range of 10 to 20 rad/WLM.

To date, there remains little consensus as to the atmospheric characteristics of radon progeny aerosols in mine and home environments (NCRP, 1984; ICRP, 1987; NRC, 1988; United Nations Scientific Committee on the Effects of Atomic Radiation, 1988). Nor is there agreement as to which choices in dosimetric model structure and parameter values are most appropriate. This difficulty arises because of a general lack of either relevant data (e.g. in upper airway morphometry, regional deposition, and mucous clearance) or understanding of the nature of the different processes that must be considered in calculating alpha-radiation doses to airway epithelial cells. As the knowledge base matures, construction of better models will be possible. Later sections of this report deal with updating knowledge of radon and radon progeny dosimetry as it applies to the different scenarios specific to exposure within a mine environment, for which epidemiological data are available, and in homes, for which human exposure-response data are being developed.

REFERENCES

Albert, R. E., M. Lippmann, J. Spiegelman, A. Liuzzi, and N. Nelson. 1967. The deposition and clearance of radioactive particles in the human lung. Arch. Environ. Health 14:10-15.

Altshuler, B. 1959. Calculation of regional deposition of aerosol in the respiratory tract. Bull. Math. Biophys. 21:257-270.

Altshuler, B., N. Nelson, and M. Kuschner. 1964. Estimation of lung tissue dose from the inhalation of radon and daughters. Health Phys. 10:1137-1161.

Altshuler, B., E. D. Palmes, and N. Nelson. 1967. Regional aerosol deposition in the human respiratory tract. Pp. 323-337 in Inhaled Particles and Vapors II, C. N. Davies, ed. Oxford: Pergamon Press.

Andersen, I., G. R. Lundquist, and D. F. Proctor. 1971. Human nasal mucosa. Arch. Environ. Health 23:408-420.

Bailey, M. R. 1984. Assessment of the dose to the nasopharyngeal region from inhaled radionuclides. Pp. 273-280 in Lung Modelling for Inhalation of Radioactive Materials, H. Smith, and G. Gerber, eds. Brussels: Commission of the European Communities.

Bailey, M. R., F. A. Fry, and A. C. James. 1985. Long-term retention of particles in the human respiratory tract. J. Aerosol Sci. 16(4):295-305.

Bang, B. G., A. L. Mukherjee, and F. B. Bang. 1967. Human nasal mucus flow rates. Johns Hopkins Med. J. 121:38-48.

Beekmans, J. M. 1965. Correction factor for size-selective sampling results, based on a new computed alveolar deposition curve. Ann. Occup. Hyg. 8:221-231.

Bowes, S. M., and D. L. Swift. 1989. Deposition of inhaled particles in the oral airway during oronasal breathing. Aerosol Sci. Tech. 11:157-167.

Brain, J. D., and P. A. Valberg. 1979. Deposition of aerosol in the respiratory tract. Am. Rev. Respir. Dis. 120:1325-1373.

Camner, P., and K. Philipson. 1978. Human alveolar deposition of 4 μm Teflon particles. Arch. Environ. Health 33:181-185.

Chamberlain, A. C., and E. D. Dyson. 1956. The dose to the trachea and bronchi from the decay products of radon and thoron. Br. J. Radiol. 29:317-325.

Chan, T. L., and M. Lippmann. 1980. Experimental measurements and empirical modelling of the regional deposition of inhaled particles in humans. Am. Ind. Hyg. Assoc. J. 41:399-409.

Cheng, Y. S., Y. Yamada, H. C. Yeh, and D. L. Swift. 1988. Diffusional deposition of ultrafine aerosols in a human nasal cast. J. Aerosol Sci. 19(6):741-751.

Chopra, S. K., G. V. Taplin, D. Elam, S. A. Carson, and D. Golde. 1979. Measurement of tracheal mucociliary transport velocity in humans—smokers versus non-smokers. Am. Rev. Respir. Dis. 119(Suppl.):205.

Ciurea, D., and J. Gil. 1989. Morphometric study of human alveolar ducts based on serial sections. J. Appl. Physiol. 67:2512-2521.

Clarke, S. W., and D. Pavia. 1984. Aerosols and the lung. Boston: Butterworths.

Cohen, B. S. 1987. Deposition of ultrafine particles in the human tracheobronchial tree: A determinant of the dose from radon daughters. Pp. 475-486 in Radon and Its Decay Products, P. H. Hopke, ed. Washington, D.C.: American Chemical Society.

Cohen, B. S., N. H. Harley, R. B. Schlesinger, and M. Lippmann. 1988. Nonuniform particle deposition on tracheobronchial airways: Implications for lung dosimetry. Ann. Occup. Hyg. 32:1045-1053.

Cuddihy, R. G., and H. C. Yeh. 1988. Respiratory tract clearance of particles and substances dissociated from particles. Pp. 169-193 in Inhalation Toxicology: The Design and Interpretation of Inhalation Studies and Their Use in Risk Assessment, D. Dungworth, G. Kimmerle, J. Lewkowski, R. McClellan, and W. Stober, eds. New York: Springer-Verlag.

Dagle, G. E., J. F. Park., E. S. Gilbert, and R. E. Weller. 1989. Risk estimates for lung tumours from inhaled ^{239}PuO$_2$, and ^{238}PuO$_2$, and ^{239}Pu(NO$_3$)$_4$ in beagle dogs. Radiat. Prot. Dosim. 26:173-176.

Dautrebande, L., and W. Walkenhorst. 1961. Deposition of NaCl microaerosols in the respiratory tract. Arch. Environ. Health 3:411-419.

Davies, C. N. 1961. A formalized anatomy of the human respiratory tract. Pp. 82-87 in Inhaled Particles and Vapors, C. N. Davies, ed. Oxford: Pergamon Press.

Egan, M. J., and W. Nixon. 1985. A model of aerosol deposition in the lung for use in inhalation dose assessments. Radiat. Prot. Dosim. 11(1):5-17.

Egan, M. J., and W. Nixon. 1988. A modelling study of regional deposition of inspired aerosols with reference to dosimetric assessments. Ann. Occup. Hyg. 32:909-918.

Emmett, P. C., and R. J. Aitken. 1982. Measurements of the total and regional deposition of inhaled particles in the human respiratory tract. J. Aerosol Sci. 13(6):549-560.

Ewert, G. 1965. On the mucus flow rate in the human nose. Acta Otolaryngol. 200 (Suppl.):7-62.

Ferron, G. A. 1977. Deposition of polydisperse aerosols in two glass models representing the upper human airways. J. Aerosol Sci. 8:409-427.

Ferron, G. A., S. Hornik, W. G. Kreyling, and B. Haider. 1985. Comparison of experimental and calculated data for the total and regional deposition in the human lung. J. Aerosol Sci. 16(2):133-143.

Findeisen, W. 1935. Über das Absetzen kleiner, in der Luft suspendierter Teilchen in der menschlichen Lunge bei der Atmung. Pflügert Arch Ges. Physiol. 236:367-379.

Foord, N., A. Black, and M. Walsh. 1978. Regional deposition of 2.5-7.5 μm diameter inhaled particles in healthy male non-smokers. J. Aerosol Sci. 9:343-357.

Foster, W. M., E. H. Bergofsky, D. E. Bohning, M. Lippmann, and R. E. Albert. 1976. Effect of adrenergic agents and their mode of action on mucociliary clearance in man. J. Appl. Physiol. 41(2):146-152.

Foster, W. M., E. Langenback, and E. H. Bergofsky. 1980. Measurement of tracheal and bronchial mucus velocities in man: Relation to lung clearance. J. Appl. Phys. Respir. Environ. Exercise Phys. 48(6):965-971.

Friedman, M., F. D. Stott, D. O. Poole, R. Dougherty, G. A. Chapman, H. Watson, M. A. Sackner. 1977. A new roentgenographic method for estimating mucous velocity in airways. Am. Rev. Respir. Dis. 115:67-72.

Fry, F. A., and A. Black. 1973. Regional deposition and clearance of particles in the human nose. Aerosol Sci. 4:113-124.

Gamsu, G., R. M. Weintraub, and J. A. Nadel. 1973. Clearance of tantalum from airways of different caliber in man evaluated by a roentgenographic method. Am. Rev. Respir. Dis. 107:214-224.

Gehr, P., M. Bachofen, and E. R. Weibel. 1978. The normal human lung: Ultrastructure and morphometric estimation of diffusion capacity. Respir. Phys. 32:121-140.

George, A., and A. J. Breslin. 1969. Deposition of radon daughters in humans exposed to uranium mine atmospheres. Health Phys. 17:115-124.

Gerrity, T. R., P. S. Lee, F. J. Hass, A. Marinelli, P. Werner, and R. V. Lourenco. 1979. Calculated deposition of inhaled particles in the airway generations of normal subjects. J. Appl. Physiol. 47:867-873.

Goodman, R. M., B. M. Yergin, J. F. Landa, M. H. Golinvaux, and M. A. Sackner. 1978. Relationship of smoking history and pulmonary function tests to tracheal mucous velocity in nonsmokers, young smokers, ex-smokers, and patients with chronic bronchitis. Am. Rev. Respir. Dis. 117:205-214.

Gradon, L., and D. Orlicki. 1990. Deposition of inhaled aerosol particles in a generation of the tracheobronchial tree. J. Aerosol Sci. 21:3-19.

Gradon, L., and C. P. Yu. 1989. Diffusional particle deposition in the human nose and mouth. Aerosol Sci. Tech. 11:213-220.

Guilmette, R. A., J. D. Wicks, and R. K. Wolff. 1989. Morphometry of human nasal airways in vivo using magnetic resonance imaging. J. Aerosol Med. 2:365-377.

Gurman, J. L. 1983. Particle deposition patterns in human upper bronchial airways undera simulated inspiratory flow. Ph. D. dissertation. Institute of Environmental Medicine, New York University, New York, N.Y.

Gurman, J. L., M. Lippmann, and R. B. Schlesinger. 1984. Particle deposition in replicate casts of the human upper tracheobronchial tree under constant and cyclic inspiratory flow. I. Experimental. Aerosol Sci. Tech. 3:245-252.

Haefeli-Bleuer, B., and E. R. Weibel. 1988. Morphometry of the human pulmonary acinus. Anatomic Rec. 220:401-414.

Haight, J. S., and P. Cole. 1983. The site and function of the nasal valve. Laryngoscope 93:49-55.

Haque, A. K. M. M., and M. J. Geary. 1981. The radiation dose to the respiratory system due to the daughter products of radon. Pp. 113-127 in Seventh Symposium on Microdosimetry, J. Booz, G. Ebert, and H. D. Hartfiel, eds. Oxford: Harwood Academic Publishers Ltd.

Harley, N. H., and B. S. Pasternack. 1972. Alpha absorption measurements applied to lung dose from radon daughters. Health Phys. 23:771-781.

Harley, N. H., and B. S. Pasternack. 1982. Environmental radon daughter alpha dose factors in a five-lobed human lung. Health Phys. 42:789-799.

Heyder, J., and G. Rudolf. 1977. Deposition of aerosol particles in the human nose. Pp. 107-126 in Inhaled Particles IV, Part 1, W. H. Walton and B. McGovern, eds. Oxford: Pergamon Press.

Heyder, J., and G. Rudolf. 1984. Mathematical models of particle deposition in the human respiratory tract. J. Aerosol Sci. 15:697-707.

Heyder, J., J. Gebhart, G. Rudolf, C. F. Schiller, and W. Stahlhofen. 1986. Deposition of particles in the human respiratory tract in the size range 0.005-15 μm. Aerosol Sci. 17:811-825.

Hofmann, W. 1982. Cellular lung dosimetry for inhaled radon decay products as a base for radiation-induced lung cancer risk assessment. 1. Calculation of mean cellular doses. Radiat. Environ. Biophys. 20:95-112.

Horsfield, K., and G. Cumming. 1968. Morphology of the bronchial tree in man. J. Appl. Physiol. 24:373-383.

Horsfield, K., G. Dart, D. E. Olson, G. F. Filley, and G. Cumming. 1971. Models of the human bronchial tree. J. Appl. Physiol. 31:207-217.

Hounam, R. F., A. Black, and M. Walsh. 1971. The deposition of aerosol particles in the nasopharyngeal region of the human respiratory tract. Pp. 71-80 in Inhaled Particles III, Volume 1, W. H. Walton, ed. Old Woking, Surrey, England: The Gresham Press.

Ingham, D. B. 1976. Diffusion of aerosols from a stream flowing through a cylindrical tube. J. Aerosol Sci. 6:125-132.

International Commission on Radiological Protection (ICRP). 1974. Report of the Task Group on Reference Man, W. S. Snyder, M. J. Cook, E. S. Nasset, L. R. Karhausen, G. P. Howells, and I. H. Tipton, eds. Oxford: Pergamon Press.

International Commission on Radiological Protection (ICRP). 1987. Lung Cancer Risk from Indoor Exposure to Radon Daughters. Publ. No. 50. Oxford: Pergamon Press.

James, A. C. 1988. Lung dosimetry. Pp. 259-310 in Radon and Its Decay Products in Indoor Air, W. W. Nazaroff and A. V. Nero, eds. New York: John Wiley & Sons.

Johnson, J. R., K. D. Isles, and D. C. F. Muir. 1977. Inertial deposition of particles in human branching airways. Pp. 61-73 in Inhaled Particles IV, Part 1, W. H. Walton and B. McGovern, eds. Oxford: Pergamon Press.

Koblinger, L., and W. Hofmann. 1985. Analysis of human lung morphometric data for stochastic aerosol deposition calculations. Phys. Med. Biol. 30:541-556.

Lambert, B. E., M. L. Phipps, P. J. Lindop, A. Black, and S. R. Moores. 1982. Induction of lung tumours in mice following the inhalation of ^{239}PuO$_2$. Pp. 370-375 in Proceedings of the 3rd International Symposium on Radiation Protection—Adv. in Theory and Practice, Vol. 1. Inverness, Scotland: Society for Radiation Protection.

Landahl, H. D. 1950. On the removal of air-borne droplets by the human respiratory tract. II. The nasal passages. Bull. Math. Biophys. 12:161-169.

Landahl, H. D., and S. Black. 1947. Penetration of air-borne particulates through the human nose. J. Ind. Hyg. Toxicol. 29:269-277.

Lee, P. S., T. R. Gerrity, F. J. Hass, and R. V. Lourenco. 1979. A model for tracheobronchial clearance of inhaled particles in man and a comparison with data. IEEE Trans. Biomed. Eng. BME-26:624-630.

Lee, W. C., and C. S. Wang. 1977. Particle deposition in systems of repeatedly bifurcating tubes. Pp. 49-59 in Inhaled Particles IV, Part 1, W. H. Walton and B. McGovern, eds. Oxford: Pergamon Press.

Lippmann, M. 1970. Deposition and clearance of inhaled particles in the human nose. Ann. Otol. Rhinol. Laryngol. 79:519-528.

Lippmann, M. 1977. Regional deposition of particles in the human respiratory tract. Pp. 213-232 in Reaction to Environmental Agents, D. H. K. Lee, H. L. Falk, S. D. Murphy, and S. R. Geiger, eds. Bethesda, Md.: American Physiological Society.

Lippmann, M., and R. E. Albert. 1969. The effect of particle size on the regional deposition of inhaled aerosols in the human respiratory tract. Am. Ind. Hyg. Assoc. J. May/June:257-275.

Lippmann, M., R. E. Albert, and H. T. Peterson. 1971. The regional deposition of inhaled aerosols in man. Pp. 105-122 in Inhaled Particles III, Volume 1, W. H. Walton, ed. Old Woking, Surrey: The Gresham Press.

Lippmann, M., D. B. Yeates, and R. E. Albert. 1980. Deposition retention and clearance of inhaled particles. Br. J. Ind. Med. 37:337.

Lundgren, D. L., F. F. Hahn, A. H. Rebar, and R. O. McClellan. 1983. Effects of the single or repeated inhalation exposure of Syrian hamsters to aerosols of ^{239}PuO$_2$. Int. J. Radiat. Biol. 43:1-18.

Martin, D., and W. Jacobi. 1972. Diffusion deposition of small-sized particles in the bronchial tree. Health Phys. 23:23-29.

Martonen, T. B. 1983. Measurement of particle dose distribution in a model of a human larynx and tracheobronchial tree. J. Aerosol Sci. 14:11-22.

Martonen, T. B., and J. Lowe. 1983. Assessment of aerosol deposition patterns in human respiratory tract casts. Pp. 151-164 in Aerosols in the Mining and Industrial Work Environments, Vol. 1, Fundamentals and Status, V. A. Marple and B. Y. H. Liu, eds. Ann Arbor, Mich.: Ann Arbor Science.

Mercer, T. T. 1975. The deposition model of the task group on lung dynamics: A comparison with recent experimental data. Health Phys. 29:673-680.

Morrow, P. E., and C. P. Yu. 1985. Models of aerosol behavior in airways. Pp. 150-191 in Aerosols in Medicine, Principles, Diagnosis and Therapy, M. F. Newhouse and M. B. Dolovich, eds. New York: Elsevier Science Publishers.

National Research Council (NRC). 1988. Health Risks of Radon and Other Internally Deposited Alpha Emitters. BEIR IV. Committee on the Biological Effects of Ionizing Radiation. Washington, D.C.: National Academy Press.

National Council on Radiation Protection and Measurments (NCRP). 1984. Evaluation of Occupational and Environmental Exposures to Radon and Radon Daughters in the United States. Bethesda, Md.: National Council on Radiation Protection and Measurements.

Nikiforov, A. I., and R. B. Schlesinger. 1985. Morphometric variability of the human upper bronchial tree. Respir. Physiol. 59:289-299.

Nuclear Energy Agency Group Experts (NEA). 1983. Dosimetry Aspects of Exposure to Radon and Thoron Daughter Products. Paris: Organization for Economic Cooperation and Development.

Olson, D. E., M. F. Sudlow, K. Horsfield, and G. F. Filley. 1973. Convective patterns of flow during inspiration. Arch. Intern. Med. 131:51-57.

Parker, H., K. Horsfield, and G. Cumming. 1971. Morphology of distal airways in the human lung. J. Appl. Physiol. 31:386-391.
Pattle, R. E. 1961. The retention of gases and particles in the human nose. Pp. 302-309 in Inhaled Particles and Vapors I, C. N. Davies, ed. Oxford: Pergamon Press.
Phalen, R. F., M. J. Oldham, C. B. Beaucage, T. T. Crocker, and J. D. Mortensen. 1985. Postnatal enlargement of human tracheobronchial airways and implications for particle deposition. Anat. Rec. 212:363-380.
Phalen, R. F., M. J. Oldham, and W. J. Mautz. 1989. Aerosol deposition in the nose as a function of body size. Health Phys. 57:299-305.
Philipson, K., R. Falk, and P. Camner. 1985. Long-term lung clearance in humans studied with teflon particles labeled with chromium-51. Exp. Lung Res. 9:31-42.
Proctor, D. F. 1973. Clearance of inhaled particles from the human nose. Arch. Intern. Med. 131:132-139.
Proctor, D. F., and H. N. Wagner, Jr. 1965. Clearance of particles from the human nose. Arch. Environ. Health 11:377-381.
Proctor, D. F., and D. L. Swift. 1971. The nose—A defence against the atmospheric environment. Pp. 59-70 in Inhaled Particles III, Vol. 1, W. H. Walton, ed. Old Woking, Surrey: The Gresham Press.
Proctor, D. F., I. Andersen, and G. Lundquist. 1977. Nasal mucociliary function in humans. Pp. 427-452 in Respiratory Defense Mechanisms, Part I, J. D. Brain, D. F. Proctor, and L. M. Reid, eds. New York: Marcel Dekker.
Quinlan, M. F., S. D. Salman, D. L. Swift, H. N. Wagner, and D. F. Proctor. 1969. Measurement of mucociliary function in man. Am. Rev. Respir. Dis. 99:13-23.
Raabe, O. G. 1982. Deposition and clearance of inhaled aerosols. Pp. 28-68 in Mechanisms in Respiratory Toxicology, Vol. 1, H. Witschi and P. Nettesheim, eds. Boca Raton, Fla.: CRC Press.
Raabe, O. G., H. C. Yeh, G. M. Schum, and R. F. Phalen. 1976. Tracheobronchial geometry: human, dog, rat, hamster. Lovelace Foundation Report, U.S. Department of Energy Document No. LF-53. Springfield, Va.: NTIS.
Raabe, O. G., H. C. Yeh, G. J. Newton, R. F. Phalen, and D. J. Velasquez. 1977. Deposition of inhaled monodisperse aerosols in small rodents. P. 3 in Inhaled Particles IV, Part 1, W. H. Walton and B. McGovern, eds. Elmsford, N.Y.: Pergamon Press.
Rudolf, G., J. Gebhart, J. Heyder, C. F. Schiller, and W. Stahlhofen. 1986. An empirical formula describing aerosol deposition in man for any particle size. J. Aerosol Sci. 17:350-355.
Sanders, C. L., G. E. Dagle, W. C. Cannon, G. J. Powers, and D. M. Meier. 1977. Inhalation carcinogenesis of high fired $^{238}PuO_2$ in rats. Radiat. Res. 71:528-546.
Santa Cruz, R., J. Landa, J. Hirsch, and M. A. Sackner. 1974. Tracheal mucous velocity in normal man and patients with obstructive lung disease: Effects of terbutaline. Am. Rev. Respir. Dis. 109:458-463.
Scherer, P. W., L. H. Shendalman, and N. M. Greene. 1972. Simultaneous diffusion and convection in single breath lung washout. Bull. Math. Biophys. 34:393-412.
Schiller, C. F., G. J. Gebhart, G. Rudolf, and W. Stahlhofen. 1988. Deposition of monodisperse insoluble aerosol particles in the 0.005 to 0.2 μm size within the human respiratory tract. Ann. Occup. Hyg. 32:41-49.
Schlesinger, R. B., and M. Lippmann. 1972. Particle deposition in casts of the human upper tracheobronchial tree. Am. Ind. Hyg. Assoc. J. 33:237-251.
Schlesinger, R. B., D. E. Bohning, T. L. Chan, and M. Lippmann. 1977. Particle deposition in a hollow cast of the human tracheobronchial tree. J. Aerosol Sci. 8:429-445.

Schlesinger, R. B., J. L. Gurman, and M. Lippmann. 1982. Particle deposition within bronchial airways: Comparisons using constant and cyclic inspiratory flows. Ann. Occup. Hyg. 26:47-64.

Schlesinger, R. B., J. Concato, and M. Lippmann. 1983. Particle deposition during exhalation: A study in replicate casts of the human upper tracheobronchial tree. In Aerosols in the Mining and Industrial Work Environments, Vol. 1, Fundamentals and Status, V. A. Marple and B. Y. H. Liu, eds. Ann Arbor, Mich.: Ann Arbor Science.

Scott, W. R., D. B. Taulbee, and C. P. Yu. 1978. Theoretical study of nasal deposition. Bull. Math. Biol. 40:581-603.

Soong, T. T., P. Nicolaides, C. P. Yu, and S. C. Soong. 1979. A statistical description of the human tracheobronchial tree geometry. Respir. Physiol. 37:161-172.

Stahlhofen, W., J. Gebhart, and J. Heyder. 1980. Experimental determination of the regional deposition of aerosol particles in the human respiratory tract. Am. Ind. Hyg. Assoc. J. 41:385-398.

Stahlhofen, W., J. Gebhart, J. Heyder, and G. Scheuch. 1983. New regional deposition data of the human respiratory tract. J. Aerosol Sci. 14:186-188.

Stahlhofen, W., J. Gebhart, G. Rudolf, and G. Scheuch. 1986. Measurement of lung clearance with pulses of radioactively-labeled aerosols. J. Aerosol. Sci. 17:333-336.

Stuart, B. O. 1984. Deposition and clearance of inhaled particles. Environ. Health Perspect. 55:369-390.

Swift, D. L. 1981. Aerosol deposition and clearance in the human upper airways. Ann. Biomed. Eng. 9:593-604.

Swift, D. L., and D. F. Proctor. 1976. Access of air to the respiratory tract. Pp. 63-93 in Respiratory Defense Mechanisms, J. D. Brain, D. F. Proctor, and L. M. Reid, eds. New York: Marcel Dekker.

Task Group on Lung Dynamics. 1966. Deposition and retention models for internal dosimetry of the human respiratory tract. Health Phys. 12:173-207.

Taulbee, D. B., and C. P. Yu. 1975. A theory of aerosol deposition in the human respiratory tract. J. Appl. Physiol. 38:77-85.

Thomson, M. L., and M. D. Short. 1969. Mucociliary function in health, chronic obstructive airway disease, and asbestosis. J. Appl. Physiol. 26:535-539.

Thurlbeck, W. M. 1967. The internal surface area of nonemphysematous lungs. Am. Rev. Respir. Dis. 95:765-773.

Tu, K. W., and E. O. Knutson. 1984. Total deposition of ultrafine hydrophobic and hygroscopic aerosols in the human respiratory system. Aerosol Sci. Tech. 3:453-465.

United Nations Scientific Committee on the Effects of Atomic Radiation. 1988. Sources, Effects and Risks from Ionizing Radiation. 1988 Report to the General Assembly. New York: United Nations.

van Hengstum, M., J. Festen, W. Buijs, W. van den Broek, and F. Corstens. 1989. Variability of tracheobronchial clearance in healthy non-smoking subjects. Respiration 56:94-102.

Weibel, E. R., ed. 1963. Morphometry of the Human Lung. New York: Academic Press.

Wood, R. E., A. Wanner, J. Hirsch, and P. M. Farrell. 1975. Tracheal mucociliary transport in patients with cystic fibrosis and its stimulation by terbutaline. Am. Rev. Repir. Dis. 111:733-738.

Yeates, D. B., and N. Aspin. 1978. A mathematical description of the airways of the human lungs. Respir. Physiol. 32:91-104.

Yeates, D. B., N. Aspin, H. Levison, M. T. Jones, and A. C. Bryan. 1975. Mucociliary tracheal transport rates in man. J. Appl. Physiol. 39:487-495.

Yeates, D. B., T. R. Gerrity, and C. S. Garrard. 1982. Characteristics of tracheobronchial deposition and clearance in man. Ann. Occup. Hyg. 26:245-257.

Yeh, H. C., and G. M. Schum. 1980. Models of human lung airways and their application to inhaled particle deposition. Bull. Math. Biol. 42:461-480.

Yergin, B. M., K. Saketkhoo, E. D. Michaelson, S. M. Serafini, K. Kovitz, and M. A. Sackner. 1978. A roentgenographic method for measuring nasal mucous velocity. J. Appl. Physiol. Respir. Environ. Exercise Physiol. 44:964-968.

Yu, C. P. 1978. Exact analysis of aerosol deposition during steady state breathing. Powder Technol. 21:55-62.

Yu, C. P., and C. K. Diu. 1982. A comparative study of aerosol deposition in different lung models. J. Am. Ind. Hyg. Assoc. 43:54-65.

Yu, C. P., P. Nicolaides, and T. T. Soong. 1979. Effect of random airway sizes on aerosol deposition. J. Am. Ind. Hyg. Assoc. 40:999-1005.

Yu C. P., C. K. Diu, and T. T. Soong. 1981. Statistical analysis of aerosol deposition in nose and mouth. J. Am. Ind. Hyg. Assoc. 42:726-733.

Yu, C. P., J. P. Hu, G. Leikauf, D. Spektor, and M. Lippmann. 1983. Mucociliary transport and particle clearance in the human tracheobronchial tree. Pp. 177-184 in Aerosols in the Mining and Industrial Work Environments, Vol. 1, Fundamentals and Status, V. A. Marple and B. Y. H. Liu, eds. Ann Arbor, Mich.: Ann Arbor Science.

6

Aerosols in Homes and Mines

The measurement methods used to determine the effective size of the radon progeny in air are generally based on the use of diffusional deposition in the sampler to remove some of the airborne activity in a well-defined manner. By making a series of such measurements and knowing the theoretical rates at which the deposition should occur, the size distribution of the radioactive aerosol can be inferred. Thus, to understand the experimental methods used for activity-weighted size distribution and/or unattached fraction measurements, it is necessary to present the theoretical considerations upon which the measurement instruments are based. In the next sections, the relationship between size and diffusivity will be presented along with the theory of diffusive deposition of particles on the inside surfaces of cylinders and on strands of wire mesh screen.

RELATIONSHIP BETWEEN PARTICLE SIZE AND DIFFUSION COEFFICIENT

The particle diffusion coefficient, D, for aerosols is commonly estimated by the equation,

$$D = \frac{kTC}{3\pi\mu d_p} \qquad (6\text{-}1)$$

where k is the Boltzmann constant (1.38×10^{-16} erg $°K^{-1}$), T is the temperature in degrees Kelvin (293 K at 20°C and 1 atm), d_p is the particle diameter in centimeters, μ is the gas viscosity (1.83×10^{-4} g cm/s for air), and C is the Cunningham correction factor given by Friedlander (1977),

$$C = 1 + \frac{\lambda}{d_p}[2.514 + 0.8 \exp(-0.55 d_p/\lambda)] \qquad (6\text{-}2)$$

where λ is the mean free path of the gas (0.65×10^{-5} cm for air at 20°C and 1 atm of pressure).

Equation 6-2 was empirically derived to fit the entire range of values of d_p/λ from the continuum to the free molecular regimes (Davies, 1945). However, Equations 6-1 and 6-2 overestimate the particle diffusion coefficient in the 0.5-1.75-nm particle diameter range. To illustrate this overestimation, consider the diffusion coefficient for a radon-222 (^{222}Rn) atom whose atomic diameter is estimated to be 0.46 nm. This diffusion coefficient was measured by Hirst and Harrison (1939) to be 0.12 cm^2/s. Equations 6-1 and 6-2 predict a diffusion coefficient of 0.20 cm^2/s for a 0.5-nm-diameter cluster. This result is further evident from a comparison (Figure 6-1) with the diffusion coefficient calculated by using the kinetic theory of gases (Loeb, 1961; Porstendörfer, 1968; Raabe, 1969). The kinetic theory equation for the diffusion coefficient of molecular clusters in a gas is

$$D = \frac{0.815 V_r}{3\pi s^2 N}[(M+m)/M]^{1/2} \qquad (6\text{-}3)$$

where V_r is the root mean square velocity of the gas (5.02×10^4 cm/s for air at 20°C), N is the number concentration of gas molecules (2.51×10^{19} cm^3 for air at 760 mm Hg and 20°C), s is the sum of the radii of the gas molecules (0.155×10^{-7} cm for air) and of the cluster, M is the molecular weight of the cluster, and m is that of the gas (28.9 for air).

In order to use a single equation for the particle diffusion coefficient over the entire size range for $d_p > 0.5$ nm, the Einstein-Cunningham Equations 6-1 and 6-2 may be fitted to the kinetic theory Equation 6-3 in the 0.5-1.75-nm size range in a manner that yields the original Einstein-Cunningham Equations 6-1 and 6-2 for $d_p > 1.75$ nm. This result may be obtained by the substitution of

$$d_p^* = d_p(1 + 3e^{-2.7 \times 10^7 d_p}) \qquad (6\text{-}4)$$

for d_p (in centimeters) in Equation 6-2 for the Cunningham constant C. The molecular weight-cluster size relationship for polonium-218 (^{218}Po)-H$_2$O clusters was used for the molecular weight factor $[(M+m)/M]^{1/2}$ in the kinetic theory equation (Equation 6-3) and to obtain the fitting parameter d^* in Equation 6-4. Figure 6-1 is a plot of the corrected and uncorrected Einstein-Cunningham equations and the kinetic theory diffusion coefficient equation versus particle diameter (the kinetic theory equation is plotted without the $[(M+m)/M]^{1/2}$ factor, which rapidly approaches unity).

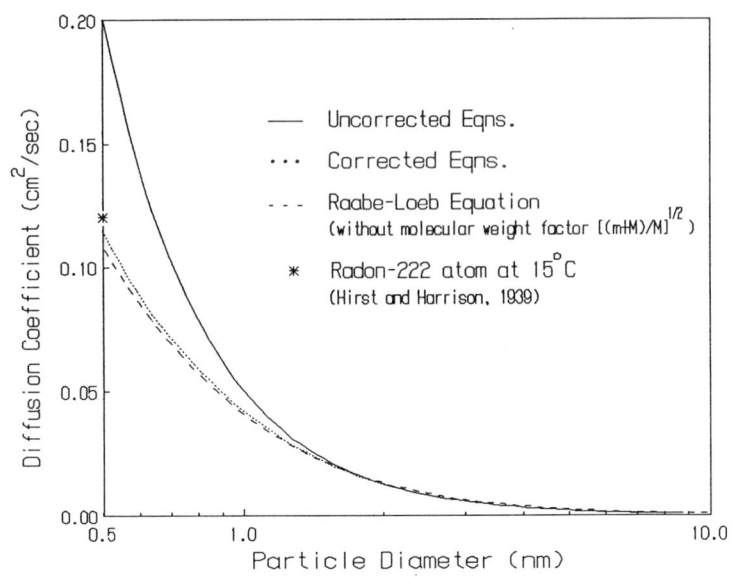

FIGURE 6-1 Diffusion coefficient for particles 0.5-10 nm in diameter.

Penetration of Aerosols Through a Tube

Theoretical equations for diffusional deposition in circular tubes are well documented in the literature (e.g., Gormley and Kennedy, 1949; Fuchs, 1964). Of particular interest when sampling highly diffusive ultrafine cluster aerosols are tube penetration equations for the prediction of diffusional deposition or wall losses in the entrances of sampling tubes. Typically, the "wall loss" lengths L in these situations are smaller than the tube length required for the development of a laminar flow profile ("entrance length"). This entrance length z beyond which a parabolic flow field is established is (Fuchs, 1964),

$$z = 0.1 R Re_t \tag{6-5}$$

where R is the radius of the tube, and $Re_t = 2U\rho R/\mu$ is the Reynolds number of the tube. Within this entrance distance, z, a developing flow exists and could enhance diffusional deposition. Theoretical studies by Tan (1969) and Chen and Comparin (1976) suggest that for highly diffusive aerosols with small Schmidt numbers ($S_c = \nu/D$) and penetration parameter, $\mu^* = DL/R^2U < 0.05$ (small entrance length L and high flow velocity U), assuming a uniform flow profile may be a suitable first-order approximation for the prediction of diffusional deposition losses.

Ingham (1975) developed an analytical, matched-asymptote solution for uniform flow penetration through a circular tube, P_t, given by,

AEROSOLS IN HOMES AND MINES 93

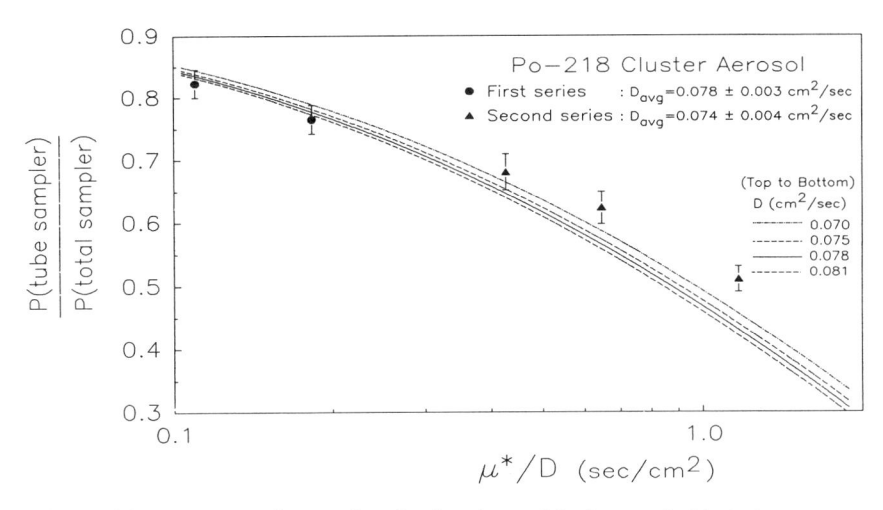

FIGURE 6-2 Assessment of penetration of radioactive particles into a cylindrical tube compared to the theoretical predictions of Ingham (1975).

$$P_t = \frac{4}{a_1^2} e^{-a_1^2 \mu^*} + \frac{4}{a_2^2} e^{-a_2^2 \mu^*} + \frac{4}{a_3^2} e^{-a_3^2 \mu^*} + \{1 - 4(1/a_1^2 + 1/a_2^2$$
$$+ 1/a_3^2)\} e^{-4\mu^{*1/2}} / [\pi^{1/2}(1 - 4(1/a_1^2 + 1/a_2^2 + 1/a_3^2))] \quad (6\text{-}6)$$

The parameters a_1^2 (=5.783186), a_2^2 (=30.471262), a_3^2 (=74.887007) are the zeros of the zero-order Bessel function of the first kind, $J_0(a_n) = 0$ (Tan, 1969). Equation 6-6 is valid for tube Reynolds number $Re_t < 1{,}200$ and tube Peclet number ($Pe_t = RU/D$) > 100.

Thomas (1955) attempted to verify the laminar flow tube penetration theory equations using gas molecules and reported agreement to within 20-30%. The discrepancies were attributed to experimental difficulties and possible entrance effects. Scheibel and Porstendörfer (1984) attempted a verification of tube penetration theory using monodisperse particles ($d_p > 4$ nm). They reported good agreement with the laminar flow theory (Gormley and Kennedy, 1949) for particle sizes greater than 15 nm and large deviations for particle sizes below 5 nm. The discrepancies were resolved by including terms for uniform flow deposition in the entrance of the tube and for deposition on the front face of the tubes. Recent studies by Ramamurthi et al. (in press) have found that tube penetration theory accurately predicts the deposition behavior of highly diffusive radioactive particles if the value of $\mu^* \leq 0.05$. For $\mu^* > 0.05$, Equation 6-6 systematically underpredicts the penetration of particles through the tube. Figure 6-2 shows the results of this test of penetration of a single tube as a function of the diffusion parameter, μ^*/D.

Deposition of Ultrafine Particles onto Wire Screens

An equation of penetration through wire screens based on the theory of fibrous filtration was derived by Cheng and Yeh (1980) and Cheng et al. (1980). The penetration equation for a wire screen, with wire diameter d_f, thickness w, volume fraction α, and sampling face velocity U, for ultrafine particles, $d_p < 0.1$ μm, is given by,

$$P = 1 - \eta = \exp[-\frac{(4)(2.7)}{\pi}(KVF)^{-2/3}(D)^{2.3}] \quad (6\text{-}7)$$

where

$$KVF = \frac{U}{[WF]^{1.5}} \text{ is the wire-velocity parameter (cm}^2\text{ s}^{-1}) \quad (6\text{-}8)$$

$$WF = \alpha \frac{W}{(1-\alpha)d_f^{5/3}} \text{ is the wire factor (cm}^{-2/3}), \quad (6\text{-}9)$$

and D is the particle diffusion coefficient. Equation 6-7 is valid for wire Reynolds numbers ($Re_f = U\rho d_f/\mu$, where ρ is the density of air) less than 1 (Emi et al., 1982). The fan model filtration theory has been applied to wire screens and experimentally verified for particle sizes $d_p > 15$ nm (0.015 μm) (Cheng and Yeh, 1980). Scheibel and Porstendörfer (1984) have further verified the fan model for particle sizes $d_p \approx 4$ nm. Recent work with charged and uncharged ^{218}Po clusters in the 0.5-1.5-nm size range has also indicated general agreement with the wire screen fan model penetration Equation 6-7 (Holub and Knutson, 1987; Ramamurthi et al., in press), although calibration studies remain necessary for $d_p < 4$ nm. The penetration characteristics for a wire screen operating in the 0.5-100 nm size range can thus be determined from Equations 6-7, 6-8, and 6-9. Typical penetration curves are shown in Figure 6-3.

Further, each screen-velocity combination can be conveniently parameterized by its $d_p(50\%)$ (Ramamurthi and Hopke, 1989), the particle diameter for 50% collection efficiency, because of the form of the penetration in Equation 6-7. This parameter can be determined (using a log-linear diffusion coefficient approximation) as,

$$d_p(50\%) \text{ (nm)} = 10^7 \exp\left[-\frac{32.193 + LN(KVF)}{1.957}\right] \quad (6\text{-}10)$$

for $0.001 < KVF < 0.325$. These $d_p(50\%)$ values are a convenient way of characterizing the particular screen parameters and face velocity of the sampling flow. However, as can be seen by the curves in Figure 6-3, there is a distinct difference between the relatively slowly changing penetration-versus-diameter behavior of screens and the sharp cutoffs that are observed with inertial collections systems such as impactors or cyclones.

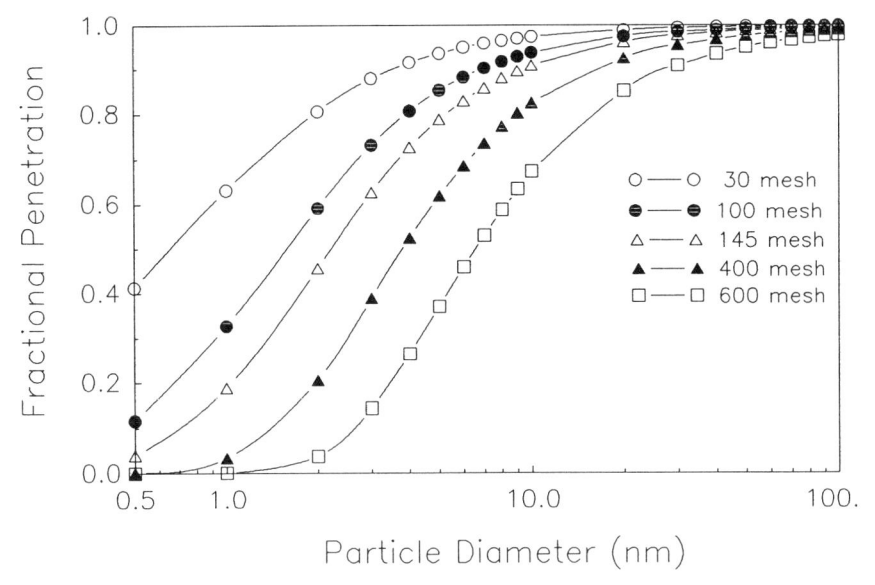

FIGURE 6-3 Fractional penetration through various mesh screens at a face velocity of 10 cm s^{-1}.

MEASUREMENT METHODS FOR UNATTACHED FRACTION

Diffusion Sampler

Most of the early work on "unattached" fraction measurements was carried out in uranium mines. In the earliest studies, including the work of Chamberlain and Dyson (1956), the unattached fraction was determined by measurement of the penetration of the activity through a right circular cylinder. Using the tube penetration theory of Gormley and Kennedy (1949), the fractional penetration can be related to $\pi DL/2Q$, where D is the diffusion coefficient of the radioisotope, L is the length of the tube, and Q is the volumetric flow rate through the tube. Details of the theory of penetration of particles through a tube are given in the previous section. The diffusion coefficients of the radon decay products are assumed to have a single, very much larger value than the diffusion coefficients of the condensation nuclei to which the radon progeny become attached. Thus, by comparing the amount of activity penetrating through a tube of given dimensions at a given flow rate with the total airborne decay product activity, the fraction of unattached activities could be estimated.

The amount of unattached radioactivity has been measured in this manner by a number of investigators (Craft et al., 1966; Fusamura et al., 1967; Duggan and Howell, 1969). A problem that tends to confuse the literature of unattached fraction measurements is the number of ways in which the unattached radioactivity is reported. The International Commission on Radiological Protection

(ICRP) has defined the unattached fraction as the fraction of the equilibrium number of ^{218}Po ions that are unattached to particles (ICRP, 1959). However, the method used to measure the attached fraction readily yields the fraction of unattached ^{218}Po atoms to the total number of ^{218}Po atoms actually present. Chamberlain and Dyson (1956) made the first report of unattached fraction to be 0.1 using what became the ICRP definition (ICRP, 1959). It is important to carefully determine in each case what the investigator means by "unattached" fraction in order to compare results.

Wire Screen Samplers

The diffusion samplers were fairly cumbersome devices to use, and therefore a simpler and more portable system was developed based on the collection of the activity on wire mesh screens. Wire mesh screens have become the most commonly used method for estimating unattached radon daughter fractions. The early development of these systems was described by James et al. (1972), Thomas and Hinchliffe (1972), and George (1972). These systems were much easier to use, although initially they suffered from the lack of a well-developed theory to relate the screen properties to their collection efficiency. The collection efficiency of wire screens for unattached ^{218}Po was therefore determined empirically from calibration experiments with fresh ^{218}Po in the absence of ambient aerosols as a function of screen parameters and face velocity. Compared with the currently accepted theory described below, the equation of Thomas and Hinchliffe (1972) overestimates wire screen collection efficiencies for particle diameters $d_p < 1.0$ to 2.0 nm and $d_p > 5.0$ to 7.0 nm for most screen-velocity combinations.

George (1972) developed a standard method for measuring unattached ^{218}Po fractions: a 60-mesh stainless steel screen is sampled simultaneously with a parallel filter, followed by counting of the alpha particles on both the screen and filter; ^{218}Po data are extracted from both the screen and the filter by the modified Tsivoglou technique, and the unattached fraction is calculated as the ratio of twice the activity on one face of the screen to the activity on the filter. This method has been widely used to obtain estimates of the unattached fraction in a variety of different environments.

The wire screen penetration theories developed by Cheng and Yeh (1980), Cheng et al. (1980), and Yeh et al. (1982) are presented earlier, along with a semi-empirically corrected diffusion coefficient equation in the molecular cluster size range to characterize unattached fraction measurements reported in earlier studies, Equations 6-1 and 6-2.

A number of wire screen measurement studies of the unattached ^{218}Po fraction are reported in the literature (George, 1972; James et al., 1972; Raghavayya and Jones, 1974; Bigu and Kirk, 1980; Stranden and Berteig, 1982; Bigu, 1985; Reineking et al., 1985). Table 6-1 provides a compilation

TABLE 6-1 Wire Screen Parameters and Face Velocities Used in Previously Published Wire Screen "Unattached" Fraction Measurements

						"Unattached" Mode Underestimation (%)	
Study	Mesh No.	d_f	U	$d_p(50\%)$	"E"	$\frac{1-C}{C}$	$\frac{1-C/E}{C/E}$
James et al. (1972)	200	50.8	12.0	2.7	—	15	—
George (1972)	60	178.0	11.5	1.7	0.76	45	10
George (1972)	80	127.0	11.5	1.9	—	36	—
Raghavayya and Jones (1974)	120	94.0	12.2	2.0	—	29	—
Bigu and Kirk (1980)	150	76.0	9.6	2.8	—	14	—
Stranden and Berteig (1982)	174	56.0	21.0	2.3	0.86	22	5
Bigu (1985)	150	76.0	27.3	1.6	0.77	48	14
Reineking et al. (1985)	188	50.0	8.5	2.0	—	31	—

NOTE: Underestimation of "unattached" cluster fraction is based on a log-normally distributed cluster mode, $0.5 < d_p < 3.0$ nm, with $d_m = 1.0$ nm and $\sigma_g = 1.5$. "C" is the fraction of activity collected by the screen, and "E" is a "collection efficiency" correction factor used in some of the studies.

of the wire screen parameters and face velocities used in these studies. Each screen-velocity combination is characterized by its $d_p(50\%)$ parameter obtained from Equation 6-1. Nominal values for commonly undocumented parameters such as w and α were taken from wire screen manufacturers' catalogs for screens with the appropriate combination of mesh number and wire diameter reported in the studies.

Figure 6-4 shows the calculated particle collection efficiency characteristics for each of the studies listed in Table 6-1, as determined by the theory of Cheng and Yeh (1980) discussed above. The collection efficiencies for the 60-mesh screens used by George (1972) and the 150-mesh screens used by Bigu (1985) are also plotted, although the wire Reynolds numbers are greater than 1. Wire screens with collection efficiencies that differ only slightly are plotted together for comparison purposes. The collection efficiencies of the screens are between 70 and 90% and 6 and 12%, respectively, for 1- and 10-nm-diameter particles and decrease rapidly in the 1- to 3-nm size range.

The concept of an unattached fraction measurement is to have a system that will collect all of the highly diffusive activity without collecting any activity attached to particles. The separation of aerosol size distributions into well-defined modes has been used to great advantage in studying the much larger sized modes (≥ 1 μm) in the ambient aerosol. However, such measurements

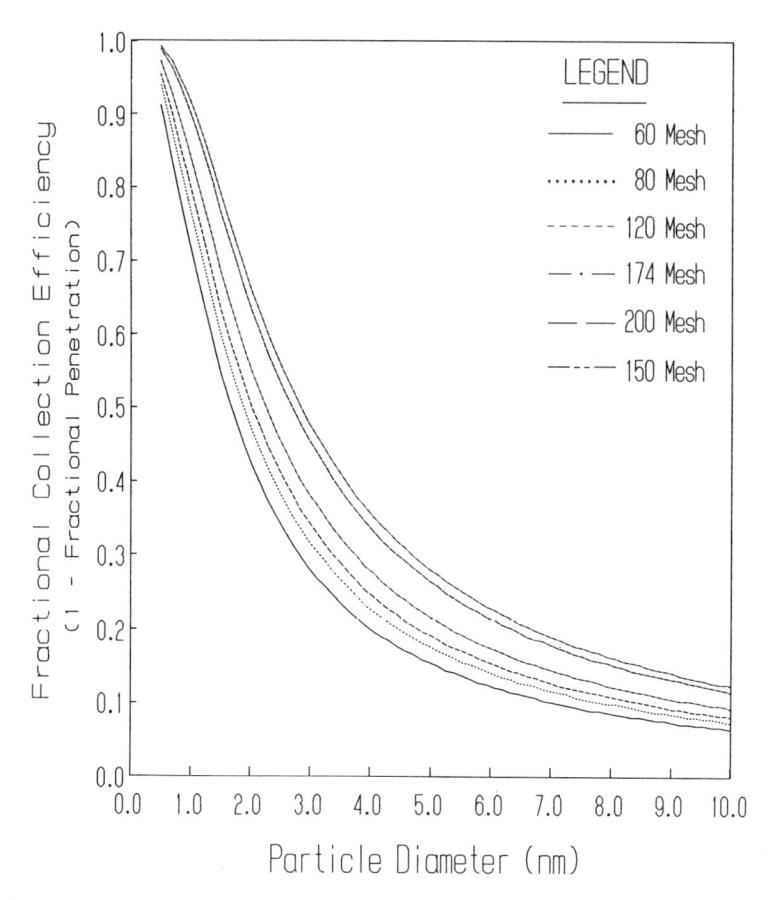

FIGURE 6-4 Collection characteristics for the wire mesh screen systems used in a number of the reported "unattached" fraction measurements.

are possible because the separation of large particles is based on their inertial properties. In this case, devices with sharp cutoff points such as cyclones or impactors can be designed (Lodge and Chan, 1986). However, because of the stochastic nature of diffusional deposition, the collection curves for wire screens or tubes are much more gradual functions of particle size. To illustrate the problem with using a single screen for making such a dichotomous measurement, Figure 6-5 shows the collection efficiency curves with $d_p(50\%)$ along with the characteristics of a log-normally distributed aerosol size distribution with a median diameter of 1.0 nm and a geometric standard deviation of 1.5 in the range suggested by Reineking and Porstendörfer (1986). Nominal values for w and α were obtained from screen manufacturers' catalogs (see Hopke et al. [in press] for details and references).

AEROSOLS IN HOMES AND MINES 99

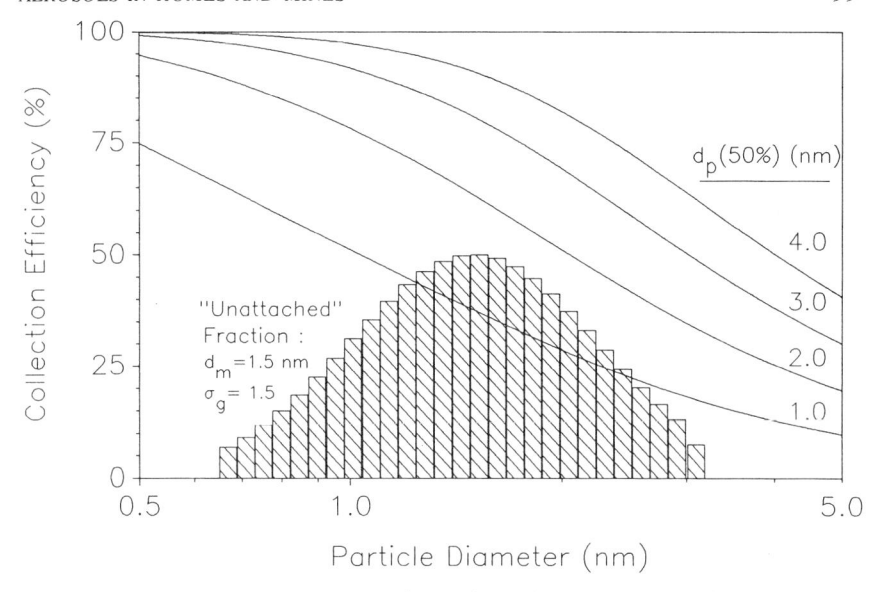

FIGURE 6-5 Collection efficiency curves for "typical" wire screens characterized by $d_p(50\%)$ values of 1, 2, 3, and 4 nm and presentation of a log-normal size distribution characteristic of the "unattached" fraction.

It can be seen that the use of wire screens underestimates the unattached fraction if it indeed consists of an ultrafine cluster mode in the 0.5- to 3-nm size range (Reineking and Porstendörfer, 1986). The cumulative collection of activity as a function of size is presented in Figure 6-6. The underestimation of the unattached cluster mode in each of the studies is presented in Table 6-1. The wire screens used in these studies would underestimate such an unattached cluster mode by 14 to 48%, depending on the $d_p(50\%)$ parameter for the particular screen-velocity combination. The choice of a single, optimized screen $d_p(50\%)$ parameter that maximizes the collection of the unattached mode while minimizing the collection of attached activity may hence be beneficial for single-screen unattached fraction measurements.

To facilitate the appropriate choice of this parameter, the collection efficiency of a log-normally distributed aerosol particle mode was determined as a function of the wire screen $d_p(50\%)$, as shown in Figure 6-6. From the data in Figure 6-6 it appears that a wire screen $d_p(50\%)$ is an optimal choice yielding about 90% collection of the unattached mode ($d_m = 1$ nm, $\sigma_g = 1.5$, $0.5 < d_p < 3$ nm), while minimizing the collection of activity in the second aerosol mode ($d_m = 25$ nm, $\sigma_g = 1.75$, $10 < d_p < 60$ nm) to less than 8%. Activity in the latter size range may be significant in indoor air following cooking, as reported by Tu and Knutson (1988a). The collection of activity attached to the larger aerosol particle mode ($d_m = 125$ nm, $\sigma_g = 2.0$, $40 < d_p < 400$ nm) is minimal for a $d_p(50\%)$ of <10 nm, and is about 1% for a $d_p(50\%)$ of 4 nm. This

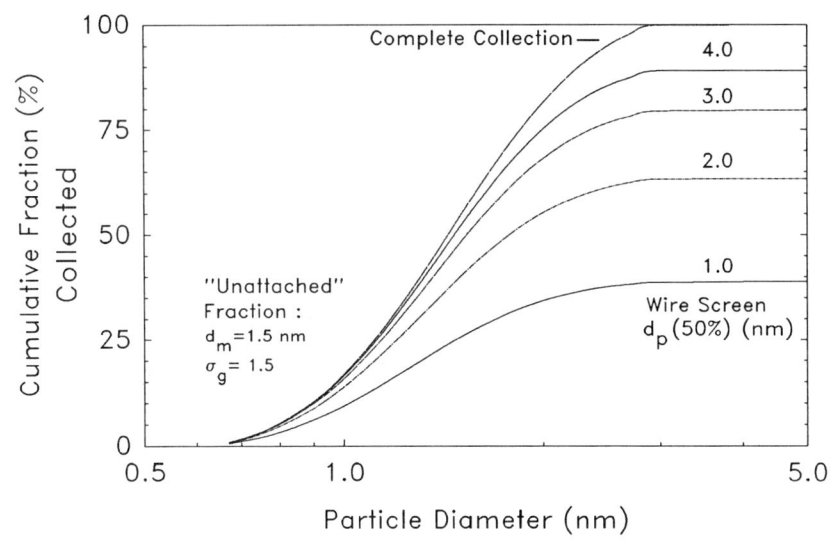

FIGURE 6-6 Cumulative collection fraction of an ultrafine activity mode.

result is consistent with the calculations by Van der Vooren et al. (1982) for the collection of attached activity in this size range by wire screens sampling the unattached mode. In the sampling of dusty atmospheres with particle sizes of a d_p of >0.5 μm (500 nm), collection by impaction and interception may become significant; but for wire screens with a d_f of >100 μm, an α of <0.3 and a U of <10.0 cm/s the wire screen collection efficiency for a 5-μm-diameter particle is less than 5%. More recent work by Reineking and Porstendörfer (1990) suggests that there can be errors in unattached fraction measurements caused by inertial collection of radioactivity-carrying particles of >100 nm in diameter.

Another consideration of wire screen systems is the activity measurements. Measurement of the activity collected by a wire screen is complicated by the deposition of activity on the front and back faces of the screen as well as alpha absorption losses in the screen weaves. The ratio of activity collected on the front and back faces of a wire screen has been found to be clearly dependent on both the screen parameters and the activity distribution sampled by the screen (Holub and Knutson, 1987). For single wire screen samplers, analysis of total or reference filter (A_t) and wire screen backup filter (A_{bf}) activities would yield more reliable estimates of the unattached fraction and would help to circumvent the problems associated with analyzing the wire screens themselves for collected activity. However, depending on the amount of activity that attaches to the screen, such a procedure may lead to low statistical precision in the activity determination on the screen backup filter.

REVIEW OF PAST UNATTACHED FRACTION MEASUREMENTS IN MINE ATMOSPHERES

Diffusion Sampler Measurements

Craft et al. (1966) used a diffusion sampler and an assumed diffusion coefficient of 0.045 cm^2/s, to find a wide range of fractions of 0 to 0.73. However, they defined their fraction as the ratio of unattached radon daughter alpha-energy concentration to the total radon decay product alpha-energy concentration. They concluded that there is so much variability from location to location and mine to mine that it is not possible to select any particular value of unattached activity as being representative of actual mine conditions. They observed that in the presence of diesel smoke, the unattached fractions were low.

Fusamura et al. (1967) also used a tube diffusion sampler. They assumed a diffusion coefficient of 0.054 cm^2/s and found fractions of ^{218}Po from 0.06 to 0.13 in active areas of mines where dust-producing activities were in progress and from 0.09 to 0.25 in inactive areas of mines according to the ICRP definition. They found that drilling operations produced substantial oil mists to which the radon progeny attached. In areas where pneumatic loaders or picks were used, the unattached fractions were in the 0.09 to 0.13 range. It was anecdotally reported that in areas where no work was being done, the unattached fraction was as high as 0.5, but no specific number was reported related to their measurements in particular locations in several mines.

George and Hinchliffe (1972) and George et al. (1975, 1977) of the Health and Safety Laboratory (HASL; now the Environmental Measurements Laboratory) described an extensive series of measurements in active uranium mines in the Grants mineral belt of New Mexico. In those studies, they used a diffusion sampler that was based on the studies of deposition of unattached radon progeny in an impactor stage (Mercer and Stowe, 1969). The collection efficiency as a function of sample flow rate is given in Figure 6-7.

However, Subba Ramu (1980) built a similar device and suggests that the device is about 90% efficient at 1 liter/min. Subba Ramu claims that the difference in efficiencies between the HASL devices described by George and Hinchliffe (1972) and his sampler were due to his "extreme precautions to minimize the presence of attached radon daughter products, which interfere with accurate calibration." Thus, there exists some uncertainty in the earlier measurements, and they may underpredict the unattached fractions.

The results of the series of measurements are summarized in Figures 6-8 and 6-9. Figure 6-9 shows the unattached fractions of ^{218}Po, f_a, as defined by the ratio of the total ^{218}Po activity collected in the sampler divided by the collection efficiency times the total airborne ^{218}Po activity as measured on an open face filter. Figure 6-8 shows the unattached fraction as defined by ICRP (1959).

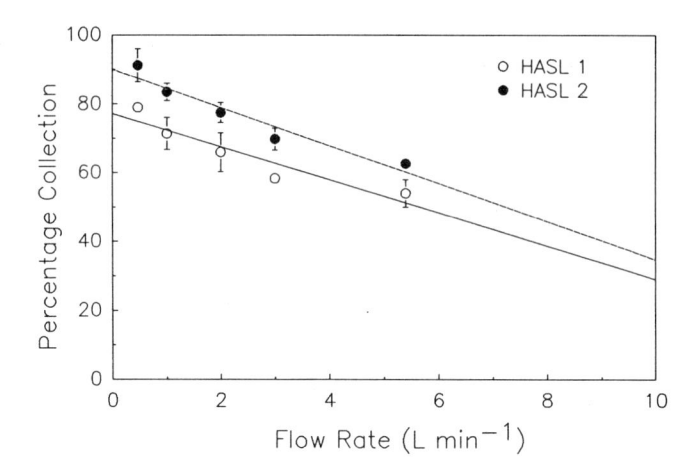

FIGURE 6-7 Collection efficiency of the HASL diffusion samplers for ^{218}Po as a function of flow rate redrawn from data in George and Hinchliffe (1972). The lines are simple first-order regression lines drawn through each set of points.

These values were obtained by multiplying the f_a values by the ratio of the total airborne ^{218}Po activity concentration (picocuries [pCi]/liter) to the radon concentration (pCi/liter). It should be noted that the ICRP unattached fraction values are substantially lower than the fraction of ^{218}Po that is unattached.

Cooper et al. (1973) also used a Mercer and Stowe (1969) diffusion sampler to measure unattached fractions in mines. They examined a mine in the Uravan Belt area of Colorado (mine A) and a mine in Ambrosia Lake, New Mexico (Grants mineral belt) (mine B). The results of these studies also included detailed analyses of the collected inorganic and organic particulate matter. Only 13 unattached fraction measurements are reported: 10 in mine A and 3 in mine B. The results of Cooper et al. (1973) are summarized in Figure 6-10. Unattached ^{218}Po compared with total ^{218}Po was determined to be 0.49 ± 0.14; for ^{214}Pb it was 0.067 ± 0.019, and for ^{214}Bi it was 0.032 ± 0.009. The unattached fraction of the potential alpha-energy concentration was 0.12, with a range of 0.09 to 0.14.

Wire Screen Measurements

Raghavayya and Jones (1974) made measurements of radon, decay product concentrations, unattached fractions of each decay product, and condensation nuclei concentration in three mines in Colorado and New Mexico. They used the approach of Thomas and Hinchliffe (1972) to calculate collection efficiency. However, since they calculated that they would obtain 92.59% collection for D of 0.06 cm^2/s activity, they assumed that there is 100% unattached activity and

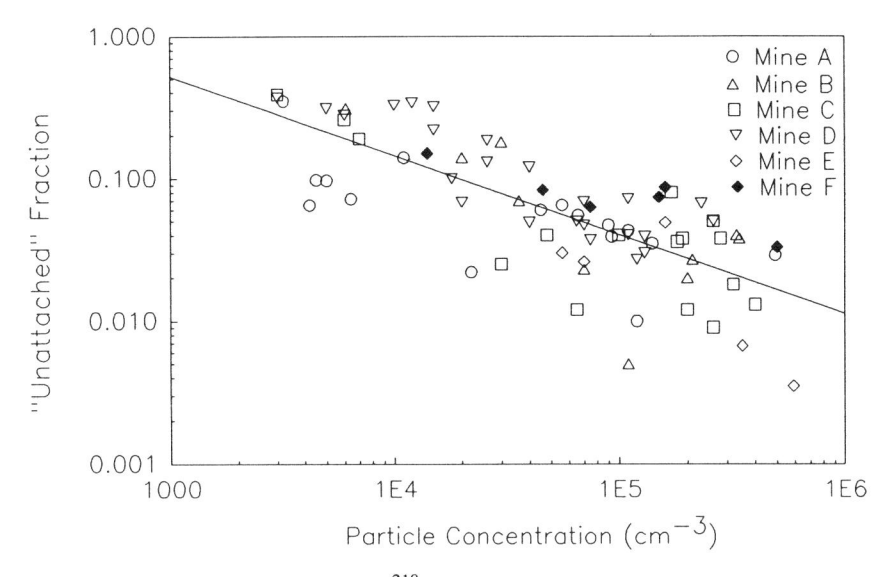

FIGURE 6-8 "Unattached" fraction of ^{218}Po as defined by ICRP (1959) measured in New Mexico uranium mines as a function of particle concentration. Data for this plot were taken from George and Hinchliffe (1972) and George et al. (1977).

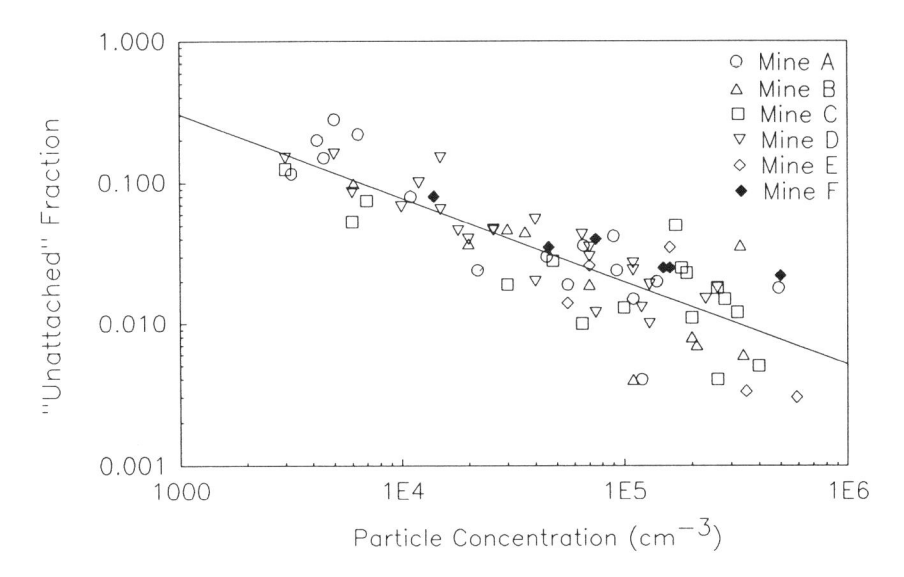

FIGURE 6-9 "Unattached" fraction of ^{218}Po as defined relative to total ^{218}Po measured in New Mexico uranium mines as a function of particle concentration. Data for this plot were taken from George and Hinchliffe (1972) and George et al. (1977).

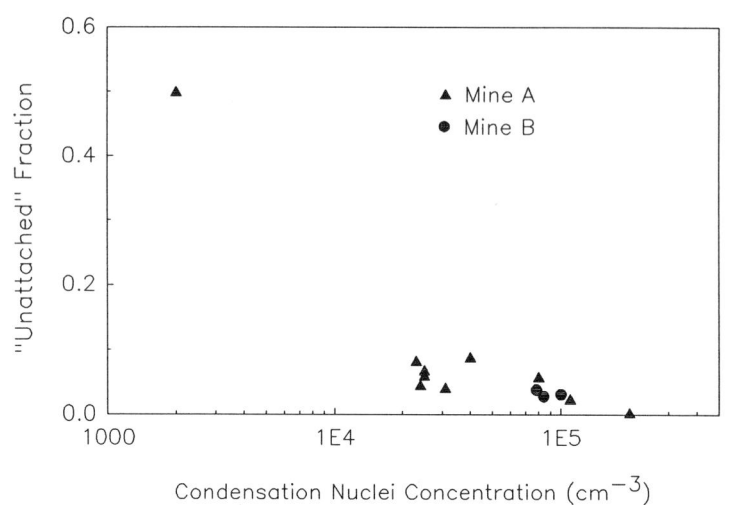

FIGURE 6-10 "Unattached" fraction measurements of Cooper et al. (1973). All of the results are the fraction of ^{218}Po activity that is "unattached."

no attached collected activity in the high-particle-concentration atmospheres in the mines. They counted the wire screens and had difficulties getting the total activity measured with a separate filter to match the sum of the activity on the screen plus that on the backup filter. They developed an empirical correction factor to obtain the values they present as the fraction of activity for each decay product that is unattached and activity for a total unattached fraction relative to the activity of radon.

Mercer (1975) noted that their correction factor produces a minimum value of the unattached fraction. He provided an alternative analysis of the calibration data and a new set of f values. However, other problems have been noted in Mercer's results. Kotrappa and Mayya (1976) examined the equations used to calculate the ^{218}Po, ^{214}Pb, and ^{214}Bi concentrations and found errors in the equations of Raghavayya and Jones (1974). Kotrappa and Mayya (1976) have recalculated the concentrations of the decay products, working levels, and a revised set of total unattached fractions that they incorrectly described as following the ICRP (1959) definition. The corrected total unattached fractions are presented as a function of the condensation nuclei concentration in Figure 6-11.

In general, the ^{218}Po results show the largest unattached fraction compared with the results for longer-lived ^{214}Pb and ^{214}Bi. The ranges of the fractions of unattached atoms to atoms present are as follows:

$$0.0013 < F_{^{218}Po} < 0.554,$$
$$0.0006 < F_{^{214}Pb} < 0.314,$$

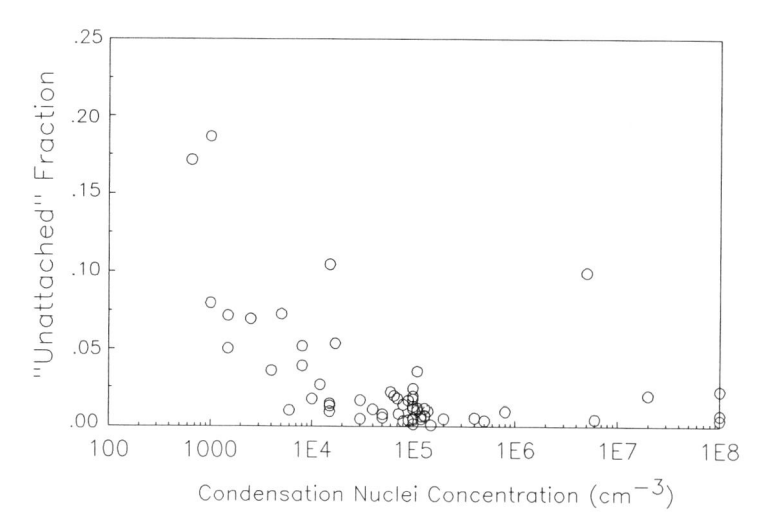

FIGURE 6-11 Total "unattached" fraction, f_t, as measured by Raghavayya and Jones (1974) with corrected values of Kotrappa and Mayya (1976).

$$0.0008 < F_{^{214}Bi} < 0.164.$$

Further examination of the results of Raghavayya and Jones (1974) still suggest problems in their measurements. For example, it would be expected that as the condensation nuclei count (c) increases, there should also be an increase in the equilibrium fraction (F), as defined by

$$F = \frac{0.105c_a + 0.516c_b + 0.379c_c}{c_{Rn}} \tag{6-11}$$

The equilibrium fraction as a function of the condensation nuclei concentration is presented in Figure 6-12. It can be seen that there is no clear trend in the data. From prior measurements of attachment coefficients (Raabe, 1969; Porstendörfer and Mercer, 1979), it would be anticipated that there should be a substantial increase in the equilibrium factor with increasing particle concentrations. These results thus suggest a substantial problem in either the radon progeny measurements or the field measurements of the number of condensation nuclei.

Bigu and Kirk (1980) measured unattached fractions in two Canadian mines using both a diffusion sampler and a wire screen system. The collection efficiencies of the diffusion and wire screen samplers were reported to be 53 and 95%, respectively, when operated at 2 liters/min. Their unattached ^{218}Po fraction results are presented in Table 6-2. In general, there is reasonable agreement between the two measurements. Unfortunately, there are only limited condensation nuclei measurements, and no radon measurements were reported. Thus, the ICRP F values cannot be calculated, nor can the equilibrium fraction

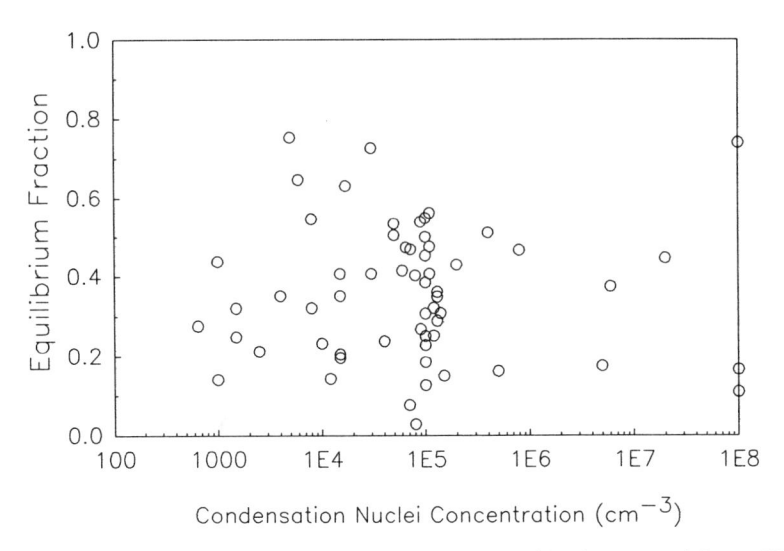

FIGURE 6-12 Equilibrium fraction, calculated from the data of Raghavayya and Jones (1974) as corrected by Kotrappa and Mayya (1976), as a function of the condensation nuclei count.

be determined. The values are similar to the other values reported for active working areas of mines.

Other measurements were performed in one Canadian mine by Busigin et al. (1981). They used parallel plate diffusion batteries. Their measurements were in active mining areas with high particle concentrations (approximately 10^6 nuclei/cm^3). Under these conditions, they found no distinct unattached fractions, with an upper limit of 1 to 2%. Subsequently, the same group made more extensive measurements in two mines using a wire screen sampler (Khan et al., 1987). Although that report followed publication of the detailed theory of wire screen collection in the early 1980s, this group continued to use the collection efficiency curve of Thomas and Hinchliffe (1972). Thus, the penetration efficiencies have some error and lead to an underestimation of the unattached fraction by approximately 10 to 15%. In subsequent measurements, Bigu (1985) also found very small values of unattached progeny (≤ 0.01) with aerosol concentrations in excess of 10^5/cm^3.

Stranden and Berteig (1982) made a series of 33 measurements in an iron ore mine in Norway. They also used the collection efficiency curve of Thomas and Hinchliffe (1972) and reported the fractions of unattached ^{218}Po, ^{214}Pb, and ^{214}Bi and the fraction of unattached potential alpha energy concentration (PAEC). The unattached PAEC is obtained from the individual concentrations as follows:

TABLE 6-2 "Unattached" Fraction Measurements Reported by Bigu and Kirk (1980)

Mine	Test No.[a]	"Unattached" Fraction D.S.[b]	W.S.[c]	Condensation Nuclei (cm^{-1})	Mining Operation	Area of Mine
A	1	.09	.07	2.5 × 10^5	Drilling and slushing	Non-Diesel
B	2	.064	—	3.0 × 10^5	Drilling	Non-Diesel
A	3	.058	—	—	Drilling	Non-Diesel
A	4	.014	.025	—	Roof bolting	Non-Diesel
A	5	.0047	.0047	—	Mucking	Diesel
A	6	.006	.0025	—	Mucking and drilling	Diesel
A	7	.0067	.007	—	Mucking, drilling, and roof bolting	Diesel

[a]Each test represents one day during which a number of measurements were made.
[b]Diffusion sampler measurements.
[c]Wire screen measurements.

$$f_p = \frac{1.05 C_a^f + 5.16 C_b^f + 3.8 C_c^f}{1.05 C_a + 5.16 C_b + 3.8 C_c} \quad (6\text{-}12)$$

The distribution of values for these four variables are presented in Figure 6-13. In this case, condensation nuclei concentrations were not measured. However, they did obtain very reasonable relationships for f_a and f_p, with the respirable dust concentration (particles of <7 μm) in mg/m^3. Their results are summarized in Table 6-3.

Summary of Mining Exposure

From the limited number of measurements of activity-size distributions made in active mines, it is necessary to estimate the typical exposure conditions in the mine. In this review, an attempt has been made to estimate the conditions both in the areas where active drilling, slushing, and other activities could produce substantial airborne particle concentrations, in the haulage drifts where the material was being transported from the active mine sites to the sites where it could be conveyed to the surface, and in areas where nonmining activities were being conducted (workshops, lunchrooms, etc.). In the report of Cooper et al.

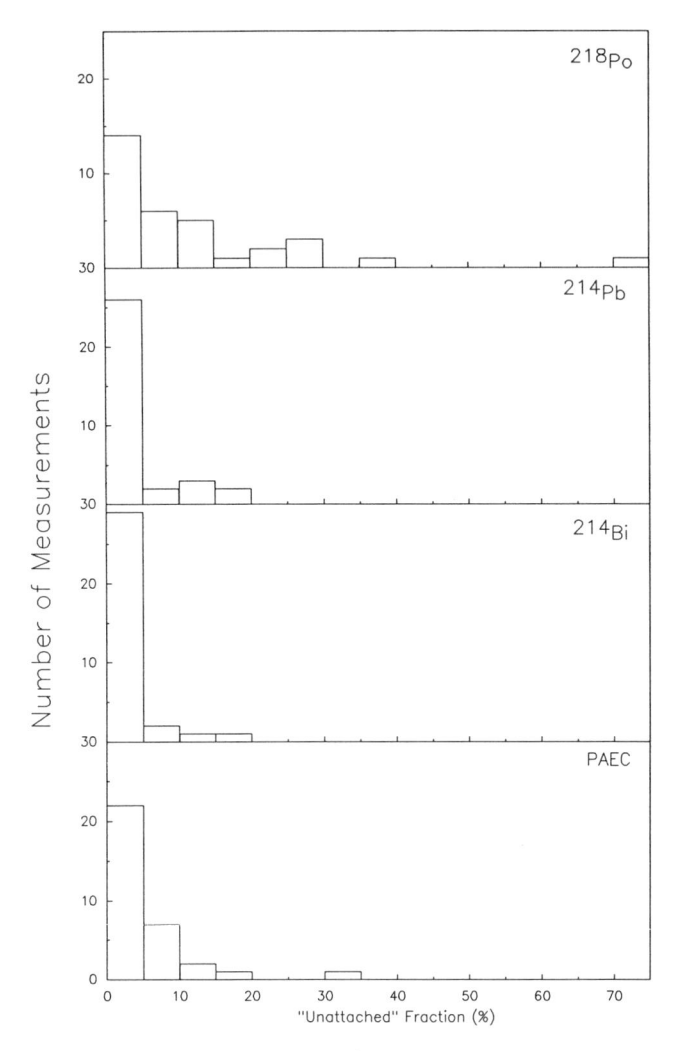

FIGURE 6-13 Distributions of the "unattached" fractions of the three radon decay products and the potential alpha-energy concentration (PAEC) measured in an iron ore mine. Redrawn from the graph in Stranden and Berteig (1982).

(1973), very detailed descriptions of the sampling locations are provided. They found AMD values between 0.20 and 0.36 μm in locations near active mining sites with an average value around 0.25 μm. Bigu and Kirk (1980) provide airborne respirable dust concentrations, AMAD and unattached fraction values specifically in working locations such as "Slusher position 1." They found very low unattached fractions and AMAD values also of the size of 0.25 μm. Thus, because of the very clear descriptions of locations and corresponding activity,

TABLE 6-3 Values for the Mean and Median "Unattached" Fractions as Measured in a Norwegian Iron Ore Mine by Stranden and Berteig (1982)

Species	Mean	Median
^{218}Po	0.123	0.063
^{214}Pb	0.057	0.032
^{214}Bi	0.032	0.018
PAEC	0.059	0.038

these results were given high weight in concluding the results for the active mining area. For the haulage drifts, the extensive values of George (1972) and George et al. (1975, 1977), particularly as reevaluated by Knutson and George (1990) were given the most credence and their average AMAD value of 0.15 μm was used for the average size in the open areas of the mine away from the actual mining activities. The values for the nonmining enclosed spaces were estimated from the indoor results presented in the next section.

INDOOR ATMOSPHERES

Diffusion Sampler Measurements

Duggan and Howell (1969) first attempted to use electrostatic collection of the decay products to determine the unattached fractions. However, as is now known (Hopke, 1989a,b), neutralization of the decay products is sufficiently rapid in normal air that only a small fraction of the highly diffusive fraction of the activity is charged. Duggan and Howell then used a rectangular channel diffusion battery to remove the unattached activity. The battery had 28 channels with cross-sections of 0.07 by 5 cm and operated at a flow rate of 80 liters/min. They measured the unattached fraction in outdoor and laboratory air at quite low radon concentrations (0.04 to 0.39 pCi/liter). In the approximately 50 measurements that were made, they obtained values in the range of 0.07 to 0.40 with no discernible relationship between the unattached fraction and the radon concentration. No measurements of the condensation nuclei concentrations were reported, nor was there any apparent attempt to obtain size distribution results from the measurements. Thus, it is difficult to use these values to assess the exposure to unattached progeny in either the ambient atmosphere or indoor air.

Shimo and Ikebe (1984) and Shimo et al. (1985) presented results of unattached fraction measurements in an underground tunnel of the Mikawa Crustal Movement Observatory of Nagoya University (Nagoya, Japan) using a diffusion tube. They then used the tube as the body of a proportional counter and followed the decay of the alpha activity in order to extract the concentrations of the three decay products. The total activity of each decay product was determined by using the Thomas-modified Tsivoglou method (Thomas, 1970). Radon concentrations were in the range of 73.5 to 251 pCi/liters. The aerosol concentrations were measured with a Pollak-type condensation nuclei counter and were found to be in the range of 2,000 to 10,000/cm.

Wire Screen Measurements

Only a limited number of measurements of unattached fractions in single-family homes, apartments, offices, and other nonmine locations have been performed. Porstendörfer and coworkers (Reineking et al., 1985; Reineking and Porstendörfer, 1986; Porstendörfer, 1987; Reineking and Porstendörfer, 1990) have made measurements of both unattached fractions and the total activity size distributions for a series of rooms in unoccupied houses. In those studies, they varied the ventilation rate and added aerosol sources to the room. From the results of the measurements on the radon progeny, radon, and particle concentrations, they then calculated the unattached fractions of ^{218}Po, ^{214}Pb, and potential alpha energy and the equilibrium fractions. The results of the unattached fraction of PAEC as a function of condensation nuclei concentration in five rooms with and without aerosol sources are shown in Figure 6-14. Figure 6-15 presents the equilibrium factor as a function of particle concentration for these same rooms. In a later report (Porstendörfer, 1987), three additional rooms were characterized with similar results. In their most recent report, Reineking and Porstendörfer (1990) also provide some results for a limited number of measurements in the ambient atmosphere. These results are presented in Table 6-4.

Vanmarcke et al. (1985, 1987, 1989) used a screen diffusion battery with a characteristic $d_p(50\%)$ of 4 nm for "unattached" fraction measurements in indoor air. In addition aerosol size distributions were made with an automated aerosol spectrometer system (Raes et al., 1984). They measured the unattached fraction of PAEC and reported that their results indicate "that the fraction of unattached radon daughters is higher than assumed in earlier studies." A sequence of unattached fraction measurements made in a single Belgian house is shown in Figure 6-16 (Vanmarcke et al., 1985). Additional studies made in the same house and in several other locations including the authors' laboratory and a railroad station as well as a second room of the house. Figure 6-17 summarizes the equilibrium and unattached fraction measurements plotted as a function of the attachment rate inferred from the aerosol size distribution. Unfortunately, none of the reports by that group provided actual values of these variables, nor summary statistics of their measurements.

The Ghent, Belgium, group (Vanmarcke and coworkers) intercompared

AEROSOLS IN HOMES AND MINES 111

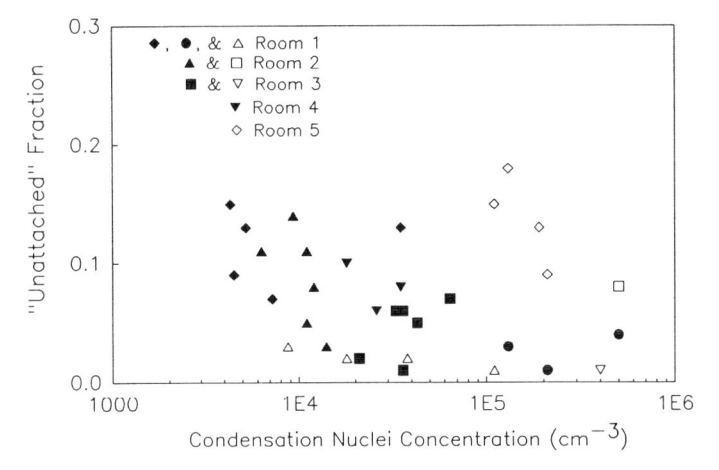

FIGURE 6-14 "Unattached" fraction of potential alpha-energy concentration, f_p, as a function of the condensation nuclei concentration. Data taken from Porstendörfer (1987).

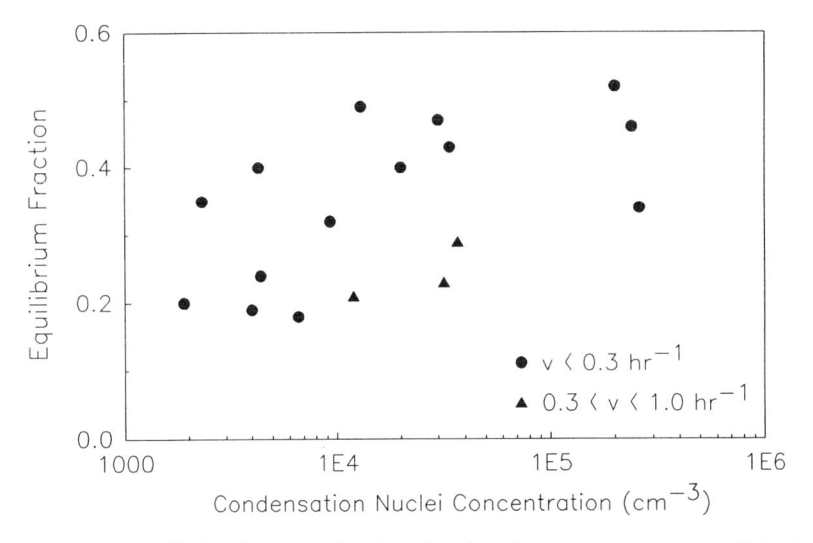

FIGURE 6-15 Equilibrium factor as a function of condensation nuclei concentration. Data taken from Porstendörfer (1987).

their measurement methodologies with those of the group from Göttingen, Germany, (Reineking and Porstendörfer). The results of this intercomparison study have been presented by Vanmarcke et al. (1988). The results of the intercomparison of radon and radon decay product concentrations are presented in Figure 6-18. It should be noted that Ghent values for ^{218}Po were consistently higher than the Göttingen values, although the radon concentrations varied

TABLE 6-4 Concentrations of Radon, "Attached" and "Unattached" Activities, f_p, F Values, and Particles (N) Measured in the Ambient Atmosphere near Göttingen (1 m Above the Ground During Daylight) by Reineking and Porstendörfer (1990)

Period	^{222}Rn (Bq m^{-3})	^{218}Po (Bq m^{-3})	^{218}Po (Bq m^{-3})	^{214}Pb (Bq m^{-3})	^{214}Pb (Bq m^{-3})	^{214}Bi (Bq m^{-3})	N (10^3 cm^{-3})	f_p	F	No. of Values
February 1986	14.6 (4.2-37.7)	6.8 (1.4-21.5)	0.8 (0.0-3.0)	7.5 (1.9-24.1)	0.1 (0.0-1.6)	8.1 (2.0-22.4)	41 (14-71)	0.012 (0-0.098)	0.51 (0.37-0.67)	8
October 1986	11.3 (2.0-23.6)	7.8 (0.8-18.2)	1.5 (0.0-5.6)	8.4 (1.1-17.6)	0.2 (0.0-1.7)	9.1 (0.8-20.6)	41 (7.5-89)	0.031 (0-0.180)	0.79 (0.51-1.15)	13
July 1988	6.9 (1.0-17.2)	3.9 (0.4-19.7)	0.6 (0.0-1.9)	4.3 (0.6-20.6)	0.1 (0.0-2.0)	4.6 (0.6-20.1)	23 (7.7-93)	0.014 (0-0.216)	0.64 (0.23-1.19)	15
Mean	10.2 (1.0-37.7)	6.0 (0.4-21.5)	1.0 (0.0-5.6)	6.5 (0.6-24.1)	0.1 (0.0-2.0)	7.0 (0.6-22.4)	34 (7.5-93)	0.020 (0-0.216)	0.67 (0.23-1.19)	36

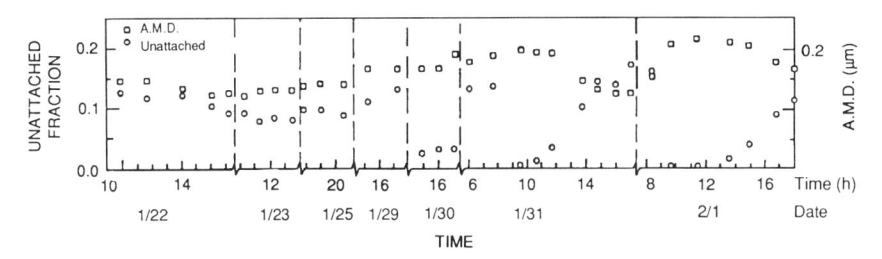

FIGURE 6-16 Time sequence of measurements in a Belgian house. Figure taken from Vanmarcke et al. (1985) and used with permission.

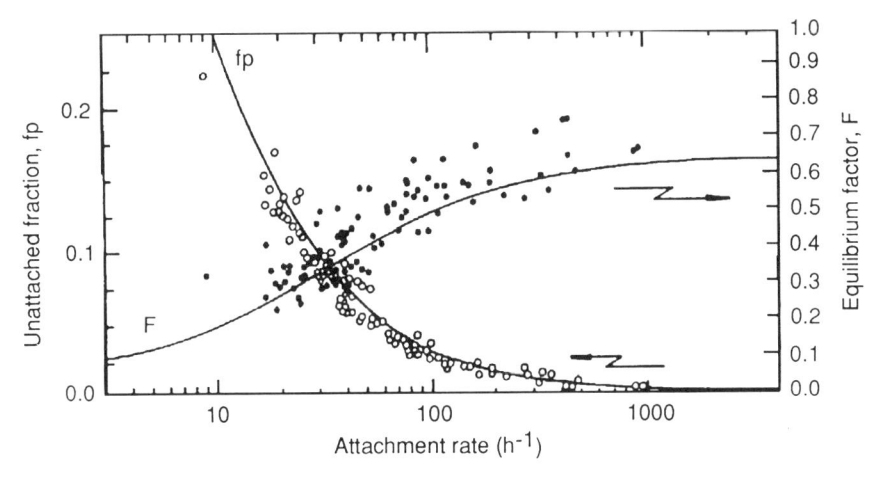

FIGURE 6-17 Equilibrium factor and the "unattached" fraction of potential alpha-energy concentration as a function of the estimated attachment rate based on measured aerosol size distributions.

between the higher and lower values. There thus appears to be a consistent measurement problem that leads to high ^{218}Po results and systematically high f_p values as shown in Figure 6-19. It can be seen that the Ghent values are generally higher than the Göttingen values, particularly in the absence of an additional aerosol source. The Göttingen unattached fraction values are generally in good agreement with those observed in other studies. Thus, although there is excellent agreement for the equilibrium fraction results, it appears that the f_p values of Vanmarcke et al. (1988) are high and may overestimate the true value by more than 50%.

Stranden and Strand (1986) measured the unattached fraction and the equilibrium factor in occupied dwellings and an underground hydroelectric generating station and compared these results with their underground mining results. There were no descriptions of the housing units, the locations in

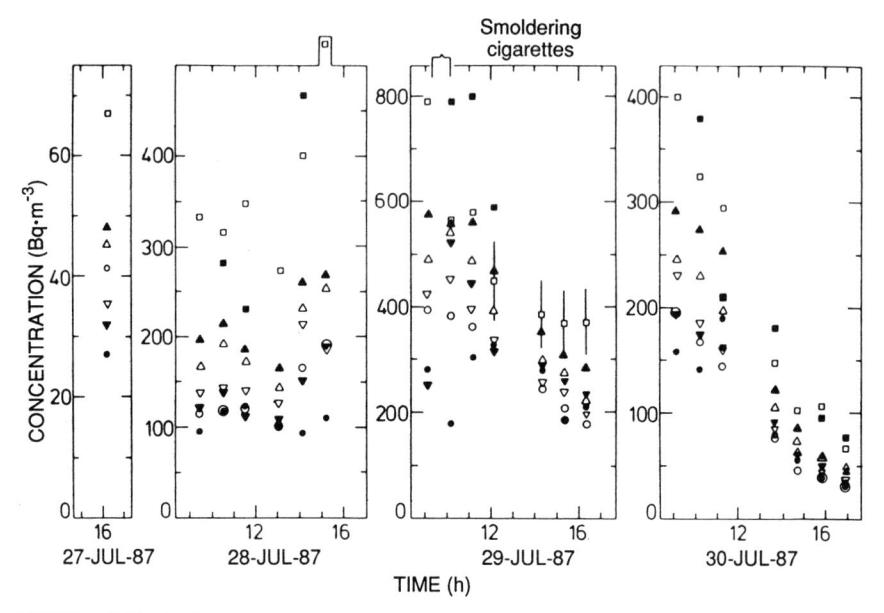

114 COMPARATIVE DOSIMETRY OF RADON IN MINES AND HOMES

FIGURE 6-18 Radon and radon decay product concentrations measured during the Ghent-Göttingen intercomparison study. Filled symbols are Ghent; open symbols are Göttingen. Figure taken from Vanmarcke et al. (1988) and used with permission.

which sampling occurred, nor the "normal activities" that were ongoing in the dwellings and the generating station. They fit their data to an equation of the form

$$f_p = aF^b \qquad (6\text{-}13)$$

where a and b are empirical coefficients. They linearized the equation using a logarithmic transformation and performed least-squares fits to the data. Such a process produces biased results because the uncertainties are not uniform and do not propagate linearly through a logarithmic transformation. They obtained the following results:

$$\text{Dwellings:} f_p = 0.0172 F^{-2.34}$$
$$\text{Hydro-electric station:} f_p = 0.0455 F^{-1.23}$$
$$\text{Mine:} f_p = 0.016 F^{-1.12}$$

If the same analysis is applied to the data obtained by Reineking and coworkers (1985, 1990), an equation of the form $f_p = 0.0177\ F^{-2.50}$ with a correlation coefficient of 0.80 is obtained. Thus, close agreement of unattached and equilibrium fractions has been obtained for aerosols in these different indoor

AEROSOLS IN HOMES AND MINES 115

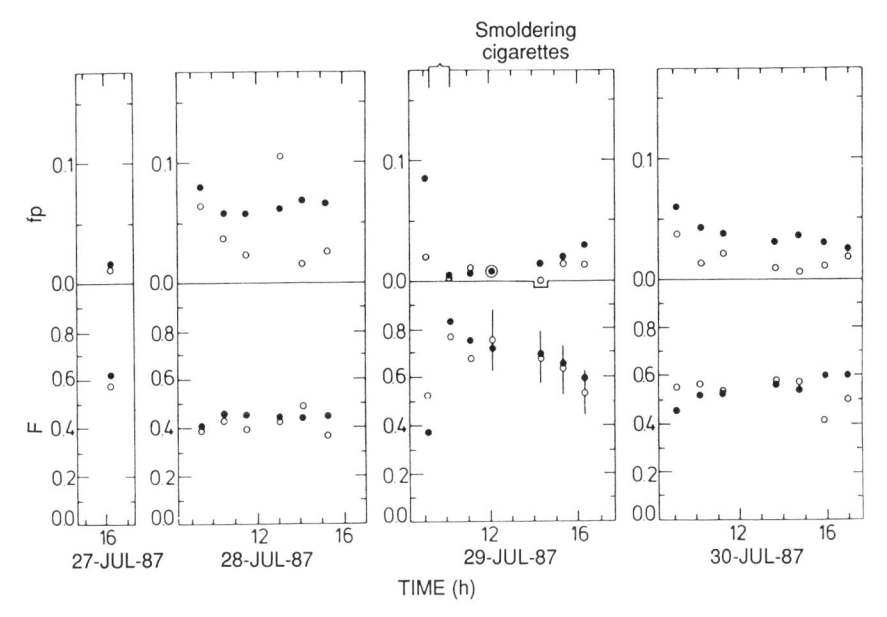

FIGURE 6-19 Evolution of the "unattached" and equilibrium fractions as measured in the Ghent-Göttingen intercomparison study. Figure taken from Vanmarcke et al. (1988) and used with permission.

spaces. These results indicate that the size distribution of the indoor aerosol must be similar in these houses, although the concentration may vary.

Kojima and Abe (1988) developed an automated tape sampler for the "alpha-energy" spectroscopic measurement of the time sequence of unattached fraction in Japanese houses. They used a 500-mesh screen at a face velocity of 29.7 cm/s. They also used the calibration curve of Thomas and Hinchliffe (1972) to estimate a collection efficiency of 99%. They calibrated the counting efficiency of activity on the screen. The correction factor for undetectable activity on the back of the screen was determined to be 0.79 ± 0.21. As in the other measurements of the front-to-back ratio (Holub and Knutson, 1987), the amount of activity on the back side of the filter is dependent on the actual size of the unattached mode. There was also a loss of activity in the sampler's inlet. The correction factor for the sampler head loss is 0.82 ± 0.10. They were able to measure 2 Bq of ^{218}Po per m^3 and 0.3 Bq of ^{214}Pb per m^3 with a relative standard deviation of 50%.

The instrument was then deployed in a detached, two-story concrete house used in the typical Japanese life-style. Figure 6-20 provides the diurnal variations of the unattached fractions of each of the three decay products and the aerosol concentration. A summary of the unattached fraction measurements made over 7 months (November 1986 to May 1987) is presented in Table 6-5.

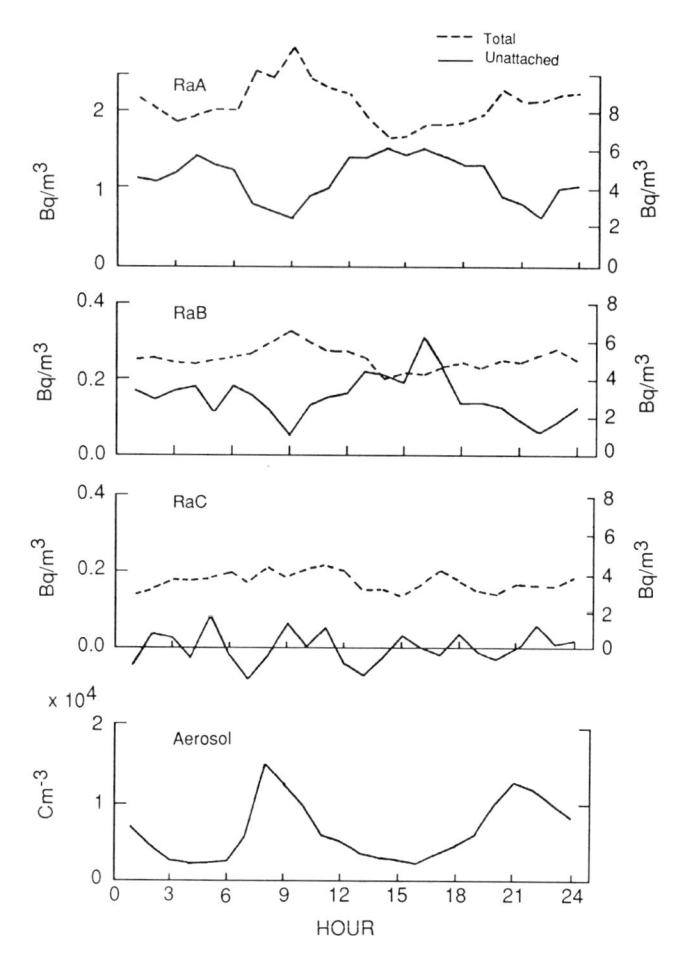

FIGURE 6-20 Diurnal variation of the "unattached" fractions of each of the radon decay products and the aerosol concentration. Left axis for unattached; right axis for total. Figure taken from Kojima and Abe (1988) and used with permission.

It should be noted that during this 7-month period the total ^{218}Po concentration dropped from 11.1 Bq/m^3 in November to 3.9 Bq/m^3 in May. Thus, concentrations in this house are very low; it should be noted that Japanese construction and life-style are quite different from those in the United States.

Unattached fraction wire screen measurements must be tempered with the recognition that (1) the unattached fraction is, in reality, an ultrafine cluster mode in the 0.5- to 3-nm size range, (2) the collection efficiency versus particle diameter characteristics for wire screens do not allow a distinct separation of the unattached and attached fractions, and (3) there can be collections of

TABLE 6-5 Summary Results of the "Unattached" Fractions Measured in a Single Japanese House (Kojima and Abe, 1988)

Period	No. of Values	"Unattached" ^{218}Po	^{214}Pb	"Unattached" PAEC
November 1986	142	0.12	0.025	0.032
December	168	0.11	0.028	0.034
January 1987	144	0.12	0.035	0.043
February	240	0.078	0.024	0.031
March	200	0.13	0.029	0.045
April	190	0.19	0.041	0.055
May	227	0.21	0.040	0.064

large particles (>100 nm) on screens through inertial impaction (Reineking and Porstendörfer, 1990). Through appropriate choice of screen operating parameters, efficient collection of the 0.5- to 3-nm activity fraction may be obtained, while minimizing the collection of attached activity. Because unattached fraction measurements are inexact representations of the particle size information needed to utilize the dosimetric model developed and presented elsewhere in this report, the next section will review the more recent measurements that have been developed and employed to characterize the indoor radioactive aerosol.

ACTIVITY-WEIGHTED SIZE DISTRIBUTIONS

Recently, methods have been developed by which the entire radioactive aerosol size distribution can be deduced from data regarding the collection of activity on or its penetration through a series of screens. From these distributions, that portion of the size range that is to be defined as "unattached" can be calculated. Early measurements of the activity-weighted size distributions were made with conventional aerosol sampling systems such as tube or screen diffusion batteries (George and Breslin, 1980). Since condensation nuclei counters are commonly used as the particle detector for the corresponding number distributions, their rapid decrease in counting efficiency below 10 nm limits their utility to particle diameters >5 nm. Thus, these systems were typically designed with a minimum-size "particle" of 10 nm in mind.

The first activity-size measurements in indoor and ambient air were made by Sinclair et al. (1977) using a specially designed high-volume flow diffusion battery. They observed bimodal distributions with activity mode diameters of 7.5 and 150 nm in indoor atmospheres and 30 and 500 nm outdoors in New York City. Similar results were reported by George and Breslin (1980). Becker et al. (1984) only observed the larger mode using a modified impactor method with a minimum detectable size of 10 nm. Their measurements were made in Göttingen, Germany. More extensive measurements in New York City by the group at the Environmental Measurements Laboratory (EML) have been reported by Knutson et al. (1984) using several different types of diffusion batteries as well as cascade impactors. They again observed modes around 10 nm and 130 nm in the PAEC-weighted size distribution measured with a low-volume screen diffusion battery. Four samples taken with a medium-volume (25-liter min^{-1}) screen diffusion battery showed a major mode at 80 to 110 nm and a minor mode containing 8 to 9% of the PAEC with a diameter <5 nm. Finally, the same group made measurements at Socorro, N. Mex. (George et al., 1984). They reported that the major mode was only slightly different from that found in New York, but the minor mode was always < 5 nm, distinctly smaller than the New York distributions.

One of the problems with the extension of screen diffusion batteries to smaller particle sizes is the substantial collection efficiency of the high-mesh-number screens typically used in diffusion batteries designed to cover the range of particle sizes from 5 to 500 nm. At normally used flow rates, a single 635-mesh screen has greater than 90% efficiency for collecting 1-nm particles, the size of "unattached" ^{218}Po having a diffusion coefficient of the order of 0.05 cm^2 s^{-1}. Thus, once the Yeh-Cheng screen penetration theory had been validated to 4 nm by Scheibel and Porstendörfer (1984) and the limitations of high-mesh-number screens were recognized, it was then possible to begin to examine alternative diffusion battery designs that could be extended to smaller particle diameters.

Reineking et al. (1985), Reineking and Porstendörfer (1986), and Reineking et al. (1988) used the high-volume-flow diffusion batteries described by Reineking and Porstendörfer (1986) to obtain activity-size distributions. They obtained their size distributions by fitting log-normal distributions using a SIM-PLEX algorithm. Size distributions of indoor air in rooms without and with additional aerosol sources are presented in Figures 6-21 and 6-22, respectively. From these results, it can be surmised that the unattached fraction is the mode with a median diameter of 1.2 nm and geometric standard deviation of 1.5 nm.

Holub and Knutson (1987) reported the development of low-flow diffusion batteries with low-mesh-number screens and extension of the EML batteries to smaller sizes. Tu and Knutson (1988a,b) used 25-liter/min screen diffusion batteries to measure the ^{218}Po-weighted size distributions in the presence of several specific aerosol sources. The results of these measurements are presented in Figures 6-23 and 6-24. The presence of a mode around 10 nm is again observed in curve 1 in Figure 6-23. Only in curve 1 (no active aerosol sources) in Figure 6-24 was a mode at 1 nm observed. In all of the other cases, the activity was attached to the aerosol present in the house. The attachment was confirmed by independently measuring the aerosol size distributions by using an electrical aerosol analyzer (Liu and Pui, 1975) and the attachment coefficients

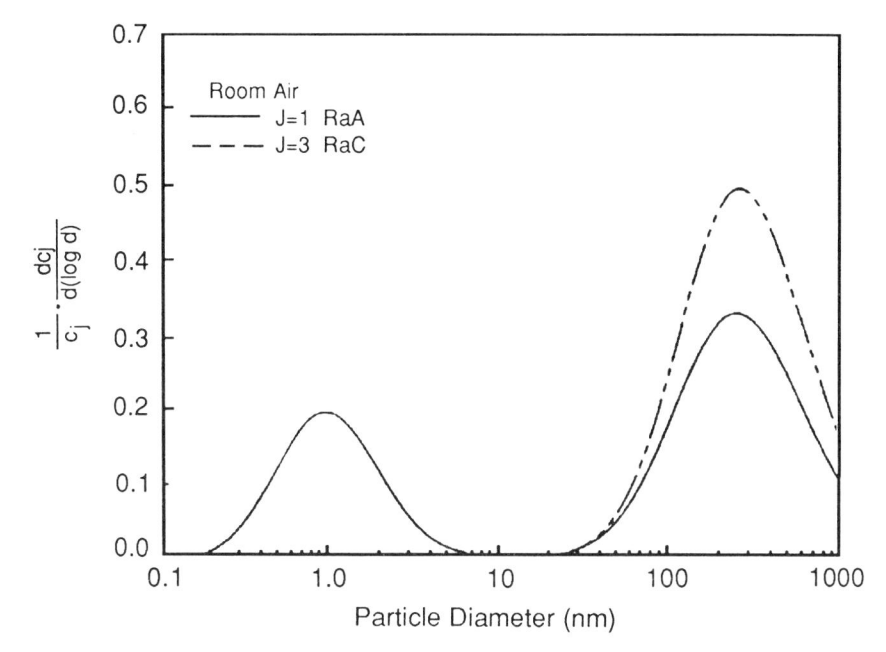

FIGURE 6-21 Activity-weighted size distribution of the indoor aerosol in a closed room without additional aerosol sources. Figure taken from Reineking and Porstendörfer (1986) and used with permission.

recommended by Porstendörfer et al. (1979). The agreement between the measured and calculated activity-weighted size distributions was excellent.

Several other groups, including the National Radiation Protection Board (NRPB) of the United Kingdom and the Australian Radiation Laboratory (ARL), have also developed these graded screen diffusion batteries for activity-size distribution measurements. An intercomparison between these three groups (EML, NRPB, and ARL) has been performed (Knutson et al., 1988). This initial comparison found difficulties for particles with sizes of >500 nm and <20 nm. For the large particles, the problem arises from impaction on the screen, and thus, there is an apparently excessive collection of activity in the ultrafine size range. For the smaller particle size range, there were several unexplainable discrepancies among the measurements.

Further tests of the single-screen methods were performed by Holub et al. (1988) in which EML, ARL, and the U.S. Bureau of Mines (BuMines) made measurements in a chamber at ARL. The results agreed well as to the size of the unattached progeny. There were differences of about a factor of 2 in the measured amounts of airborne activity, but these differences were attributed to differences in sampling location rather than difficulties with the various screen configurations. The single-screen measurements showed significant differences

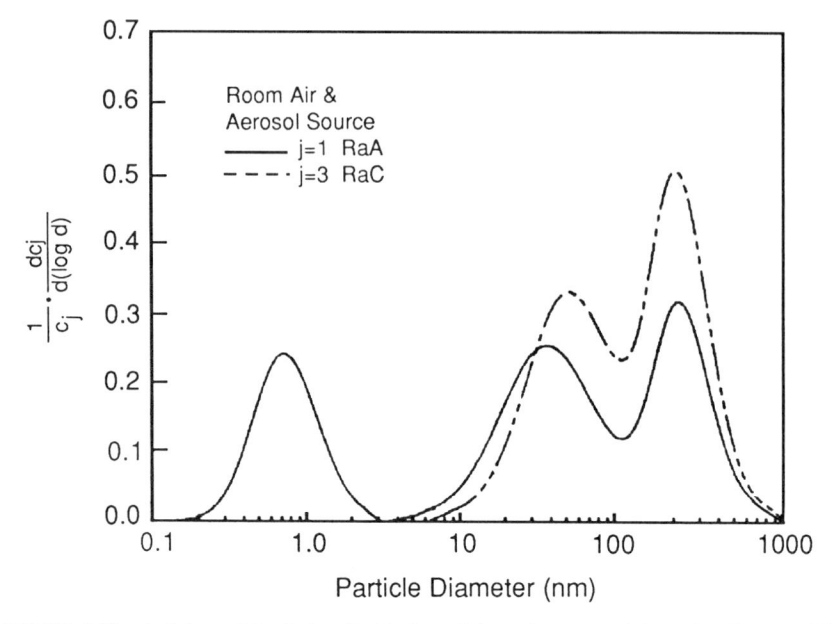

FIGURE 6-22 Activity-weighted size distribution of the indoor aerosol in a closed room with an additional aerosol source. Figure taken from Reineking and Porstendörfer (1986) and used with permission.

from the conventional screen diffusion battery measurements for the unattached fraction size, but were in good agreement for the attached mode sizes. These results can be anticipated because of the lack of resolution for the diffusion battery below 5 nm.

Several more recent intercomparison studies involving these three groups, the U.S. Bureau of Mines, Denver Research Center (BuMines), P. K. Hopke's group (then at the University of Illinois, Urbana), and the Inhalation Toxicology Research Institute (ITRI) at Albuquerque, N. Mex., found excellent agreement among the various activity-weighted size measurements as well as with the size distributions inferred from measurements of the particle size distributions by using a differential mobility analyzer (Ramamurthi et al., 1989). Thus, the results of these studies suggest that it is now possible to measure activity-weighted size distributions from 0.5 to 500 nm.

Several automated systems that make use of this methodology have been developed. Strong (1988) developed a system with six sampling heads containing 0, 1, 3, 7, 18, and 45 stainless steel, 400-mesh wire screens. He measured the size distributions in several rooms in two houses at two times of the year. The size distributions observed in the kitchen are presented in Figure 6-25. These results are summarized in Table 6-6. It should be noted that in the kitchen curve in Figure 6-25, a trimodal distribution is observed: a true unattached fraction

AEROSOLS IN HOMES AND MINES 121

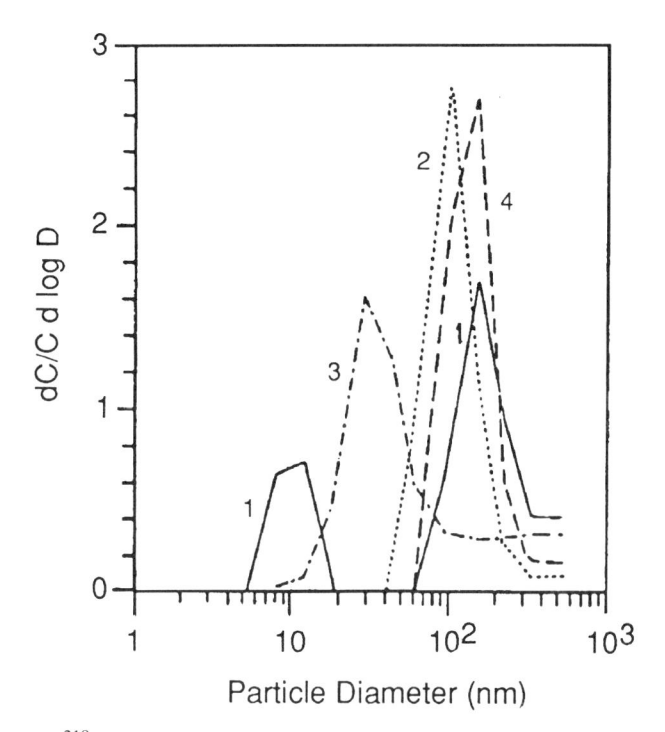

FIGURE 6-23 ^{218}Po-weighted size distributions measured in house I. 1. cooking (5 min); 2. frying food; 3. cooking soup; 4. cigarette smoldering. Figure taken from Tu and Knutson (1988a) and used with permission.

at 1 nm, a nuclei mode at 10 nm, and an accumulation mode at 100 to 130 nm. In Table 6-6, the unattached fractions presented are the integrated values from the size distributions. A problem then arises as to what unattached means since Strong integrated the distribution up to >10 nm to obtain that fraction that he attributes as being unattached. For the distribution in the kitchen with cooking, the activity median diameter for the unattached fraction was given as 11 nm. Such attribution is a clear departure from the original purpose for defining an unattached fraction. The advent of these measurement systems requires a more precise definition of the meaning of the unattached fraction.

Subsequent to these original measurements, Strong (1989) modified his system by changing the screens to 1,200-mesh screens and 1, 4, 14, and 45 400-mesh screens as well as the open channel. This modification provides a stage such that there is better resolution at the smallest-sized particles and the range of the system can be extended to 0.5 nm. The effective resolution cutoff of the original battery was about 2 nm. With the new battery, trimodal distributions were clearly observed (Figure 6-26). Although these measurements were made in the living room, the kitchen was adjacent to the living room, and

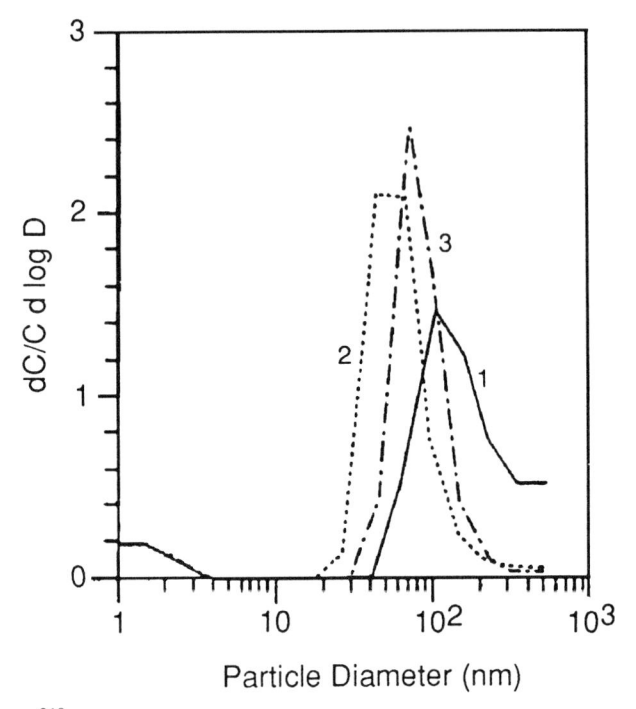

FIGURE 6-24 ^{218}Po-weighted size distributions measured in house II after a kerosene heater was operated for: 1. 0 min; 2. 80 min; 3. 200 min. Figure taken from Tu and Knutson (1988a) and used with permission.

cooking with a gas stove was being performed at the time these measurements were made. These results show the advantage of being able to measure the size distribution and determine the actual exposure of individuals to airborne radon progeny activity.

A similar system has been developed at the Australian Radiation Laboratory by Solomon (personal communication, 1989). It was designed for measurements in the particle size range of 2 to 600 nm. The measurements were extended to a smaller size range (0.5 to 100 nm) by using a manual, serial, single-screen array sampling at 1 to 6 liters/min and to lower concentrations in the same size range by using larger screens (9.5-cm diameter) and a 100-liter/min flow rate. Solomon examined both the Twomey (1975) and expectation-maximization (Maher and Laird, 1985) algorithms for deconvoluting the size distributions from the screen penetration data. In both cases, Solomon has developed a Monte Carlo method for determining the stability of the inferred size distributions. New input values for the concentrations found on each stage were chosen from a normal distribution by using the measured radon decay product activity as the mean value and the measured uncertainty as the standard deviation of the distribution.

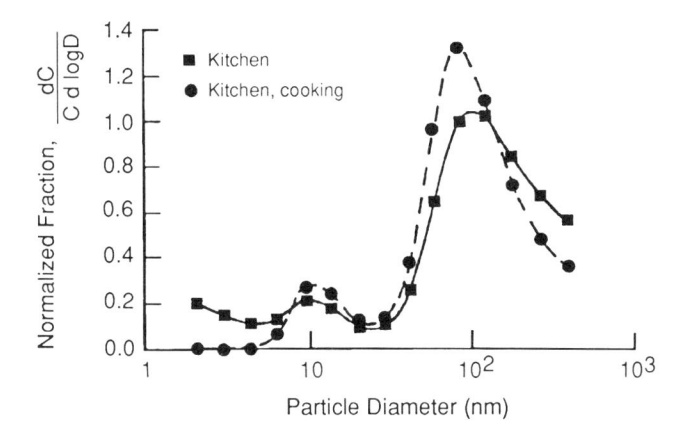

FIGURE 6-25 Activity-size distributions measured in a rural house kitchen by Strong (1988). Used with permission.

TABLE 6-6 Summary of Activity-Size Measurements Made by Strong (1988) in Two Houses in the United Kingdom

	Ambient Aerosol			Attached		Unattached	
Site	Median (nm)	GSD[a]	N (cm^{-3})	AMD (nm)	GSD	f_p (%)	AMD (nm)
Rural (summer)							
Bedroom	42	2.0	5,000	130	2.4	17	2.0
Living room	30	2.0	5,100	150	2.1	17	2.0
Kitchen	33	1.7	11,000	130	2.0	18	6.0
Kitchen (cooking)	30	1.7	470,000	110	1.9	11	11.0
Rural (winter)							
Living room	32	1.7	4,700	130	2.1	20	2.0
Urban (winter)							
Living room	30	2.1	15,000	110	2.1	20	3.5
Mean (all sites)	33	2.0	8,200	130	2.1	18	3.1

[a]GSD = Geometric standard deviation.

This process can be repeated a number of times to provide a measure of the precision and robustness of the estimated size distributions.

A semicontinuous automated system has been developed at the University of Illinois by Ramamurthi and Hopke (1991). This system consists of six sampling heads with various combinations of screens, described in Table 6-7 along

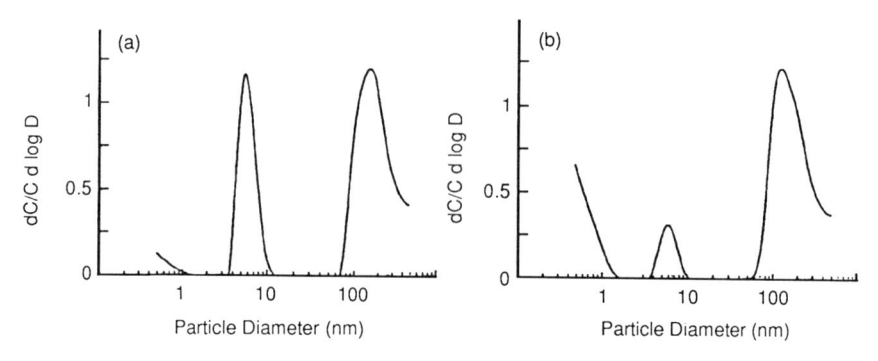

FIGURE 6-26 Trimodal activity-size distribution measured by Strong (1989) under conditions of (a) $F = 0.36$, CN = 10,000 cm^{-3}; (b) $F = 0.26$, CN = 5,000 cm^{-3}.

with their corresponding $d_p(50\%)$ values. Activity-size distributions were determined in a single house by using this system. The measurements were conducted in a three-level residence in Princeton, N.J. The house (PU-22) is instrumented by the Center for Energy and Environmental Studies, Princeton University, for continuous measurements of the radon concentration, temperature, humidity, and differential pressures. Activity-size distributions were measured in the basement and first floor of the house over a 1-week period (September 13 to 20, 1989). Grab samples for decay product activity concentrations were taken intermittently and analyzed by the gross alpha-photomultiplier tube method for comparison with the total concentration estimates from the automated system. The detectable particle number concentration in the sampling environment was continuously monitored by an Environment One model 100 condensation nuclei counter (CNC).

Initial activity-size distribution measurements were made in the basement of the house. Number concentrations of particles in the basement varied between 2,000 and 8,000/cm³ as detected by the CNC. The lack of windows or other major openings to the outside was thought to be responsible for the low concentrations. Radon concentrations varied between 5 and 500 pCi/liter during the 7-day period with the time period for fluctuations being much longer than the 15-min sampling interval. A total of 15 measurements were made during the 7-day period, with a remarkable degree of consistency in the shape of the measured activity-size distributions. Figure 6-27 shows the typical ^{218}Po, ^{214}Pb, and ^{214}Bi distributions observed in the basement of the house at a radon concentration of 55 pCi/liter and a particle number concentration of 3,000/cm³. The ^{218}Po distribution in Figure 6-27 is plotted as a histogram to illustrate the nature of the distribution, whereas the ^{214}Pb and ^{214}Bi distributions are shown by curves connecting the midpoint diameter values.

The ^{218}Po distribution showed that ≈54% of the ^{218}Po activity was in the

TABLE 6-7 Characteristics of the Stages in the Graded Screen Array System of Ramamurthi and Hopke (1990)

Unit	Sampler Slit Width (cm)	Diameter (cm)	Wire Screen Mesh × Turns[a]	$d_p(50\%)$ (0.5-350-nm range)
1	0.5	5.3	—	—
2	0.5	5.3	145	1.0
3	0.5	5.3	145 × 3	3.5
4	0.5	5.3	400 × 12	13.5
5	1.0	12.5	635 × 7	40.0
6	1.0	12.5	635 × 20	98.0

[a] Wire screen parameters given by Yeh et al. (1982).

NOTE: Sampling flow rate = 15 liters/min (each unit). Detector-Filter separation ≈ 0.8 cm (all units).

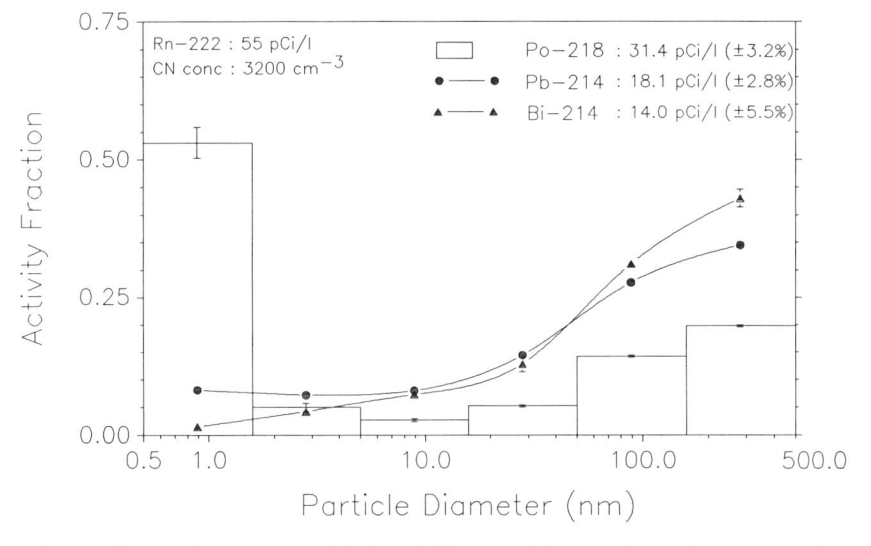

FIGURE 6-27 Typical Po-218, Pb-214, and Bi-214 activity-size distributions observed in a house basement in Princeton, N.J.

smallest inferred size interval, with a midpoint diameter of ≈0.9 nm (diffusion coefficient, $D \approx 0.04$ cm^2/s). This fraction closely resembles the classical, highly diffusive unattached fraction. The magnitude of the cluster fraction agreed well with theoretical predictions from attachment rate calculations at the observed particle number concentration (Porstendörfer et al., 1979). Very little of the ^{218}Po activity existed in the range from 1.6 to 16 nm, with the remainder of the activity attached to the larger ambient aerosol particles (diameter, >50 nm). The corresponding ^{214}Pb and ^{214}Bi distributions showed much smaller activity fractions in the 0.9-nm size range. The longer lifetime of these decay products permits a greater fraction of activity to become attached to the ambient aerosol. For all three distributions, the attached mode peaked in the 160-500-nm size range. However, this measurement system cannot be used to determine particle sizes greater than 500 nm in diameter. The activity distributions obtained are in general agreement, with respect to both the ^{218}Po cluster fraction and the size range of attached activity, with the distributions measured by Tu et al. (1989) in the basement of this house under similar conditions at an earlier date.

Several measurements of activity-size distributions were also made in the kitchen on the first level of the house. The initial measurements were performed under typical conditions of 20,000 particles/cm^3 and a radon concentration of ≈3 pCi/liter. The results of the measurement are shown in Figure 6-28. The ^{218}Po size distribution showed a large fraction of the ^{218}Po activity (≈44%) with a diffusivity similar to that of the classical unattached fraction. However, a significant fraction (≈10%) was in the 1.6-5.0-nm size interval. The corresponding ^{214}Pb and ^{214}Bi distributions indicate insignificant activity fractions in the 0.5-1.6-nm size interval, but a significant mode between 1.6 and 5.0 nm. The differences in the size distributions obtained in the basement and in the kitchen area related primarily to the 1.6- to 5.0-nm size interval, with the attached activity modes remaining in the 160- to 500-nm size range. This result suggests the presence of condensable constituents leading to the formation of particles in the 1.6- to 5.0-nm size interval or a source of very fine primary particles. This process may then allow the radon decay products to become associated with the 1.6- to 5.0-nm size mode in varying fractions depending upon the relative lifetimes. The presence of six large gas range pilot lights may be related to the formation of this mode, and similar effects are believed to have been observed by Tu et al. (1989) in other houses.

In a final experiment, activity-size distributions were measured following the continuous addition of aerosols generated in the kitchen from lighting the gas stove burners of the kitchen range. Figure 6-29 shows the distributions measured with particle number concentrations of ≈150,000/cm^3 and a radon concentration of ≈2 pCi/liter. The large concentrations of particles generated could be presumed to be rapidly coagulating soot cluster aggregates. The ^{218}Po and ^{214}Pb activity distributions measured under these conditions (Figure 6-29) were dramatically different from those measured in the basement and background kitchen conditions. The ^{218}Po distribution revealed very little activity in the 0.5- to 1.6-nm size interval (unattached fraction), with most of the activity spread out over the size interval range from 1.6 to 50 nm. The fraction of ^{218}Po attached to particles of >100 nm in diameter was reduced to a negligible level, probably because of the very large number of smaller particles produced by the gas burners. The distribution of ^{214}Pb revealed that the activity

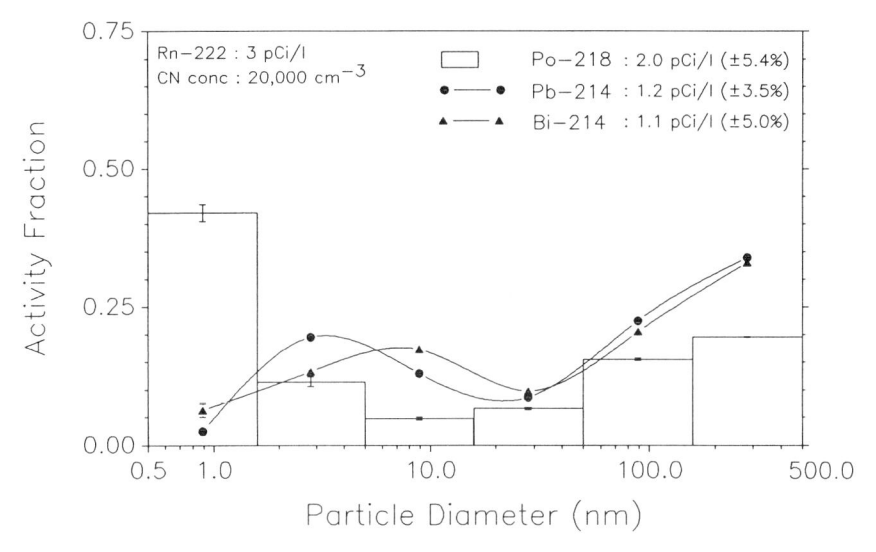

FIGURE 6-28 Po-218, Pb-214, and Bi-214 activity-size distributions measured under typical conditions in the kitchen of the test house.

was spread out over the size spectrum $d_p > 1.6$ nm, while the ^{214}Bi distribution remained similar to those measured prior to the addition of external aerosols. However, these latter results may be due to the timing of the sampling interval, which was between 20 and 35 min after the start of continuous addition of the external aerosols. Consequently, steady-state ^{214}Pb and ^{214}Bi distributions may not have been attained.

A stability analysis was performed for each of the size distributions shown in Figures 6-27, 6-28, and 6-29. This analysis provides an estimate of the stability of the inferred solutions with respect to errors in the input penetration data, and the results are represented by the error bars indicated in the figures. The size distributions obtained in the experiments were found to be stable and relatively insensitive to perturbations in the input data of the order of the associated measurement errors. The errors in the size interval fractions estimated from this procedure are too small to be seen in these figures and thus were not included.

Additional measurements were made in another one-story residence with living room, dining room, kitchen, two bedrooms, a study room, two bathrooms, and basement in the Princeton, N.J., area (Li, 1990; Hopke et al., 1990a,b). Activity-size distributions were measured in the living room and one of the bedrooms over a 2-week period (January 16 to 31, 1990). A total of about 10 measurements were made in the living room, and more than 100 measurements were made in the bedroom with different types of particle generation. Aerosols were generated from candle burning, cigarette smoking, vacuuming (electric

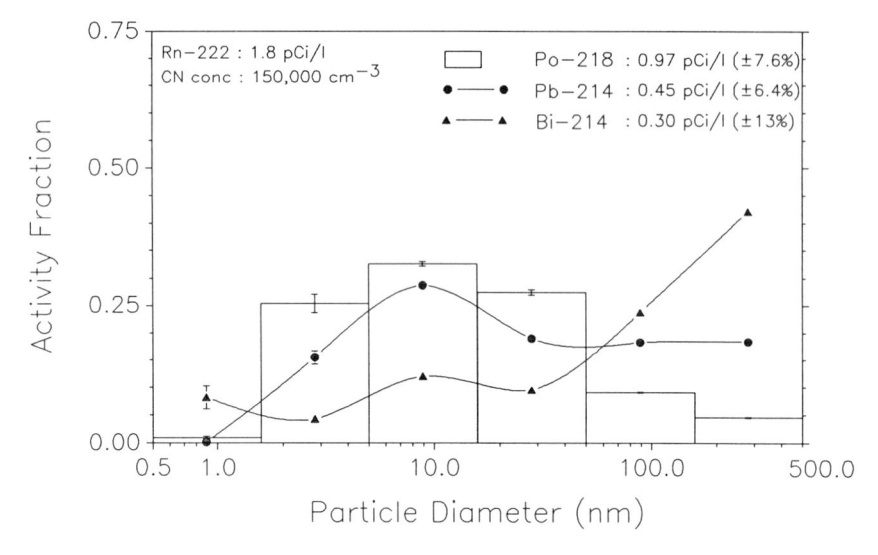

FIGURE 6-29 Po-218, Pb-214, and Bi-214 activity-size distributions measured during the continuous generation of aerosols from the kitchen gas stove burners.

motor), cooking, and opening doors from normal activities in the domestic environments. The particle concentrations were measured by using a Gardner manual condensation nucleus counter. The concentration and size distribution of radon progeny were determined by a semicontinuous graded screen array system. A sequence (0-15, 15-35, 35-75 min) with 75-min sampling was chosen because the radon concentration was in the range of 5 to 50 pCi/liter.

The influence of cigarette smoking (20 min) on the radon progeny size distributions in a closed bedroom are shown in Figure 6-30. The measurements were made 5 min after lighting the cigarette (5-20 min), 80 min later (80-95 min), and 155 min later (155-170 min). The fraction of ^{218}Po in the 0.9-nm size range changed from 60 to 8%. The fraction of ^{214}Pb and ^{214}Bi in the 0.9-nm size range was about 10% and essentially became zero. The fraction of three distributions in the 1.5- to 15-nm size range stayed the same. There was a large increase (from 40 to 80%) of ^{218}Po in the attached mode (50- to 500-nm size range), with insignificant changes (from 35 to 40%) in ^{214}Pb and ^{214}Bi fractions in this mode.

The influence of cooking on the radon progeny size distributions with an open bedroom door is shown in Figure 6-31. A steak was pan fried for 20 min (0-20 min) by using a gas stove burner in the kitchen. The measurements were made 5 min later (5-20 min), 80 min later (80-95 min), and 155 min later (155-170 min). The fraction of ^{218}Po in the 0.9-nm size range changed from 60 to 15%. The fractions of ^{214}Pb and ^{214}Bi in the 0.9-nm size range changed from 15 to 10%. There was a very low fraction of activity in the 1.5- to 15-nm

129

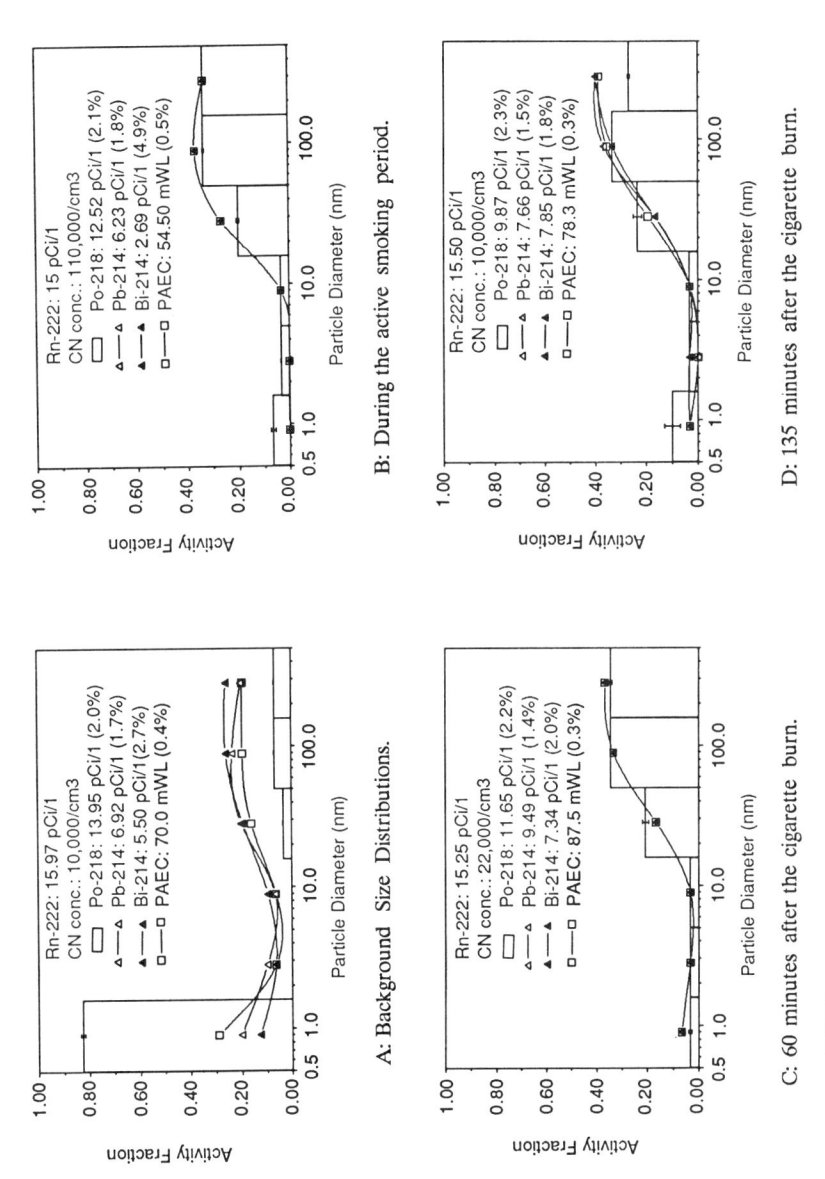

FIGURE 6-30 Activity-size distributions in a bedroom before, during, and after smoking a cigarette.

130

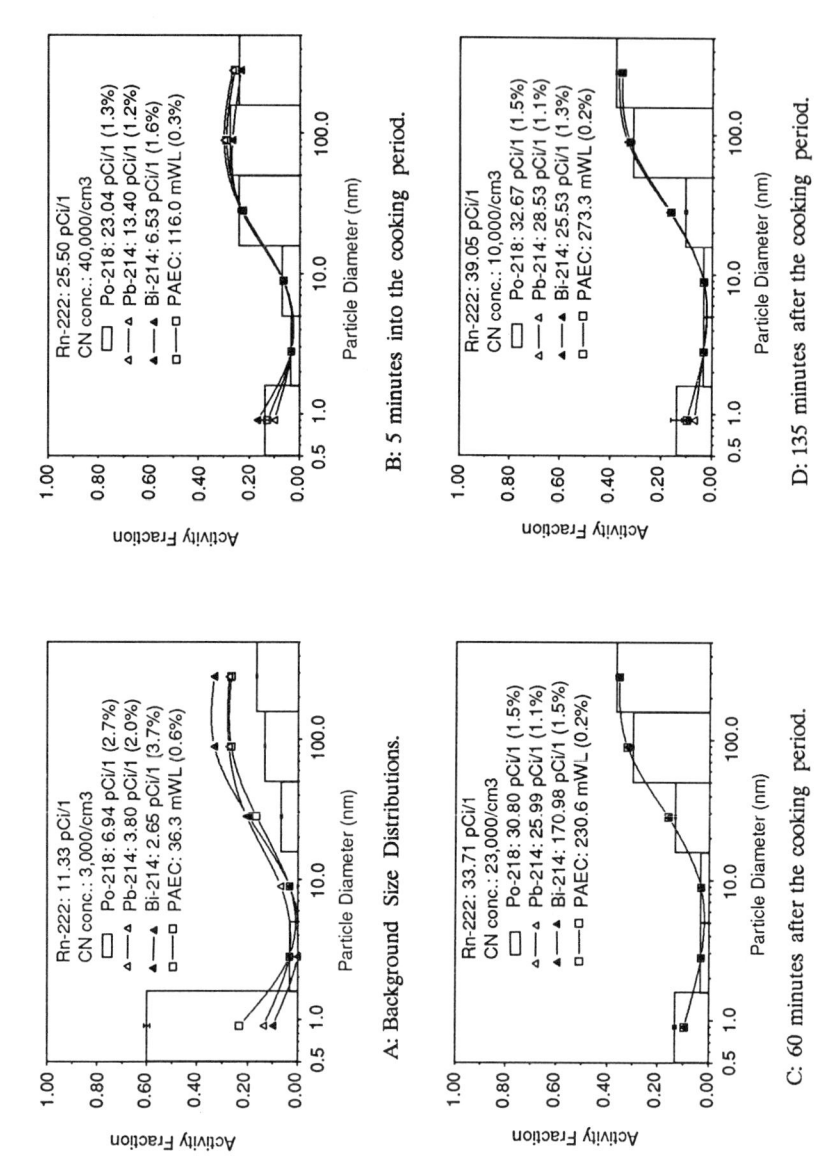

A: Background Size Distributions.
Rn-222: 11.33 pCi/l
CN conc.: 3,000/cm3
Po-218: 6.94 pCi/l (2.7%)
Pb-214: 3.80 pCi/l (2.0%)
Bi-214: 2.65 pCi/l (3.7%)
PAEC: 36.3 mWL (0.6%)

B: 5 minutes into the cooking period.
Rn-222: 25.50 pCi/l
CN conc.: 40,000/cm3
Po-218: 23.04 pCi/l (1.3%)
Pb-214: 13.40 pCi/l (1.2%)
Bi-214: 6.53 pCi/l (1.6%)
PAEC: 116.0 mWL (0.3%)

C: 60 minutes after the cooking period.
Rn-222: 33.71 pCi/l
CN conc.: 23,000/cm3
Po-218: 30.80 pCi/l (1.5%)
Pb-214: 25.99 pCi/l (1.1%)
Bi-214: 170.98 pCi/l (1.5%)
PAEC: 230.6 mWL (0.2%)

D: 135 minutes after the cooking period.
Rn-222: 39.05 pCi/l
CN conc.: 10,000/cm3
Po-218: 32.67 pCi/l (1.5%)
Pb-214: 28.53 pCi/l (1.3%)
Bi-214: 25.53 pCi/l (1.3%)
PAEC: 273.3 mWL (0.2%)

FIGURE 6-31   Activity-size distributions measured in a bedroom during and after cooking in the kitchen.

size range for background, and it increased to 10% because of cooking. A large increase (from 35 to 70%) of ^{218}Po was observed in the attached mode and peaked at the 50- to 500-nm size range, with only small changes (from 40 to 50%) in ^{214}Pb and ^{214}Bi distributions.

Because of the large number of particles generated by normal activities in the domestic environment, the working level increases for a period of time, while the unattached fraction decreases. The particles generated from cigarette smoke and cooking dramatically shifted almost all of the radon progeny to the attached fraction and remained for a long period of time. The particles produced from candle burning and vacuuming were much smaller, with an average attachment diameter around 15 nm. The candle and vacuuming particles did decrease the unattached fraction, but returned to the original background distributions about 150 min later.

Summary of Indoor Exposure

In evaluating the information on the aerodynamic size of the particles carrying the radioactivity in the indoor environment, the results of Reineking and Porstendörfer (1986, 1990), Tu and Knutson (1988a,b), Ramamurthi and Hopke (1990), Li (1990), and Hopke et al. (1990a,b) were reviewed. From these results, the unattached activity appears to have a diffusion equivalent diameter of 0.0011 μm and typically represents about 8% of the airborne alpha activity energy. The typical indoor radioactive aerosol has a mode with an AMD of 0.15 μm with a geometric standard deviation of 2.0. The presence of sidestream cigarette smoke provides a substantial number of larger sized particles so that the AMD increases to 0.25 μm with a geometric standard deviation of 2.5. During periods of active smoking, the unattached fraction diminishes to 0.1, and on average, the unattached activity represents about 3% of the total activity in houses with smokers.

Other activities can produce particles with small diameters so that during vacuuming or cooking, an additional mode in the activity-size distribution with an average diameter of 0.02 μm and containing 15% of the airborne alpha energy can be observed. Because of the high mobility of these particles, this mode is quite transient and will disappear in a few hours time.

Finally, there are times in closed rooms such as bedrooms with relatively low air exchange rates that the particle concentration can be sufficiently low that a much higher fraction of the activity is in the unattached mode. It is estimated that a typical value for the unattached fraction under these circumstances is 16%.

REFERENCES

Becker, K. H., A. Reineking, H. G. Scheibel, and J. Porstendörfer. 1984. Radon daughter activity size distributions. Radiat. Prot. Dosim. 7:147-150.

Bigu, J. 1985. Radon daughter and thoron daughter deposition velocity and unattached fraction under laboratory conditions in underground uranium mines. J. Aerosol Sci. 16:157-165.

Bigu, J., and B. Kirk. 1980. Determination of the unattached radon daughter fractions in some uranium mines. Presented at the Workshop on Attachment of Radon Daughters, Measurement Techniques and Related Topics, October 30, 1980, University of Toronto. (Report available from CANMET, P.O. Box 100, Elliot Lake, Ontario, Canada.)

Busigin, A., A. W. Van der Vooren, and C. R. Phillips. 1981. Measurement of the total and radioactive aerosol size distributions in a Canadian uranium mine. Am. Ind. Hyg. Assoc. J. 42:310-314.

Chamberlain, A. C., and E. D. Dyson. 1956. The dose to the trachea and bronchi from the decay products of radon and thoron. Br. J. Radiol. 29:317-325.

Chen, R. Y., and R. A. Comparin. 1976. Deposition of aerosols in the entrance of a tube. J. Aerosol Sci. 7:335-341.

Cheng, Y. S. 1989. Deposition of thoron daughters in human head airways. Technical Exchange Meeting, Grand Junction, Colo., September 18-19, CONF 8909190. NTIS.

Cheng, Y. S., and H. C. Yeh. 1980. Theory of screen type diffusion battery. J. Aerosol Sci. 11:313-319.

Cheng, Y. S., J. A. Keating, and G. M. Kanapilly. 1980. Theory and calibration of a screen-type diffusion battery. J. Aerosol Sci. 11:549-556.

Cooper, J. A., P. O. Jackson, J. C. Langford, M. R. Petersen, and B. O. Stuart. 1973. Characteristics of Attached Radon-222 Daughters under both Laboratory and Field Conditions with Particular Emphasis upon Underground Mine Environments. Report to the U.S. Bureau of Mines, contract H0220029. Richland, Wash.: Battelle Pacific Northwest Laboratories.

Craft, B. F., J. L. Oser, and F. W. Norris. 1966. A method for determining relative amounts of combined and uncombined radon daughter activity in underground uranium mines. Am. Ind. Hyg. Assoc. J. 27:154-159.

Davies, C. N. 1945. Definitive equations for the fluid resistance of spheres. Proc. Phys. Soc. 57:259-270.

Duggan, M. J., and D. M. Howell. 1969. The measurement of the unattached fraction of airborne RaA. Health Phys. 17:423-427.

Emi, H., C. Kanaoka, and Y. Kuhabara. 1982. The diffusion collection efficiency of fibers for aerosol over a wide range of Reynolds numbers. J. Aerosol Sci. 13:403-413.

Friedlander, S. K. 1977. Smoke, Dust and Haze. New York: John Wiley & Sons.

Fuchs, N. A. 1964. The Mechanics of Aerosols. New York: MacMillan Press.

Fusamura, N., R. Kurosawa, and M. Maruyama. 1967. Determination of f-value in uranium mine air. Pp. 213-227 in Symposium on Instruments and Techniques for the Assessment of Airborne Radioactivity in Nuclear Operations. Vienna: International Atomic Energy Agency.

George A. C. 1972. Measurement of the uncombined fraction of radon daughters with wire screens. Health Phys. 23:390-392.

George, A. C., and A. J. Breslin. 1980. The Distribution of Ambient Radon and Radon Daughters in Residential Buildings in the New Jersey-New York Area, National Radiation Environment III, Vol. 2. CONF-780422. Oak Ridge, Tenn.: Technical Information Center, U.S. Department of Energy.

George, A. C., and L. Hinchliffe. 1972. Measurements of uncombined radon daughters in uranium mines. Health Phys. 23:791-803.

George, A. C., L. Hinchliffe, and R. Sladowski. 1975. Size distribution of radon daughter particles in uranium mine atmospheres. Am. Ind. Hyg. Assoc. J. 36:484-490.

George, A. C., L. Hinchliffe, and R. Sladowski. 1977. Size Distribution of Radon Daughter Particles in Uranium Mine Atmospheres. HASL-326. New York: Health and Safety Laboratory.

George, A. C., M. H. Wilkening, E. O. Knutson, D. Sinclair, and L. Andrews. 1984. Measurements of radon and radon daughter aerosols in Socorro, New Mexico. Aerosol Sci. Technol. 3:277-281.

Gormley, P., and M. Kennedy. 1949. Diffusion for a stream flowing through a cylindrical tube. Proc. R. Irish Acad. 52A:163-167.

Hirst, B. W., and G. E. Harrison. 1939. The diffusion of Rn gas mixtures. Proc. R. Soc. Lond. Ser. A 169:573-586.

Holub, R. F., and E. O. Knutson. 1987. Measuring polonium-218 diffusion-coefficient spectra using multiple wire screens. Pp. 340-356 in Radon and Its Decay Products: Occurrence, Properties and Health Effects, P. K. Hopke, ed. Symposium Series 331. Washington D.C.: American Chemical Society.

Holub, R. F., E. O. Knutson, and S. Solomon. 1988. Tests of the graded wire screen technique for measuring the amount and size distribution of unattached radon progeny. Radiat. Prot. Dosim. 24:265-268.

Hopke, P. K. 1989a. Use of electrostatic collection of ^{218}Po for measuring Rn. Health Phys. 57:39-42.

Hopke, P. K. 1989b. The initial behavior of ^{218}Po in indoor air. Environ. Int. 15:299-308.

Hopke, P. K., M. Ramamurthi, and C. S. Li. 1990a. Measurements of the size distributions of radon progeny in indoor air. In Aerosols: Science, Industry, Health, and Environment, Vol. 2, S. Masuda and K. Takahashi, eds. Oxford: Pergamon Press.

Hopke, P. K., M. Ramamurthi, and C. S. Li. 1990b. Measurements of size distributions of indoor radioactive aerosol. Presented at the 29th Hanford Symposium on Health and the Environment. Richland, Washington, October 1990.

Hopke, P. K., M. Ramamurthi, and E. O. Knutson. In press. A measurement system for Rn decay product lung deposition based on respiratory models. Health Phys.

Ingham, D. B. 1975. Diffusion of aerosols for a stream flowing through a cylindrical tube. J. Aerosol Sci. 6:125-132.

International Commission on Radiological Protection (ICRP). 1959. Report of Committee II on Permissible Dose for Internal Radiation, ICRP Publication 2. Oxford: Pergamon Press.

James A. C., G. F. Bradford and D. M. Howell. 1972. Collection of unattached RaA atoms using wire gauze. J. Aerosol Sci. 3:243-254.

Khan, A., C. R. Phillips, and P. Duport. 1987. Analysis of errors in the measurement of unattached fractions of radon and thoron progeny in a Canadian uranium mine using wire screen methods. Radiat. Prot. Dosim. 18:197-208.

Knutson, E. O., A. C. George, R. H. Knuth, and B. R. Koh. 1984. Measurements of radon daughter particle size. Radiat. Prot. Dosim. 7:121-125.

Knutson, E. O., K. W. Tu, S. B. Solomon, and J. Strong. 1988. Intercomparison of three diffusion batteries for the measurement of radon decay product particle size distributions. Radiat. Prot. Dosim. 24:261-264.

Kojima, H., and S. Abe. 1988. Measurement of the total and unattached radon daughters in a house. Radiat. Prot. Dosim. 24:241-244.

Kotrappa, P., and Y. S. Mayya. 1976. Revision of Raghavayya and Jones' data on the radon decay in mine air. Health Phys. 31:380-382.

Li, C. S. 1990. Field Evaluation and Health Assessment of Air Cleaners in Removing Radon Decay Products in Domestic Environments. Ph.D. thesis. University of Illinois, Urbana, Ill. Department of Energy Report DOE ER61029-2.

Liu, B. Y. H., and D. Y. H. Pui. 1975. On the performance of the electrical aerosol analyzer. J. Aerosol Sci. 6:249-264.

Lodge, J. P., Jr., and T. Chan. 1986. Cascade Impactor: Sampling and Data Analysis. Akron, Ohio: American Industrial Hygiene Association,

Loeb, L. B. 1961. The Kinetic Theory of Gases, 3rd ed. New York: Dover Publications.

Maher, E. F., and N. M. Laird. 1985. Algorithm reconstruction of particle size distribution from diffusion battery data. J. Aerosol Sci. 16:557-570.

Mercer, T. T. 1975. Unattached radon decay products in mine air. Health Phys. 28:158-161.

Mercer, T. T., and W. A. Stowe. 1969. Deposition of unattached radon decay products in an impactor stage. Health Phys. 17:259-264.

Porstendörfer, J. 1968. Die Diffusionkoeffizienten und mittleren freien Weglangen der Geladenen und neutral radon-folge Produkte in Luft. Z. Physik. 213:384-396.

Porstendörfer, J. 1987. Free-fractions, attachment rates, and plate-out rates of radon daughters in houses. Pp. 285-300 in Radon and Its Decay Products: Occurrence, Properties and Health Effects, P. K. Hopke, ed. Symposium Series 331. Washington D.C.: American Chemical Society.

Porstendörfer, J., and T. T. Mercer. 1979. Influence of electric charge and humidity upon the diffusion coefficient of radon decay products. Health Phys. 15:191-199.

Porstendörfer, J., G. Röbig, and A. Ahmed. 1979. Experimental determination of the attachment coefficients of atoms and ions on monodisperse particles. J. Aerosol Sci. 10:21-28.

Raabe, O. G. 1969. Concerning the interactions that occur between radon decay products and aerosols. Health Phys. 17:177-185.

Raes, F., A. Janssens, A. DeClercq, and H. Vanmarcke. 1984. Investigation of the indoor aerosol and its effect on the attachment of radon daughters. Radiat. Prot. Dosim. 7:127-131.

Raghavayya, M., and J. H. Jones. 1974. A wire screen-filter paper combination for the measurement of fractions of unattached daughter atoms in uranium mines. Health Phys. 26:417-430.

Ramamurthi, M., and P. K. Hopke. 1989. On improving the validity of wire screen "unattached" fraction Rn daughter measurements. Health Phys. 56:189-194.

Ramamurthi, M., and P. K. Hopke. 1990. Simulation studies of reconstruction algorithms for the determination of optimum operating parameters and resolution of graded screen array systems (non-conventional diffusion batteries). Aerosol Sci. Technol. 12:700-710.

Ramamurthi, M., and P. K. Hopke. 1991. An automated, semi-continuous system for measuring indoor radon progeny activity-weighted size distributions, d_p: 0.5-500 nm. Aerosol Sci. Technol. 14:82-92.

Ramamurthi, M., P. K. Hopke, R. Strydom, K. W. Tu, E. O. Knutson, R. F. Holub, W. Winklmayr, W. Marlow, and S. C. Yoon. 1989. Radon decay product activity size distribution measurement methods—A laboratory intercomparison. Presented at the American Association for Aerosol Research Meeting, Reno, Nev., October 1989.

Ramamurthi, M., R. Strydom, and P. K. Hopke. In press. Assessment of wire and tube penetration theories using a ^{218}PoO$_x$ cluster aerosol. J. Aerosol Sci.

Reineking, A., and J. Porstendörfer. 1986. High-volume screen diffusion batteries and α-spectroscopy for measurement of the radon daughter activity size distributions in the environment. J. Aerosol Sci. 17:873-879.

Reineking, A., and J. Porstendörfer. 1990. "Unattached" fraction of short-lived decay products in the indoor and outdoor environment: An improved single screen method and results. Health Phys. 58:717-727.

Reineking, A., K. H. Becker, and J. Porstendörfer. 1985. Measurements of the unattached fractions of radon daughters in houses. Sci. Total Environ. 45:261-270.

Reineking, A., K. H. Becker, and J. Porstendörfer. 1988. Measurement of activity size distributions of the short-lived radon daughters in the indoor and outdoor environment. Radiat. Prot. Dosim. 24:245-250.

Scheibel, H. G., and J. Porstendörfer. 1984. Penetration measurements for tube and screen type diffusion batteries in the ultrafine particle size range. J. Aerosol Sci. 15:673-682.

Shimo, M., and Y. Ikebe. 1984. Measurements of radon and its short-lived decay products and unattached fraction in air. Radiat. Prot. Dosim. 8:209-214.

Shimo, M., Y. Yoshihiro, K. Hayashi, and Y. Ikebe. 1985. On some properties of ^{222}Rn short-lived decay products in air. Health Phys. 48:75-86.

Sinclair, D., A. C. George, and E. O. Knutson. 1977. Application of diffusion batteries to measurement of submicron radioactive aerosols. Pp. 103-114 in Airborne Radioactivity. La Grange Park, Ill.: American Nuclear Society.

Stranden, E., and L. Berteig. 1982. Radon daughter equilibrium and unattached fraction in mine atmospheres. Health Phys. 42:479-487.

Stranden, E., and T. Strand. 1986. A dosimetric discussion based on measurements of radon daughter equilibrium and unattached fraction in different atmospheres. Rad. Prot. Dosim. 16:313-318.

Strong, J. C. 1988. The size of attached and unattached radon daughters in room air. J. Aerosol Sci. 19:1327-1330.

Strong, J. C. 1989. Design of the NRPB activity size measurement system and results. Presented at the Workshop on "Unattached" Fraction Measurements, University of Illinois, Urbana, Ill., April 1989.

Subba Ramu, M. C. 1980. Calibration of a diffusion sampler used for the measurement of unattached radon daughter products. Atmosph. Environ. 14:145-147.

Tan, C. W. 1969. Diffusion of disintegration products of inert gases in cylindrical tubes. Int. J. Heat Mass Trans. 12:471-478.

Thomas, J. W. 1955. The diffusion battery method for aerosol particle size determination. J. Colloid Sci. 10:246-255.

Thomas, J. W. 1970. Modification of the Tsivoglou method for radon daughters in air. Health Phys. 19:691-693.

Thomas, J. W., and L. E. Hinchliffe. 1972. Filtration of 0.001 μm particles with wire screens. J. Aerosol Sci. 3:387-393.

Tu, K. W., and E. O. Knutson. 1988a. Indoor radon progeny particle size distribution measurements made with two different methods. Radiat. Prot. Dosim. 24:251-255.

Tu, K. W, and E. O. Knutson. 1988b. Indoor outdoor aerosol measurements for two residential buildings in New Jersey. Aerosol Sci. Technol. 9:71-82.

Tu, K. W., A. C. George, and E. O. Knutson. 1989. Summary of results of radon progeny particle size in indoor air. Presented at the American Association for Aerosol Research Meeting, Reno, Nev., October 1989.

Twomey, S. 1975. Comparison of constrained linear inversion and an iterative nonlinear algorithm applied to the indirect estimation of the particle size distribution. J. Comp. Phys. 18:188-200.

Van der Vooren, A. W., A. Busigin, and C. R. Phillips. 1982. An evaluation of unattached radon (and thoron) daughter measurement techniques. Health Phys. 42:801-808.

Vanmarcke, H., A. Janssens, and F. Raes. 1985. The equilibrium of attached and unattached radon daughters in the domestic environment. Sci. Tot. Environ. 45:251-260.

Vanmarcke, H., A. Janssens, F. Raes, A. Poffijn, P. Perkvens, and R. Van Dingenen. 1987. The behavior of radon daughters in the domestic environment. Pp. 301-323 in Radon and Its Decay Products: Occurrence, Properties and Health Effects, P. K. Hopke, ed. Symposium Series 331. Washington, D.C.: American Chemical Society.

Vanmarcke, H., A. Reineking, J. Porstendörfer, and F. Raes. 1988. Comparison of two methods for investigating indoor radon daughters. Radiat. Prot. Dosim. 24:281-284.

Vanmarcke, H., P. Berkvens, and A. Poffijn. 1989. Radon versus Rn daughters. Health Phys. 56:229-231.

Yeh, H. C., Y. S. Cheng, and M. M. Orman. 1982. Evaluation of various types of wire screens as diffusion battery cells. J. Colloid Interface Sci. 86:12-16.

7

Breathing, Deposition, and Clearance

INTRODUCTION

Although epidemiological and animal studies provide information on the risks of lung cancer in relation to exposure, linkage of exposure to dose is essential for extrapolating the risk from the mining to the indoor environment. Since exposure does not equal absorbed dose, it is important to describe adequately the variables that determine exposure-dose relationships.

This chapter summarizes aerosol deposition and clearance mechanisms and discusses how both of these influence the amount of alpha energy from radon progeny delivered to target sites. The chapter also discusses approaches that can be used to describe the amount and distribution of doses and, finally, some factors that are known to influence the amount and distribution of retained aerosols.

A major goal of this chapter is to understand and predict differences in response to similar concentrations of inhaled radon progeny among different groups. However, exposure-dose relationships can vary in different individuals (e.g., children versus adults). Thus, even if the inspired concentration of radon progeny were similar, the dose deposited in the lungs may vary. Such factors as metabolic rate, breathing pattern, and lung structure determine the deposition of radon progeny and may differ among individuals. The impact of changes in breathing pattern and the effects of chronic lung disease on particle retention are also discussed in this chapter.

Many aspects of the deposition of aerosols in mammalian lungs have captured the energy and imagination of many investigators, and only a brief summary is provided here. Published symposia serve as excellent sources of information in this area, for example, the First (Oxford), Second (Cambridge), Third (London), Fourth (Edinburgh), Fifth (Cardiff), and Sixth (Cambridge) International Symposia on Inhaled Particles (Davies, 1961, 1964; Walton, 1971, 1977, 1982; Dodgson and McCallum, 1988). In addition, many papers and books reviewing deposition and clearance processes are available (Altshuler et al., 1957; Hatch and Gross, 1964; Aharonson et al., 1976; Brain et al., 1977; Lippmann, 1977; Raabe et al., 1977; Brain and Valberg, 1979, 1985; Heyder et al., 1980; Lippmann et al., 1980; Clarke and Pavia, 1984; Stuart, 1984; Morén et al., 1985).

The distinction between retention (the amount of an aerosol present in the lungs at any time) and deposition (the initial attachment of suspended particles to a surface) should be kept in mind. Retention, but not deposition, is influenced by clearance and translocation. Even during a brief exposure to radon progeny (30 to 60 min), there may be loss and redistribution of deposited particles, especially in the large ciliated airways. To acknowledge the possibility of particle redistribution during a measured exposure period, the term *retention* is used to refer to the amount and distribution of particles in lungs at any time after an exposure to an aerosol.

Many parameters that influence aerosol deposition have been studied. Total deposition in the lungs depends on particle size (Morrow, 1966), tidal volume (Taulbee et al., 1978), breathing frequency (Altshuler, 1961; Muir and Davies, 1967; Valberg et al., 1982), and lung volume (Davies et al., 1972). To date, little is known about the actual anatomic distribution of deposited radon progeny within the lungs, especially at the level of small airways and parenchyma. Since the rates and pathways of clearance are determined by the sites of aerosol deposition, it is necessary to study factors such as exercise and disease that strongly influence the distribution of the retained aerosol in the lungs.

DEPOSITION OF RADON PROGENY: GENERAL PRINCIPLES

Deposition is the process that determines what fraction of the inspired particles is caught in the respiratory tract and, thus, fails to exit with expired air. It is likely that all particles that touch a wet surface are deposited; thus, the site of contact is the site of initial deposition. Distinct physical mechanisms operate on inspired particles to move them toward respiratory tract surfaces. Major mechanisms are inertial forces, gravitational sedimentation, Brownian diffusion, interception, and electrostatic forces. The extent to which each mechanism contributes to the deposition of a specific particle depends on the particle's physical characteristics, the subject's breathing pattern, and the geometry of the respiratory tract. Radon progeny are unusual in that they tend to be smaller than most aerosols of concern to health (see Chapters 2 and 6), and thus, Brownian diffusion and sedimentation dominate, as discussed below.

Detailed treatments of particle deposition have been given by Brain and Valberg (1979), Lippmann et al. (1980), Heyder et al. (1980, 1986), Morgan et al. (1983), Raabe (1982), Stuart (1984), and Agnew (1984). Comprehensive treatises on aerosol behavior are also available (Fuchs, 1964; Davies, 1967; Mercer, 1981; Hinds, 1982; Reist, 1984).

The behavior of a particle in the respiratory system is largely determined by its size and density. Particles of varying shape and density may be compared by their aerodynamic equivalent diameter (D_{ae}). Aerodynamic diameter is the diameter of the unit density (1 g/cm^3) sphere that has the same gravitational settling velocity as that of the particle in question. Aerodynamic diameter is proportional to the product of the geometric diameter and the square root of density.

FACTORS ACTING TO DEPOSIT PARTICLES IN THE LUNGS

Diffusional

Radon progeny undergo Brownian diffusion—a random motion caused by their collisions with gas molecules; this motion can lead to contact and deposition on respiratory surfaces. Diffusion is significant for particles with diameters of less than 1 μm; only then does their size approach the mean free path of gas molecules. Thus, this is probably the dominant mechanism for radon progeny since most of the alpha activity resides on such small particles. Unlike inertial or gravitational displacement, diffusion is independent of particle density; however, it is affected by particle shape (Heyder and Scheuch, 1983). For these particles, size is best expressed in terms of a thermodynamic equivalent diameter, D_t, the diameter of a sphere that has the same diffusional displacement as that of the particle. The probability that a particle would be deposited by diffusion increases with an increase in the quotient $(t/D_t)^{1/2}$, where t is the residence time. Deposition is also independent of respired flow rate. Diffusion, like sedimentation, is most important in the peripheral airways and alveoli, where dimensions are smaller. However, as particle size becomes very small, diffusion may become an important mechanism even in the upper airways.

Gravitational

Gravity accelerates falling bodies downward, and terminal settling velocity is reached when viscous resistive forces of the air are equal and opposite in direction to gravitational forces. Respirable particles reach this constant terminal sedimentation velocity in less than 0.1 ms. Then, particles can be removed if their settling causes them to strike airway walls or alveolar surfaces. The probability that a particle will deposit by gravitational settling is proportional to the product of the square of the aerodynamic diameter (D_{ae}^2) and the residence time. Thus, breathholding enhances deposition by sedimentation. Sedimentation is most important for particles larger than about 0.2 μm and within the peripheral airways and alveoli, where airflow rates are slow and residence times are long (Heyder et al., 1986).

Inertia

Inertia is the tendency of a moving particle to resist changes in direction and speed. It is related to momentum: the product of the particle's mass and velocity. High linear velocities and abrupt changes in the direction of airflow occur in the nose and oropharynx and at central airway bifurcations. Inertia causes a particle entering bends at these sites to continue in its original direction instead of following the curvature of the airflow. If the particle has sufficient mass and velocity, it will cross airflow streamlines and impact on the airway wall. The probability that a particle will deposit by inertial impaction, therefore, increases with increasing product of D_{ae}^2 and respired flow rate. Generally, inertial impaction is an important deposition mechanism for particles with aerodynamic diameters larger than 2 μm. Thus, it is probably unimportant for radon progeny in indoor air. It can occur during both inspiration and expiration in the extrathoracic airways (oropharynx, nasopharynx, and larynx) and central airways.

As particle size decreases, inertia and sedimentation become less important, but diffusion becomes more important. For example, a 2-μm-unit-density spherical particle is displaced by diffusion (Brownian displacement) by only about 9 μm in 1 s. It is important to know that this displacement varies with the square root of time. It settles by gravity by about 125 μm in the same period. However, as the particle size drops to 0.2 μm, the diffusional displacement in 1 s increases to 37 μm whereas gravitational displacement drops to only 2.1 μm. At 0.02 μm, gravitational displacement is only 0.013 μm/s while diffusional displacement in 1 s has soared to 290 μm. A comparison of settling and diffusion displacements for a range of particle sizes is shown in Table 7-1.

Other Forces

Deposition can also occur in the lungs when particles have dimensions that are significant relative to those of the air spaces. As aerosols move into smaller and smaller air structures, some particles may reach a point where the distance from their center to a surface is less than their radius. The resulting contact is called interception. Interception is most important for the deposition of fibers, but it is probably insignificant for most radon progeny unless they are attached to larger particles.

TABLE 7-1 Root Mean Square Brownian Displacement in 1 Second Compared with the Distance Fallen in Air in 1 Second for Unit-Density Particles of Different Diameters

	Particle Diameter (μm)	Brownian Displacement (μm)	Distance Fallen (μm)
Settling greater in 1 s	50	1.7	70,000
	20	2.7	11,500
	10	3.8	2,900
	5	5.5	740
	2	8.8	125
	1	13	33
Diffusion greater in 1 s	0.5	20	9.5
	0.2	37	2.1
	0.1	64	0.81
	0.05	120	0.35
	0.02	290	0.013
	0.01	570	0.0063

NOTE: Temperature = 37°C; gas viscosity = 0.19×10^{-3} poise. Appropriate correction factors were applied for motion outside the range of validity of Stokes' law.

Electrical forces may cause charged particles to deposit in the respiratory tract by the creation of image charges on airway walls. Image charges form within the airways when charged particles attract ions that have the opposite charge and repel airway ions of the same charge. Generally, electrical forces are a minor mechanism of deposition, unless the inspired particles are highly charged. When they are, electrical forces may cause significant losses within the device used to generate the aerosol. Deposition of 1.0-μm charged particles is increased over that of uncharged particles once there are about 40 charges on the particle; for 0.6-μm particles, an increase is seen once there are 30 charges, and for 0.3-μm particles only 10 charges are needed (Melandri et al., 1983). Ambient aerosols that have been airborne for several hours tend to be uncharged. However, fresh aerosols 1 to 10 μm in diameter produced by a grinding or cutting process may have hundreds of electrostatic charges per particle, which are sufficient to enhance deposition. The extent to which charge influences deposition of radon progeny is largely unknown. Other forces acting to affect deposition such as acoustic forces, magnetic forces, or thermal forces, are normally not significant in the respiratory tract.

The effectiveness of these deposition mechanisms depends on (1) the effective aerodynamic diameters of the particles, (2) the pattern of breathing, and (3) the anatomy of the respiratory tract. These factors determine the fraction of the inhaled particles that is deposited as well as the site of deposition.

Aerosol Characteristics

A major factor governing the effectiveness of the deposition mechanisms is the size of the inspired particles. The effective aerodynamic diameter is a function of the size, shape, and density of the particles and affects the magnitude of forces acting on them. For example, while inertial and gravitational effects increase with increasing particle size, the displacements produced by diffusion decrease. The importance of particle size cannot be overemphasized. It is featured prominently in most discussions of aerosol deposition in the respiratory tract (Altshuler et al., 1957; Brain et al., 1977; Lippmann, 1977; Raabe et al., 1977; Brain and Valberg, 1979; Heyder et al., 1980, 1986; Lippman et al., 1980; Agnew, 1984; Stuart, 1984).

Radon decay products have a particular size as molecular species but then attach to particles with a wide range of sizes. Since size helps to determine the site of deposition within the lungs, it is important to quantify the mass of particles within the size range that can penetrate the oropharynx. Aerosol mass distributions are characterized by two values, the mass median aerodynamic diameter (MMAD) and the geometric standard deviation (GSD). The MMAD denotes the particle size at which half of the total aerosol mass is contained in larger particles and half is contained in smaller particles. Since the MMAD is expressed as an aerodynamic diameter, it describes how the aerosol behaves in the respired air and can be used to estimate where and by what processes the aerosol deposits in the respiratory tract. The GSD denotes the spread of particle sizes. Most aerosols have sizes that are distributed log-normally; that is, on a plot of the frequency distribution versus particle diameter, the distribution looks Gaussian. An aerosol that is composed of particles of the same size would have a GSD equal to 1.0 and would be termed monodisperse. An aerosol with a GSD of 1.22 or larger is called polydisperse (Fuchs, 1964). Almost all naturally occurring aerosols are polydisperse. An aerosol with a MMAD of 2.0 μm and a GSD of 2.0 would have 1 GSD or 68% of its mass contained in particles between 1.0 and 4.0 μm in aerodynamic diameter. An important implication of a log-normal distribution is that much of the aerosol mass can be contained in the large particles, since mass is proportional to the cube of the diameter. Where these large particles deposit in the respiratory tract governs, to a large extent, where much of the dose is deposited.

Mass can be measured gravimetrically by pulling air containing the aerosol through an absolute filter at a known volumetric flow rate. Particle size can be estimated by using light or electron microscopy to examine collected particles. Optical methods utilizing light scattering can give continuous estimates of particle size and/or concentration. These methods, however, describe only the cross-sectional or geometric diameter of the aerosol. Aerodynamic diameter is a more meaningful predictor of deposition site, since it accounts for particle size, shape, and density. Several devices can provide aerodynamic diameter directly, including cascade impactors and aerosol centrifuges. Real-time aerodynamic size measurements are possible using laser Doppler velocimeters (Hiller et al., 1978; Bouchikhi et al., 1988) or laser diffraction particle sizers (Clay et al., 1983).

Accurate measurement of aerosol size distributions is complex because particle size is frequently dynamic. Evaporation, hygroscopicity, and agglomeration may cause rapid changes in particle size. Once hygroscopic particles are inhaled and mix with the warm humid air in the respiratory tract, they stop shrinking and start to adsorb water and grow in size. The relative humidity in the lungs beyond the major bronchi at resting inspiratory rates is about 99.5% (Ferron et al., 1988a); this is sufficient to cause dry salt particles to increase 3 to 4.5 times in diameter (Ferron and Gebhart, 1988). Thus, aerosol size measurements of hygroscopic particles made after drying or at a low relative humidity drastically underestimate their size in the lungs. In such cases it is possible to estimate the size attained in the lungs by taking into consideration the relative humidity, the molecular weight of the salt, and the number of ions into which it dissociates in water, among other factors. Such particle growth equations are presented by Ferron (1977, 1987) and Persons et al. (1987).

Particle size may not be constant as a generated pharmacological aerosol moves through a delivery system and the respiratory tract. Volatile aerosols composed of Freon propellant or water become smaller through evaporation (Mercer, 1973), whereas hygroscopic aerosols such as sodium chloride particles may grow dramatically, especially as the relative humidity nears 100% (Cinkotai, 1971; Ferron, 1977; Ferron and Gebhart, 1988).

The effect of hygroscopicity on deposition has been studied both experimentally (Blanchard and Willeke, 1983; Tu and Knutson, 1984) and theoretically (Ferron, 1977; Martonen et al., 1985; Xu and Yu, 1985; Persons et al., 1987; Ferron et al., 1988a,b), and has recently been reviewed (Morrow, 1986). The overall effect borne out in these studies is that, with increasing hygroscopicity and relative humidity, deposition fraction as a function of particle size is shifted to smaller particle sizes. That is, whereas the deposition minimum for nonhygroscopic particles is about 0.5 μm, the minimum for dry NaCl particles that grow in the 99.5% relative humidity of the lungs is about 0.1 μm (Xu and Yu, 1985). Consequently, the deposition of hygroscopic particles with diameters of greater than 0.1 μm when inhaled exceeds the deposition of nonhygroscopic particles of the same size. For example, a 1-μm NaCl particle has a deposition fraction or collection efficiency 3.8 times greater than that of a 1-μm nonhygroscopic particle (Xu and Yu, 1985). Many pharmacological aerosols are hygroscopic; the hygroscopic growth and deposition characteristics of histamine, for example, are similar to those of NaCl (Ferron, 1987).

Breathing Pattern

Another important factor affecting deposition is the breathing pattern. Minute volume defines the average flow velocity of the aerosol-containing air in the lung and the total number of particulates to which the lung is exposed. Respiratory frequency affects the residence time of aerosols in the lungs and, hence, the probability of deposition by gravitational and diffusional forces. A change in the lung volume alters the dimensions of the airways and parenchyma.

Anatomy of the Respiratory System

The anatomy of the respiratory tract is important since it is necessary to know the diameters of the airways, the frequency and angles of branching, and the average distances to the alveolar walls. Furthermore, along with the inspiratory flow rate, airway anatomy specifies the local linear velocity of the airstream and the character of the flow. A significant change in the effective anatomy of the respiratory tract occurs when there is a switch between nose and mouth breathing. There are inter- and intraspecies differences in lung morphometry; even within the same individual, the dimensions of the respiratory tract vary with changing lung volume, with aging, and with pathological processes.

CLEARANCE OF RADON PROGENY: GENERAL PRINCIPLES

The response to the alpha energy produced by radon progeny depends not only on the amount of aerosol deposited but also on the amount retained in the lungs over time. Retention is the amount of material present in the lungs at any time and equals deposition minus clearance. An equilibrium concentration is reached during continuous exposure to radon progeny when the rate of deposition equals the rate of clearance. The amount of particles retained within a specific lung region over time is a key determinant of dose.

As discussed elsewhere in this chapter, such factors as particle size, hygroscopicity, and breathing pattern affect the site of deposition within the respiratory tract. In turn, where particles deposit in the lungs determines which mechanisms are used to clear them and how fast they are cleared. This influences the amount retained over time. Examples of the implications of particle characteristics on integrated retention were given by Brain and Valberg (1974) using a model developed by the Task Force on Lung Dynamics. They showed that the total amount as well as the distribution of retained dose among nose and pharynx, trachea and bronchi, and the pulmonary and lymphatic compartments were dramatically altered by particle size and solubility.

Particles that deposit on the ciliated airways are cleared primarily by the mucociliary escalator. Those particles that penetrate to and deposit in the peripheral, nonciliated lung can be cleared by many mechanisms, including translocation by alveolar macrophages, particle dissolution, or movement of free particles or particle-containing cells into the interstitium and/or the lymphatics. These pathways are of great importance for materials that have a long biological half-life, such as silica, asbestos, or plutonium. However, one must realize that these fates are of relatively little importance for short-lived radon progeny. This is because the majority of the energy produced by alpha-particle emission is dissipated during the first hour. Thus, the movement of radon progeny very soon after deposition is relevant to dosimetry. What happens weeks or months later is irrelevant. Reviews of clearance processes have been prepared by Kilburn (1977), Pavia (1984), and Schlesinger (1985a).

Mucociliary Transport

Less soluble particles that deposit on the mucous blanket covering pulmonary airways and the nasal passages are moved toward the pharynx by cilia. Also present in this moving carpet of mucus are cells and particles that have been transported from the nonciliated alveoli to the ciliated airways. At the pharynx, mucus, cells, and debris coming from the nasal cavities and the lungs meet, mix with salivary secretions, and enter the gastrointestinal tract after being swallowed. In humans, the ciliated epithelium extends from the trachea down to the terminal bronchioles. The particles are removed with half-times of minutes to hours; the rate depends on the speed of the mucous blanket. The speed is faster in the trachea than it is in the small airways (Serafini et al., 1976). There is little time for solubilization of slowly dissolving materials. In contrast, particles deposited in the nonciliated compartments have much longer residence times; there, small differences in in vivo solubility can have great significance. The speed of mucous flow can be affected by factors influencing either the cilia or the amount and quality of the mucus.

Ciliary action may be affected by the number of strokes per minute, the amplitude of each stroke, the time course and form of each stroke, the length of the cilia, the ratio of ciliated to nonciliated area, and the susceptibility of the cilia to intrinsic and extrinsic agents that modify their rate and quality of motion. The characteristics of the mucus are critically important. The thickness of the mucous layer and its rheological properties may undergo wide variations. Wanner (1977) and Camner and Mossberg (1988) reviewed many of these factors that influence clearance, including clinical implications.

Mucociliary transport has been studied by a variety of techniques, such as monitoring the movement of inert or radiolabeled particles deposited on the tracheal mucus via a bronchoscope or as an inhaled bolus. Tracheal mucus velocity (TMV) can then be estimated from the distance the particles moved over time, as observed with either movies taken through a bronchoscope or by a gamma camera (Yeates et al., 1975; Chopra et al., 1979). Bronchoscopic techniques yield higher numbers for TMV (15-21 mm/min) than the noninvasive bolus techniques do (4.4 mm/min). These values are only characteristic for the trachea and large central airways. Transport in the small peripheral airways is slower probably due to the discontinuous mucous layer (Van As, 1977).

Many investigators have estimated mucociliary transport from whole-lung clearance curves. These curves are generated by monitoring the amount of radioactivity in the lungs over time (hours to days) following the inhalation of a radiolabeled aerosol.

Albert and Arnett (1955) first used this method and noted that the clearance curve can be divided into two phases: a fast and a slow phase. The fast phase was complete within 24 to 48 h and has generally been attributed to tracheobronchial clearance; the slow phase has been attributed to alveolar clearance (Booker et al., 1967; Morrow et al., 1968; Lippmann and Albert, 1969). This interpretation has been widely accepted and has been used to study clearance in normal and abnormal individuals (Lourenco et al., 1971; Sanchis et al., 1972; Poe et al., 1977; Camner and Philipson, 1978; Stahlhofen et al., 1980). However, evidence indicates that clearance from the airways might not be complete in the first 24 h. This may be even more pronounced in patients with lung disease. Gore and Patrick (1982) noted that particles instilled into the trachea can be sequestered in epithelial cells. Stahlhofen et al. (1986) noted fast and slow phases of clearance even in humans given a bolus of particles delivered only 45 cm^3 beyond the larynx. Another approach is to examine deposition and clearance of particles in central versus peripheral regions (Smaldone et al., 1988). The uses of this type of analysis in the interpretation of clearance curves have recently been noted in an editorial by Foster (1988).

More studies are needed to elucidate the best methodology to model mucociliary clearance and to understand its role and importance in patients with pulmonary disease. Clearly, it is also essential to understand the initial deposition pattern of an inhaled aerosol in order to assess the importance of mucociliary clearance on the disappearance of the aerosol from the lungs.

Nonciliated Regions

Particles deposited in the nonciliated portion of the lungs either are moved toward the ciliated region, primarily within alveolar macrophages, or they enter the lung connective tissue either as free particles or within macrophages. Macrophages are credited with keeping the alveolar surfaces clean and sterile. These cells rest on the continuous epithelial layer of the lung. It is their phagocytic and lytic potentials that provide most of the bactericidal properties of the lungs. Rapid endocytosis of insoluble particles by macrophages prevents particle penetration through the alveolar epithelia and facilitates alveolar-bronchiolar transport. Particles in connective tissue may slowly dissolve or may be transported to new sites through lymphatic pathways. Particles remaining on alveolar surfaces are cleared with biological half-times estimated to be days to weeks in humans, whereas particles that have penetrated into fixed tissues are cleared with half-times ranging from a few days to thousands of days depending on their solubility. Brain (1985) has summarized the biology of lung macrophages and in a recent review (Brain, 1988) has emphasized that many kinds of lung macrophages exist. These include alveolar, airway, connective tissue, pleural, and intravascular macrophages.

Cough

The function of cough is the removal of material from the respiratory tract when mucociliary transport is overwhelmed. Thus, there would probably be little or no transport of radon progeny in the airways produced by cough in individuals with normal amounts of secretions. However, in chronic smokers and other subjects with chronic bronchitis, cough can produce significant and rapid mucus transport. Whether cough is initiated by cough receptors or voluntarily, the subject's first action is usually inspiration. Following the inspiratory phase, the glottis is closed. At about the same time, expiratory muscles in the abdomen and chest wall contract, producing pleural and alveolar pressures of 100 mm Hg or more. When the glottis is opened abruptly, the result is extremely high linear velocities which may be hundreds of miles per hour. The kinetic velocity of this airflow is coupled to mucus and causes it to be propelled along the airways. Several reviews of cough are available (Brain et al., 1985; Leith et al., 1986).

DIFFERENCES BETWEEN WORKERS AND THE PUBLIC

The mechanisms responsible for deposition and clearance have already been discussed. The effectiveness of these mechanisms depends on various factors. Some major influences on the fraction of the inspired aerosol that deposits within the respiratory tract and its distribution are discussed here.

Anatomic Variations

The configuration of the lungs and airways is important since the efficiency of deposition depends, in part, upon the diameters of the airways, their angles of branching, and the average distances to alveolar walls. Furthermore, along with the inspiratory volumetric flow rate, airway anatomy specifies the local linear velocity of the airstream and, thus, whether the flow is laminar or turbulent. There are inter- and intraspecies differences in lung morphometry (Soong et al., 1979; Schlesinger and McFadden, 1981; Phalen, 1984; Nikiforov and Schlesinger, 1985); even within the same individual the dimensions of the respiratory tract vary with changing lung volume, with aging, and with pathological processes. Among individuals who breathe in the same manner, total deposition has been found to have a coefficient of variation of as large as 27%, much of which is believed to be due to intersubject differences in airway geometry (Heyder et al., 1982). This finding has been supported by other theoretical and experimental studies (Yeates et al., 1982; Yu and Diu, 1982). Within one person, a decrease in lung volume from 4,800 to 2,400 ml not only increases deposition but also causes the major site of particle deposition to shift from the lung periphery to the central airways (Agnew, 1984). At low lung volumes, central airways have smaller cross-sectional areas and, thus, higher linear velocities. For a given flow rate, this enhances deposition by impaction in more central intrapulmonary airways. As a person ages, anatomical changes of the respiratory tract also appear to affect deposition. It has been predicted theoretically that total deposition in children may be as much as 1.5 times higher than that in adults (Xu and Yu, 1986). Children have smaller tidal volumes and residence times, and these compensate in part for their smaller lung dimensions.

The dimensions of the airways and alveoli vary among individuals because of age, body size, genetics, and disease. However, the overall pattern of conducting airways changes little since it is established at birth. It is true that terminal bronchioles can develop into respiratory bronchioles and become more alveolated, but in general, the newborn's airways simply enlarge into the adult tracheobronchial tree (Hogg et al., 1970; Hislop et al., 1972; Burri, 1985). Between birth and adulthood, airways enlarge in length and diameter by a factor of around 3. Lung volume would thus be expected to grow 27-fold during this period of time (Dunill, 1962).

However, ventilation per gram of body weight is clearly higher in children than it is in adults. In addition, some of the more peripheral airways appear to grow less during childhood and adolescence than do more central airways. Phalen et al. (1985) have predicted that deposition is highest in newborns and decreases with increasing age to adulthood. Phalen and colleagues have also suggested that smaller individuals receive more particle deposition within the conductive zone than do larger individuals at the same level of ventilation.

There are substantial changes in the gas-exchange region during growth that reflect an enormous increase in the number and surface area of alveoli. However, these changes should not influence the deposition or clearance of radon progeny in airways, the primary region of concern.

In addition to age, gender and ethnicity may also influence airway size. Up to age 14 yr, boys were found to have lungs larger than those of girls (Thurlbeck, 1982). There are also dramatic differences in the sizes, dimensions, and structural proportions of the larynx between males and females.

Breathing Pattern

The way each underground miner or member of the general population breathes also affects deposition. Minute volume defines the average flow velocity of air in the lungs and the total number of particles to which the lungs are exposed. Respiratory frequency, tidal volume, and lung volume will affect the residence time of aerosols in the lungs and, hence, the probability of deposition by gravitational and diffusional forces. Flow rate governs the degree and extent of turbulent flow in the upper airways that enhances particle deposition. A change in lung volume also alters the dimensions of the airways and parenchyma. High levels of ventilation and breath holding represent extremes of breathing patterns that give rise to markedly different deposition patterns. Valberg et al. (1978) exposed excised dog lungs to a submicrometric radioactive aerosol while the dogs were using different breathing patterns. Even though the same particle size was used throughout, when the pattern was rapid and shallow, airway deposition was predominant, but when the pattern was slow and deep, deposition on alveolar surfaces was greater.

Activity Level and Exercise

There are compelling reasons why exercise and the resulting changes in breathing pattern should influence particle deposition in the respiratory tract. First, the level of ventilation is a major determinant of the mechanism of deposition. Inertial impaction depends on the velocity of the airstream, whereas the importance of diffusion and settling depends on the residence time in the lungs and the distance the particles must travel to reach lung surfaces. Inertial impaction is more important in central airways, where linear velocities are high; diffusion and settling are more important in peripheral airways and alveoli, where residence times are longer and distances are smaller. Second, minute volume (respiratory frequency × tidal volume) determines the total particle mass that enters the respiratory tract. Hence, even if the deposition fraction or collection efficiency is constant, an increase in the minute volume increases particle deposition in the lungs. Dennis (1971) has reported that during exercise, the deposition fraction increases with minute volume in some human subjects, particularly for larger particles (1.0 to 3.0 μm); thus, the amount of aerosol deposited in the lung per unit time increases in two ways. Both the amount entering the respiratory tract as well as the percentage deposited were elevated.

Landahl and associates (1952) demonstrated that the percentage of airborne particles deposited in the respiratory tract increases with increasing minute volume. In exercising hamsters (Harbison and Brain, 1983), the deposition of a 0.4-μm ^{99m}Tc-labeled aerosol increased as their oxygen consumption increased. Exercising animals consumed twice as much oxygen as sedentary animals did but they retained 2.5 to 3 times as many particles in their lungs. Enhanced retention probably reflects both an increase in ventilation and an increase in collection efficiency. It is likely that different relationships may pertain with larger particles. Then, increased flow rates may be more effective at depositing particles in the nose, pharynx, larynx, and large airways, leading to diminished particle deposition deep in the lungs. Recently, Zeltner et al. (1988) found the same absolute number of 1.0-μm particles deposited in the parenchyma of exercising hamsters compared with that in resting animals, whereas deposition in the upper airways increased in exercising hamsters. It is possible that for still larger particles, exercise could actually lower the particle dose to the parenchyma.

In humans who are exercising vigorously, minute volumes can exceed 120 liters/min, greatly increasing the amount of aerosol inspired. Even at rest, inspiratory flows are elevated during speech (Bouhuys et al., 1966; Bunn and Mead, 1971). The high velocities (2 m/s) achieved in the main bronchi enhance inertial impaction and turbulence. Experimenting with 0.5-μm monodisperse particles, Muir and Davies (1967) found that the percentage of particles deposited increased linearly with tidal volume and decreased with the square root of breathing frequency. Expiratory reserve volume also influenced this percentage (Davies et al., 1972).

Oral Versus Nasal Breathing

A highly significant change in the effective anatomy of the respiratory tract occurs when there is a switch between nose and mouth breathing or when the nose is bypassed by a tracheostomy or by an endotracheal tube. The nose has a central role as a collector of inhaled aerosol particles and as a conditioner that warms and humidifies inspired air. The combination of a small cross section for airflow, sharp curves, and interior nasal hairs helps to maximize particle impaction. Abundant evidence indicates that significant fractions of inhaled particles and gases can be deposited in the nose and pharynx. Excellent reviews of particle deposition in that region are available (Hounam and Morgan, 1977). Deposition and clearance of particles in the head during nose breathing have been studied extensively by Pattle (1961), Hounam et al. (1969), Lippmann (1970, 1977), Märtens and Jacobi (1974), Swift and Proctor (1982), and Heyder et al. (1986). Collectively, these reports indicate that although the nose is an inefficient filter for submicrometric particles, large quantities of large particles (>5.0 μm) are deposited in the nasopharyngeal region. The probability for particle deposition increases with increasing particle diameter, flow rate, and nasal flow resistance. Almost all particles larger than 10 μm are trapped in the nose. Large individual variations are apparent, possibly because of intersubject differences in the distensibility of the nasal passages that affect nasal resistance.

The loss of the filtering capacity of the nose may be important. With rising levels of exercise-enhanced metabolism and increasing ventilation, the high-resistance nasal pathway is progressively abandoned in favor of the low-resistance oral pathway. As the nose is bypassed, more and more particles penetrate to and deposit in the airways of the lungs. With exercise, the amount of aerosol depositing in the lungs may increase in excess of that predicted by the increased ventilation.

The surface area of the human nose is small relative to the 140 m^2 of the total respiratory tract, but it has a rich vasculature that is well designed for heating or cooling and for rapid exchange of dissolved substances between the mucus, tissues, and blood vessels. Most capillaries have fenestrated endothelial linings enclosed by porous basement membranes.

Proctor (1973) estimates that the nasal surface area between the bony opening into the nasal cavity and the posterior end of the turbinates is 160 cm^2, or 0.016 m^2. The volume of air within the human nasal passages is about 20 ml. At an inspiratory flow rate of 0.4 liters/s, the residence time of the inspired gas in the nose is only 0.05 s. Despite this brief contact time with the nasal mucosa, special anatomic features help to condition the inspired gas. At modest flow rates, inspired gas is warmed and humidified before it reaches the subglottal area. However, the upper airways are less effective at higher flow rates, especially when the inspired air is cold and dry. The nose also has a major role as a collector of inhaled aerosol particles. A small cross section for airflow and the resulting high linear velocities, sharp curves, and interior nasal hairs all help to promote particle impaction.

In both animals and humans, particle deposition and clearance rates are markedly affected by the choice of pathway to the trachea. Shifts from the nasal to the oral route are of major significance. An excellent review of particle deposition in the nose is that of Hounam and Morgan (1977). Deposition and clearance of particles in the head during nose breathing have been extensively studied by Pattle (1961), Hounam and coworkers (1969), Lippmann (1970), and Märtens and Jacobi (1974). Collectively, these reports indicate that substantial quantities of particles larger than 5.0 μm are deposited in the nasopharyngeal region. There are large individual variations, depending on the structure and aerodynamic properties of the nose. Hounam et al. (1969) showed an increase in the percentage of inspired particles deposited in the nose as the resistance to airflow increased. Forsyth and associates (1983) have demonstrated that nasal resistance decreases dramatically with exercise, probably because of mucosal shrinkage mediated by the effects of the sympathetic nervous system on the nasal microcirculation. Lower nasal resistance would decrease the particle collection efficiency in the nose and would increase particle penetration to the pulmonary airways and parenchyma. The nose would become a less efficient filter.

The nasal route is more efficient at removing large particles from inspired air than is the oral route. In bypassing the nose, mouth breathers deposit more particles in their lungs and may concomitantly increase the risk of pulmonary damage. Mouth breathing is common under conditions of heavy exercise, during speech or singing (Bouhuys et al., 1966; Bunn and Mead, 1971), and in those who have decreased nasal conductance caused by infection or allergy. This may result in markedly increased exposure of the airways and alveoli to large particles.

Some inhaled toxins and the diseases they produce may induce changes that in turn affect where radon progeny deposit. For example, sulfur dioxide and sulfate particles alter the dimensions of the upper and lower airways in both humans and experimental animals. The principal effect in acute experiments is to increase pulmonary flow resistance; this reflects the bronchoconstrictive action of the pollutant. Chronic exposure may lead to chronic bronchitis, and again, airways may be narrowed. In either case, the reduction in airway cross section increases linear flow velocities and turbulence, with the result that more particles deposit by impaction. Thus, chronic exposure to air pollutants or tobacco smoke is likely to be associated with changes in the uptake of radon progeny.

OTHER MODIFYING FACTORS INFLUENCING DEPOSITION AND/OR CLEARANCE OF RADON PROGENY

Preexisting Disease

Respiratory diseases can influence the distribution of inspired radon progeny. Bronchoconstriction or obstruction of airways leads to diversion of flow to less obstructed airways. With advancing disease, the remaining healthy airways and alveoli may be increasingly exposed to inspired particles. Narrowing of airways by mucus, inflammation, or bronchial constriction can increase the linear velocities and turbulence of airflow, enhance inertial deposition, and cause more central deposition patterns (Albert et al., 1973; Goldberg and Lourenco, 1973; Taplin et al., 1977; Kim et al., 1983a). In very sick patients with chronic obstructive pulmonary disease (COPD), there may be increases in aerosol deposition that may be associated with flow limitation (Smaldone and Messina, 1985).

One factor that determines where radon progeny are deposited in the lungs is the distribution of ventilation. Particles are carried into the lungs with the inhaled air and therefore reach only areas that are ventilated. It is known that nonventilated regions of the lungs exist in patients with a variety of pathological conditions (Milic-Emili, 1974). Because diseased lungs are nonuniformly affected, they are more likely than normal lungs to have large alterations in the distribution of ventilation. Disease may impair ventilation in some regions because of airway closure, atelectasis, or obstructing mucous plugs (Macklem, 1971), and fewer particles reach these regions. Conversely, the remaining ventilated areas of the lungs may be exposed to increased toxic loads when a significant amount of the lung is not ventilated. Maintaining patent airways and open alveoli requires a balance between surface tension and tissue forces (Macklem, 1971; Weibel and Gil, 1977; Hoppin et al., 1986). Alterations in these forces can cause alveoli to collapse, especially when breathing at low lung volumes (Bachofen et al., 1979; Gil et al., 1979). Thus, airway and alveolar patency during breathing depends upon surface and tissue forces, and any disruption in lung structure can lead to the creation of nonventilated areas of the lung.

Chronic pulmonary disease alters the architecture of the lungs. Emphysema involves alveolar wall destruction, which leads to enlarged air spaces. When radial traction forces of the alveolar septa are reduced and surface tension is diminished because of loss of alveolar surface area, intrapulmonary bronchi are poorly supported and tend to collapse during expiration (American Lung Association, 1981). Similarly, the increase of secretions in patients with chronic bronchitis favors airway obstruction by mucous plugs and the regional thickening of alveolar walls in interstitial fibrosis makes alveoli less compliant and thus favors local hypoventilation. In each of these cases, the changes in airway and alveolar properties cause changes in the distribution of ventilation. Although airway closure has been noted in patients with chronic lung disease (Macklem, 1971), little is known about the anatomic distribution of poorly ventilated regions of the lungs and how this distribution changes during the disease process.

Radioactive aerosols have been used in humans to detect airway obstruction in COPD (Lourenco et al., 1972; Goldberg and Lourenco, 1973; Ramanna et al., 1975; Dolovich et al., 1976; Pavia et al., 1977; Taplin et al., 1977; Itoh et al., 1981; Kim et al., 1983a,b). Scintigraphic analysis of particle deposition shows that deposition is greater in the central airways of patients with airway obstruction than in airways of normal patients (Lourenco et al., 1972; Ramanna et al., 1975; Taplin et al., 1977; Itoh et al., 1981). Investigators have quantified this phenomenon by calculating a penetration index to assess how deeply the aerosol enters the lungs (Thomson and Pavia, 1973; Dolovich et al., 1976; Agnew et al., 1981). Penetration of aerosol has been found to correlate inversely with airway obstruction. Other investigators have used aerosol rebreathing techniques to measure total deposition in normal individuals versus that in patients with obstructed airways (Kim et al., 1983a). Theoretical analyses of these data suggest that aerosol deposition may be a more sensitive indicator of airway abnormalities than is measurement of overall airway resistance (Kim et al., 1983b). Total deposition in the lungs of humans with airway obstruction is enhanced compared with that in the lungs of normal patients (Kim et al., 1983a).

Since human disease is progressive, experiments such as those described above are complicated and difficult to interpret. Nevertheless, some conclusions can be made from these studies. Two types of airway obstruction can occur in patients with COPD: physical obstruction of the airways with mucous plugs, as in those with chronic bronchitis, and obstruction caused by too-small bronchioles that collapse during expiration. Airway collapse or closure does not occur in normal individuals who are breathing at rest, but can occur during normal breathing in humans with emphysema (Macklem, 1971).

Obstruction of small airways has three effects on aerosol deposition. First, particles deposit more centrally because they fail to reach the terminal air units. Second, deposition is enhanced at the site of obstruction in both obstructed airways and in those that collapse during expiration. Smaldone and associates (1979) examined the latter type of obstruction and showed that deposition is enhanced around the site of closure because of changes in airflow patterns. Turbulent eddies enhance deposition at the constriction site; collapse brings particles in close contact with airway walls.

Third, Macklem and associates (1973) have postulated that obstructed airways cause inspired gas to be diverted to the healthier regions of the lungs; thus, particles preferentially deposit in healthy regions. Because of the increased linear velocity of the gas in these healthy regions and enhanced penetration to alveoli, deposition would be enhanced (Macklem et al., 1973).

Animal models have been used to explore how aerosol deposition is influenced by lung pathology. In these studies, variable factors in humans like age, environmental exposure, and extent of disease can be controlled. In 1979, Hahn and Hobbs studied both papain- and elastase induced emphysema in hamsters. Intratracheal instillation of elastase produced focal destruction and enlargement of alveoli accompanied by loss of lung elastic recoil (Hayes et al., 1975; Snider et al., 1977, 1986).

Hahn and Hobbs (1979) exposed hamsters to an aerosol of ^{137}Cs-labeled aluminum silicate particles 21 days after elastase or papain administration. The emphysematous hamsters, examined 3 h after the aerosol exposure, retained 45 to 65% fewer particles in their lungs than did the control hamsters. This decrease in initial dose of aerosol in emphysematous lungs was also observed by Damon and associates (1983) in rats exposed to iron oxide particles and by Lundgren and associates (1981) in rats exposed to ^{169}Yb-labeled plutonium oxide particles.

Sweeney et al. (1983a, 1985, 1987) have studied the progressive influence of emphysema, chronic bronchitis, and fibrosis on the distribution of deposited submicrometric particles throughout rodent lungs. One common result in all three diseases is that the presence of detectable pulmonary disease always results in less uniform patterns of particle retention throughout the lungs. This pattern could not be explained by differences in breathing pattern (Sweeney et al., 1983b, 1986). With a restrictive lung disease, such as fibrosis, deposition was more heterogeneous in the early stages when the fibrotic lesions were focal. As the fibrosis became more uniformly distributed throughout the lungs, particle retention also became more uniformly distributed. These changes were inversely correlated with the presence of pulmonary fibrosis; fibrotic lung regions retained fewer particles than the more normal regions of the lung did (Sweeney et al., 1983a).

For pulmonary emphysema, the heterogeneity of deposition increased with increasing severity of the disease, as measured by changes in the mean linear intercept of the parenchyma (Sweeney et al., 1987). Variations in local deposition could not be explained by the anatomic distribution of emphysema throughout the lungs. In animals with emphysema, although the parenchyma is the primary site of the lesion, airways are also influenced in the later stages of the disease. Loss of elastic recoil caused by tissue destruction would tend to reduce airway caliber. With the progressive loss of supportive parenchymal tissue as emphysema progresses, the decreasing airway caliber would favor enhanced airway deposition and increased overall heterogeneity of deposition. Airway changes caused by physical obstruction of the airways because of mucus plugging (as in animals with chronic bronchitis) also produce increased heterogeneity with enhanced deposition in the affected airways (Sweeney et al., 1985).

Nevertheless, extrapolation of these findings in animals to humans with chronic obstructive pulmonary disease or idiopathic pneumonitis with fibrosis must be done cautiously. These data in animals can be used to identify some factors influencing deposition, and thus risk, but they are limited in predicting causal relationships in human disease.

Smoking

There are many reasons to believe that smokers may have altered exposure-dose relationships as well as altered responses to radon progeny. First, and probably most important, there are mechanistic interactions among the carcinogens and irritants in tobacco smoke and radon progeny. However, in addition, smoking can alter the structure of the lung and thus alter both deposition and clearance kinetics. Cigarette smoking frequently leads to chronic bronchitis, a disease characterized by proliferation of mucus-secreting cells and glands. Thus, chronic bronchitis is often characterized by narrowing of airways, and this can enhance airway deposition. This should lead to increased deposition of radon progeny. Also enhancing the dose from radon progeny would be diminished clearance rates from airways. Counteracting this would be the protective effect of an increased mucous layer thickness.

OTHER CHARACTERISTICS OF WORKERS AND THE PUBLIC

Occupancy Data

If one wants to compare exposure-dose relationships between miners and the general population, one factor that must be considered is where people spend their time. For early miners, the number of hours spent underground on a weekly basis has not been well described in published literature. Contemporary underground miners spend about 35 or 40 h/week in mining environments where radon progeny are encountered. The actual time spent there is probably less than that, since some time is spent preparing for work and descending into and ascending the mine. Clearly, the general population spends a much greater portion of its time in indoor environments. In 1955, Stewart et al. reported that the general population spent approximately 90% of its time indoors in the United Kingdom. In 1982, the United Nations Scientific Committee on the Effects of Atomic Radiation used a value of 80%. More recently, Francis (1987) produced a report describing indoor occupancy in the United Kingdom. On average, 92% of time was spent indoors. That was divided into 77% spent in their own residence and 15% spent in other indoor locations. Of the 77% spent indoors in their own residence, 42% was in the bedroom, 18% was in the living area, and the remaining 17% was in the kitchen, dining room, and bathroom. Similar data have been reported for adults in the United States; in a survey of 44 U.S. cities, participants spent less than 10% of their time outdoors (Szalai, 1972).

The data suggested that occupancy patterns are relatively independent of season, but differences among different days of the week were noted. For example, Sundays in the winter yielded the highest indoor occupancy in residences (about 90%). They also encountered some variation among various population subgroups. Housewives spent approximately 97% of their time indoors; 88% was spent in their own residence, while 9% was spent in other indoor locations. Furthermore, of the remaining 3% that was spent outdoors, about half of that was taken up with traveling.

Ethnic Differences

Another issue that has been raised in relation to comparisons between miners and the general population are ethnic differences. Is it possible that different ethnic groups have different lung geometries, lung volumes, or breathing patterns? Recently, Roy and Courtay (in press) reviewed the existing literature in relation to ventilation rates and lung volumes. They assessed more than a dozen papers providing information in relation to black, Chinese, Japanese, Indians, Caucasian Americans, and other ethnic groups in the United States. The general conclusion was that there are some differences in lung size among different ethnic groups. American Caucasians tended to have greater portal capacities than did American blacks or Americans of Asian descent. Nevertheless, variability in ventilation among these groups was relatively small, especially when compared with variations in activity and therefore ventilation within individuals. A compelling conclusion is that ethnicity is not a major factor influencing exposure-dose relationships for inhaled particles.

Altitude

Frequently, underground miners, especially in the United States, work at elevated altitudes. Many of the uranium miners in the Colorado Plateau worked at altitudes of between 5,000 and 8,000 feet above sea level. The related effects caused by changes in oxygen concentration and gas density on breathing are slight in relation to work-ventilation relationships and other concerns addressed in this report. It is true that changes in gas density can affect deposition mechanisms such as sedimentation, but again, these effects are likely to be relatively small.

Species Differences

Animal models of radon deposition and retention have been used (Cross, 1988). However, even when the same aerosol size distribution is breathed by different species, very disparate lung doses may still result. There are both systematic and unusual variations in ventilation, collection efficiency, and lung anatomy among species that influence the amount and distribution of deposited aerosols. Thus, the problem of extrapolating findings from one species to another and from animals to humans is difficult. Several scientists (Palm et al., 1956; Friedlander, 1964; Kliment, 1973; Stauffer, 1975; McMahon et al., 1977; Raabe et al., 1977; Schlesinger, 1980; Phalen, 1984; Xu and Yu, 1986) have made either theoretical or experimental contributions, but the problem is far from solved. How well aerosol deposition in humans and animals compares has been reviewed recently by Schlesinger (1985b) and by Brain et al. (1988). It is essential to realize that different rates of clearance among species can also influence retention and thus the total dose retained by the lung.

CONCLUSIONS

Determination of the distribution of inhaled radon daughters within the respiratory tract is one dimension of the more general problem of determining exposure-dose and dose-response relations. It is likely that some of the variability in response among different individuals and various animal species may result from differences in the concentration of radon progeny at the site of action as well as variations in the inherent responsiveness of specific active tissues. Furthermore, the properties of the bronchial epithelium are important, since many radon daughters may breach epithelial barriers and come into contact with more reactive cells.

REFERENCES

Agnew, J. E. 1984. Physical properties and mechanisms of deposition of aerosols. P. 49 in Aerosols and the Lung: Clinical and Experimental Aspects, S. W. Clarke and D. Pavia, eds. Boston: Butterworths.

Agnew, J. E., D. Pavia, and S. W. Clark. 1981. Airways penetration of inhaled radioaerosol: An index to small airways function? Eur. J. Respir. Dis. 62:239-255.

Aharonson, E. F., A. Ben-David, and M. A. Klingberg, eds. 1976. Air Pollution and the Lung. New York: Halsted Press-Wiley.

Albert, R. E., and L. C. Arnett. 1955. Clearance of radioactive dust from the lung. Arch. Environ. Health 12:99.

Albert, R. E., M. Lippmann, H. Peterson, J. Berger, K. Sanborn, and D. Bohning. 1973. Bronchial deposition and clearance of aerosols. Arch. Intern. Med. 131:115.

Altshuler, B. 1961. The role of the mixing of intrapulmonary gas flow in the deposition of aerosol. Pp. 47-54 in Inhaled Particles and Vapours, C. N. Davies, ed. Oxford: Pergamon Press.

Altshuler, B., L. Yarmus, E. D. Palmes, and N. Nelson. 1957. Aerosol deposition in the human respiratory tract. I. Experimental procedures and total deposition, A.M.A. Arch. Ind. Health 15:293.

American Lung Association. 1981. Chronic Obstructive Lung Disease, 5th ed. New York: American Lung Association.

Bachofen, H., P. Gehr, and E. R. Weiber. 1979. Alterations of mechanical properties and morphology in excised rabbit lungs rinsed with a detergent. J. Appl. Physiol. 47:1002-1010.

Blanchard, J. D., and K. Willeke. 1983. Total deposition of ultrafine sodium chloride particles in human lungs. J. Appl. Physiol. 57:1850.

Booker, D. V., A. C. Chamberlain, J. Rudo, D. C. Muir, and M. L. Thompson. 1967. Elimination of 5 micron particles from the human lung. Nature 215:30.

Bouchikhi, A., M. H. Becquemin, J. Bignon, M. Roy, and A. Teillac. 1988. Particle size study of nine metered dose inhalers, and their deposition probabilities in the airways. Eur. Respir. J. 1:547.

Bouhuys, A., D. F. Proctor, and J. Mead. 1966. Kinetic aspects of singing. J. Appl. Physiol. 21:2-10.

Brain, J. D. 1985. Macrophages in the respiratory tract. Pp. 447-471 in Handbook of Physiology, Vol. 1, Circulation and Nonrespiratory Functions, A. P. Fishman and A. B. Fisher, eds. Bethesda, Md.: American Physiological Society.

Brain, J. D. 1988. Lung macrophages—How many kinds are there? What do they do? Am. Rev. Respir. Dis. 137:507.

Brain, J. D., and P. A. Valberg. 1974. Models of lung retention based on the Report of the ICRP Task Group. Arch. Environ. Health 28:1-11.

Brain, J. D., and P. A. Valberg. 1979. Deposition of aerosols in the respiratory tract. Am. Rev. Respir. Dis. 120:1325-1373.

Brain, J. D., and P. A. Valberg. 1985. Aerosols: basics and clinical considerations. Pp. 594-603 in Bronchial Asthma: Mechanisms and Therapeutics, 2nd ed., E. B. Weiss, M. S. Segal, and M. Stein, eds. Boston: Little, Brown and Co.

Brain, J. D., D. F. Proctor, and L. M. Reid, eds. 1977. Respiratory defense mechanisms. In Lung Biology in Health and Disease, Vol. 5. New York: Marcel Dekker.

Brain, J. D., P. A. Valberg, and S. Sneddon. 1985. Mechanisms of aerosol deposition and clearance. Pp. 123-147 in Aerosols in Medicine. Principles, Diagnosis and Therapy, F. Morén, M. T. Newhouse, and M. B. Dolovich, eds. Amsterdam: Elsevier Science Publishers.

Brain, J. D., P. A. Valberg, and G. A. Mensah. 1988. Species differences. Pp. 89-103 in Variations in Susceptibility to Inhaled Pollutants, J. D. Brain, B. D. Beck, A. J. Warren, and R. A. Shaikh, eds. Baltimore: Johns Hopkins Press.

Bunn, J. D., and J. Mead. 1971. Control of ventilation during speech. J. Appl. Physiol. 31:870-872.

Burri, P. H. 1985. Lung development and growth. Pp. 1-46 in Handbook of Physiology, S. R. Geiger, ed. Sect. 3: The Respiratory System, A. P. Fishman and A. B. Fisher, eds. Bethesda, Md.: American Physiological Society.

Camner, P., and B. Mossberg. 1988. Mucociliary disorders: A review. J. Aerosol Med. 1:21-28.

Camner, P., and M. S. Philipson. 1978. Human alveolar deposition of 4 micron teflon particles. Arch. Environ. Health 33:181.

Chopra, S. K., G. V. Taplin, D. Elam, S. W. Carson, and D. Golde. 1979. Measurement of tracheal mucociliary transport velocity in humans—smokers versus non-smokers (preliminary findings). Am. Rev. Respir. Dis. 119(Suppl.):205.

Cinkotai, F. F. 1971. The behavior of sodium chloride particles in moist air. J. Aerosol Sci. 2:325.

Clarke, S. W., and D. Pavia, eds. 1984. Aerosols and the Lung. Boston: Butterworths.

Clay, M. M., D. Pavia, S. P. Newman, and S. W. Clarke. 1983. Factors influencing the size distribution of aerosols from jet nebulizers. Thorax 38:755.

Cross, F. T. 1988. Radon inhalation studies in animals. Radiat. Prot. Dosim. 24:463-466.

Damon, E. G., B. V. Mokler, and R. K. Jones. 1983. Influence of elastase-induced emphysema and the inhalation of an irritant aerosol on deposition and retention of an inhaled insoluble aerosol in Fischer-344 rats. Toxicol. Appl. Pharmacol. 67:322-330.

Davies, C. N., ed. 1961. Inhaled Particles and Vapours. Elmsford, N.Y.: Pergamon Press.

Davies, C. N., ed. 1964. Inhaled Particles and Vapours II. Elmsford, N.Y.: Pergamon Press.

Davies, C. N., ed. 1967. Aerosol Science. New York: Academic Press.

Davies, C. N., J. Heyder, and M. C. S. Ramu. 1972. Breathing of half-micron aerosols. I. Experimental. J. Appl. Physiol. 32:591-600.

Dennis, W. L. 1971. The effect of breathing rate on the deposition of particles in the human respiratory system. Pp. 91-102 in Inhaled Particles III, W. H. Walton, ed., Surrey, England: Unwin Brothers.

Dodgson, J., and R. I. McCallum, eds. 1988. Inhaled Particles VI. New York: Pergamon Press.

Dolovich, M. B., J. Sanchis, C. Rossman, and M. T. Newhouse. 1976. Aerosol penetrance: A sensitive index of peripheral airways obstruction. J. Appl. Physiol. 40:468-471.

Dunill, M. S. 1962. Postnatal growth of the lung. Thorax 17:329-333.

Ferron, G. A. 1977. The size of soluble aerosol particles as a function of the humidity of the air: Application to the human respiratory tract. J. Aerosol Sci. 8:251.

Ferron, G. A. 1987. A method for the calculation of aerosol particle growth and deposition in the human respiratory tract. Pp. 105-110 in Deposition and Clearance of Aerosols in the Human Respiratory Tract, W. Hoffman, ed. Vienna: Facultas.

Ferron, G. A., and J. Gebhart. 1988. Estimation of the lung deposition of aerosol particles produced with medical nebulizers. J. Aerosol Sci. (in press).

Ferron, G. A., B. Haider, and W.G. Kreyling. 1988a. Inhalation of salt aerosol particles. I. Estimation of the temperature and relative humidity of the air in the human upper airways. J. Aerosol Sci. 19:343.

Ferron, G. A., W. G. Kreyling, and B. Haider. 1988b. Inhalation of salt aerosol particles. II. Growth and deposition in the human respiratory tract. J. Aerosol Sci. 19:611-631.

Forsyth, R. D., P. Cole, and R. J. Shephard. 1983. Exercise and nasal patency. J. Appl. Physiol. 55:860-865.

Foster, W. M. 1988. Editorial: Is 24 hour lung retention an index of alveolar deposition? J. Aerosol Med. 1:1.

Francis, E. A. 1987. Patterns of building occupancy for the general public. NRPB-M129. National Radiation Protection Board.

Friedlander, S. K. 1964. Particle deposition by diffusion in the lower lung: Application of dimensional analysis. Am. Ind. Hyg. Assoc. J. 25:37.

Fuchs, N. A. 1964. The Mechanics of Aerosols. Elmsford, N.Y.: Pergamon Press.

Gil, J., H. Bachofen, P. Gehr, and E. R. Weibel. 1979. Alveolar volume-surface area relation in air- and saline-filled lungs fixed by vascular perfusion. J. Appl. Physiol. 47:990-1001.

Goldberg, I. S., and R. V. Lourenco. 1973. Deposition of aerosols in pulmonary disease. Arch. Intern. Med. 131:88-91.

Gore, D. J., and G. Patrick. 1982. A quantitative study of the penetration of insoluble particles into the tissues of the conducting airways. Ann. Occup. Hyg. 26:149.

Hahn, F. F., and C. H. Hobbs. 1979. The effect of enzyme-induced pulmonary emphysema in Syrian hamsters on the deposition and long-term retention of inhaled particles. Arch. Environ. Health 34:203-211.

Harbison, M. L., and J. D. Brain. 1983. Effects of exercise on particle deposition in Syrian golden hamsters. Am. Rev. Respir. Dis. 128:904-908.

Hatch, T. F., and P. Gross. 1964. Pulmonary Deposition and Retention of Inhaled Aerosols. New York: Academic Press.

Hayes, J. A., A. Korthy, and G. L. Snider. 1975. The pathology of elastase-induced panacinar emphysema in hamsters. J. Pathol. 117:1-14.

Heyder, J., and G. Scheuch. 1983. Diffusional transport of nonspherical aerosol particles. Aerosol Sci. Technol. 2:41.

Heyder, J., J. Gebhart, W. Stalhofen, and B. Stuck. 1982. Biological variability of particle deposition in the human respiratory tract during controlled and spontaneous mouth-breathing. Ann. Occup. Hyg. 26:137.

Heyder, J., J. Gebhart, G. Rudolf, C. F. Schiller, and W. Stahlhofen. 1986. Deposition of particles in the human respiratory tract in the size range 0.005-15 μm. J. Aerosol Sci. 17:811.

Heyder, J., J. D. Blanchard, H. A. Feldman, and J. D. Brain. 1988. Convective mixing in human respiratory tract: Estimates with aerosol boli. J. Appl. Physiol. 64:1273.

Hiller, F. C., M. K. Mazumder, J. D. Wilson, and R. C. Bone. 1978. Aerodynamic size distribution of metered dose bronchodilator aerosols. Am. Rev. Respir. Dis. 118:311.

Hinds, W. C. 1982. Aerosol Technology—Properties, Behavior, and Measurement of Airborne Particles. New York: Wiley-Interscience.

Hislop, A., C. F. Muir, M. Jacobson, G. Simon, and L. Reid. 1972. Postnatal growth and function of the pre-acinar airways. Thorax 27:265-274.

Hogg, J. C., J. Williams, J. B. Richardson, P. T. Macklem, and W. M. Thurlbeck. 1970. Age as a factor in the distribution of lower-airway conductance in the pathologic anatomy of obstructive lung disease. N. Engl. J. Med. 282:1283-1287.

Hoppin, F. G., J. C. Stothert, I. A. Greaves, Y.-L. Lai, and J. Hildebrandt. 1986. Lung recoil: Elastic and rheological properties. Pp. 195-215 in Handbook of Physiology, Section 3, The Respiratory System, Volume III, The Mechanics of Breathing, Part 1, P. T. Macklem and J. Mead, eds. Bethesda, Md.: American Physiological Society.

Hounam, R. F., and A. Morgan. 1977. Particle deposition. Pp. 125-156 in Respiratory Defense Mechanisms, J. D. Brain, D. F. Proctor, and L. M. Reid, eds. New York: Marcel Dekker.

Hounam, R. F., A. Black, and M. Walsh. 1969. Deposition of aerosol particles in the nasopharyngeal region of the human respiratory tract. Nature 221:1254-1255.

Itoh, H., Y. Ishii, H. Maeda, G. Todo, K. Torizuka, and G. C. Smaldone. 1981. Clinical observations of aerosol deposition in patients with airways obstruction. Chest 80(Suppl.):837-839.

Kilburn, K. H. 1977. Clearance mechanisms in the respiratory tract. In Handbook of Physiology, Section 9, Reactions to Environmental Agents, D. H. K. Lee, H. L. Falk, and S. D. Murphy, eds. Bethesda, Md.: American Physiological Society.

Kim, C. S., L. K. Brown, G. G. Lewars, and M. A. Sackner. 1983a. Aerosol rebreathing method for assessment of airway abnormalities: Theoretical analysis and validation. Am. Ind. Hyg. Assoc. J. 44:349-357.

Kim, C. S., L. K. Brown, G. G. Lewars, and M. A. Sackner. 1983b. Deposition of aerosol particles and flow resistance in mathematical and experimental airway models. J. Appl. Physiol. 55:154-163.

Kliment, V. 1973. Similarity and dimensional analysis, evaluation of aerosol deposition in the lungs of laboratory animals and man. Folia Morphol. 21:59.

Landahl, H. D., T. N. Tracewell, and W. H. Lassen. 1952. Retention of airborne particulates in the human lung. Arch. Ind. Hyg. Occup. Med. 6:508-511.

Leith, D. E., J. P. Butler, S. L. Sneddon, and J. D. Brain. 1986. Cough. Pp. 315-336 in Handbook of Physiology, The Respiratory System, Section 3, Volume III, Part 1, P. T. Macklem and J. Mead, eds. Bethesda, Md.: American Physiological Society.

Lippmann, M. 1970. Deposition and clearance of inhaled particles in the human nose. Ann. Otol. Rhinol. Laryngol. 79:519-528.

Lippmann, M. 1977. Regional deposition of particles in the human respiratory tract. Pp. 213-232 in Handbook of Physiology, Section 9, Reactions to Environmental Agents, D. H. K. Lee, H. L. Falk, and S. D. Murphy, eds. Bethesda, Md.: American Physiological Society.

Lippmann, M., and R. E. Albert. 1969. The effect of particle size on the regional deposition of inhaled aerosols in the human respiratory tract. Am. Ind. Hyg. Assoc. J. 30:257.

Lippmann, M., D. B. Yeates, and R. E. Albert. 1980. Deposition, retention, and clearance of inhaled particles. Br. J. Ind. Med. 37:337.

Lourenco, R. V., M. F. Klimek, and C. J. Borowski. 1971. Deposition and clearance of 2 micron particles in the tracheo-bronchial tree of normal subjects—smokers and nonsmokers. J. Clin. Invest. 50:1411.

Lourenco, R. V., R. Loddenkemper, and R. W. Carton. 1972. Patterns of distribution and clearance of aerosols in patients with bronchiectasis. Am. Rev. Respir. Dis. 106:857-866.

Lundgren, D. L., E. G. Damon, J. H. Diel, and F. F. Hahn. 1981. The deposition, distribution, and retention of inhaled ^{239}PuO$_2$ in the lungs of rats with pulmonary emphysema. Health Phys. 40:231-235.

Macklem, P. T. 1971. Airway obstruction and collateral ventilation. Physiol. Rev. 51:368-436.

Macklem, P. T., W. E. Hogg, and J. Brunton. 1973. Peripheral airway obstruction and particulate deposition in the lung. Arch. Intern. Med. 131:93-97.

Märtens, A., and W. Jacobi. 1974. Die in vivo Bestimmung der Aerosolteilchendeposition im Atemtrakt bei Mund-bzw. Nasenatmung. Pp. 117-121 in Aerosole in Physik, Medizin und Technik, W. Stahlhofen, ed. Bad Soden, West Germany: Gesellschaft für Aerosolforschung.

Martoner, T. B., A. E. Barnett, and F. J. Miller. 1985. Ambient sulfate aerosol deposition in man: Modeling the influence of hygroscopicity. Environ. Health Perspect. 63:11.

McMahon, T. A., J. D. Brain, and S. R. LeMott. 1977. Species differences in aerosol deposition. In Inhaled Particles IV, W. H. Walton, ed. Elmsford, N.Y.: Pergamon Press.

Melandri, C. V., G. Tarroni, V. Prodi, T. DeZaiacomo, M. Formignani, and C. C. Lombardi. 1983. Deposition of charged particles in the human airways. J. Aerosol Sci. 14:657.

Mercer, T. T. 1973. Aerosol Technology in Hazard Evaluation. New York: Academic Press.

Mercer, T. T. 1981. Production of therapeutic aerosols; principles and techniques. Chest 80(Suppl.):813.

Milic-Emili, J. 1974. Small airway closure and its physiological significance. Scand. J. Respir. Dis. 85(Suppl.):181-189.

Morén, F., M. T. Newhouse, and M. B. Dolovich, eds. 1985. Aerosols in Medicine—Principles, Diagnosis, and Therapy. New York: Elsevier.

Morgan, W. K. C., H. W. Clague, and S. Vinitski. 1983. On paradigms, paradoxes, and particles. Lung 161:195.

Morrow, P. E. 1986. Factors determining hygroscopic aerosol deposition in airways. Physiol. Rev. 66:330.

Morrow, P. E., Chairman, Task Group on Lung Dynamics. 1966. Deposition and retention models for internal dosimetry of the human respiratory tract. Health Phys. 12:173-188.

Morrow, P. E., F. R. Gibb, and K. M. Gazioglu. 1968. A study of particulate clearance from the human lungs. Am. Rev. Respir. Dis. 96:1209.

Muir, D. C. F., and C. N. Davies. 1967. The deposition of 0.5 μm diameter aerosols in the lungs of man. Ann. Occup. Hyg. 10:161-174.

Nikiforov, A. T., and R. B. Schlesinger. 1985. Morphometric variability of the human upper bronchial tree. Respir. Physiol. 59:289.

Palm, P. E., J. M. McNerney, and T. Hatch. 1956. Respiratory dust retention in small animals: A comparison with man. A.M.A. Arch. Ind. Health 13:355.

Pattle, R. E. 1961. The retention of gases and particles in the human nose. Pp. 302-309 in Inhaled Particles and Vapours, Vol. I, C. N. Davies, ed., Oxford: Pergamon Press.

Pavia, D. 1984. Lung mucociliary clearance. In Aerosols and the Lung: Clinical and Experimental Aspects, S. W. Clarke and D. Pavia, eds. London: Butterworths.

Pavia, D., M. Thomson, and H. S. Shannon. 1977. Aerosol inhalation and depth of deposition in the human lung. Arch. Environ. Health 32:131-137.

Persons, D. D., G. D. Hess, W. J. Muller, and P. W. Scherer. 1987. Airway deposition of hygroscopic heterodispersed aerosols: Results of computer calculation. J. Appl. Physiol. 63:1195.

Phalen, R. F., ed. 1984. Inhalation Studies: Foundations and Techniques. Boca Raton, Fla.: CRC Press.

Phalen, R. F., M. J. Oldham, C. B. Beucage, T. D. Crocker, and J. D. Mortensen. 1985. Tracheobronchial airways and implications for particle deposition. Anat. Rec. 212:368-380.

Poe, N. D., M. B. Cohen, and R. L. Yanda. 1977. Application of delayed lung imaging following radioaerosol inhalation. Radiology 122:739.

Proctor, D. F. 1973. The upper respiratory tract and the ambient air. Clin. Notes Respir. Dis. 12:2-10.

Raabe, O. G. 1982. Deposition and clearance of inhaled aerosols. In Mechanisms in Respiratory Toxicology, Vol. I, H. Witschi and P. Nettesheim, eds. Boca Raton, Fla.: CRC Press.

Raabe, O. G., H.-C. Yeh, G. J. Newton, R. F. Phalen, and D. J. Velasquez. 1977. Deposition of inhaled monodisperse aerosols in small rodents. In Inhaled Particles IV, W. H. Walton, ed. Elmsford, N.Y.: Pergamon Press.

Ramanna, L., D. P. Tashkin, G. V. Taplan, D. Elam, R. Detels, A. Coulson, and S. N. Rokaw. 1975. Radioaerosol lung imaging in chronic obstructive pulmonary disease. Chest 68:634-640.

Reist, P. C. 1984. Introduction to Aerosol Science. New York: McMillan.

Roy, M., and C. Courtay. In press. Daily activities and breathing parameters for use in respiratory tract dosimetry. Radiat. Protect. Dosim.

Sanchis, J., M. Dolovich, R. Chalmers, and M. Newhouse. 1972. Quantitation of regional aerosol clearance in the normal human lung. J. Appl. Physiol. 33:757.

Schlesinger, R. B. 1980. Particle deposition in model systems of human and experimental animal airways. In Generation of Aerosols and Facilities for Exposure Experiments, K. Willeke, ed. Ann Arbor, Mich.: Ann Arbor Science Publishers.

Schlesinger, R. B. 1985a. Clearance from the respiratory tract. Fund. Appl. Toxicol. 5:435.

Schlesinger, R. B. 1985b. Comparative deposition of inhaled aerosols in experimental animals and humans: A review. J. Toxicol. Environ. Health. 15:197.

Schlesinger, R. B., and L. McFadden. 1981. Comparative morphometry of the upper bronchial tree in six mammalian species. Anat. Rec. 199:99.

Serafini, S. M., A. Wanner, and E. D. Michaelson. 1976. Mucociliary transport in central and intermediate size airways and effect of aminophyllin. Bull. Eur. Physiopathol. Respir. 12:415.

Smaldone, G. C., and M. S. Messina. 1985. Flow limitation, cough, and patterns of aerosol deposition in humans. J. Appl. Physiol. 59:515.

Smaldone, G. C., H. Itoh, D. L. Swift, and H. N. Wagner. 1979. Effect of flow-limiting segments and cough on particle deposition and mucociliary clearance in the lung. Am. Rev. Respir. Dis. 120:747-758.

Smaldone, G. C., R. J. Perry, W. D. Bennett, M. S. Messina, J. Zwang, and J. Ilowite. 1988. Interpretation of "24 hour lung retention" in studies of mucociliary clearance. J. Aerosol. Med. 1:11.

Snider, G. L., C. B. Sherter, K. W. Koo, J. B. Karlinsky, J. A. Hayes, and C. Franzblau. 1977. Respiratory mechanics in hamsters following treatment with endotracheal elastase or collagenase. J. Appl. Physiol. 42:206-215.

Snider, G. L., E. C. Lucey, and P. J. Stone. 1986. Animal models of emphysema. Am. Rev. Respir. Dis. 133:149-169.

Soong, T. T., P. Nicolaides, C. P. Yu, and S. C. Soong. 1979. A statistical description of the human tracheobronchial tree geometry. Respir. Physiol. 37:161.

Stahlhofen, W., J. Gebhart, and J. Heyder. 1980. Experimental determination of the regional deposition of aerosol particles in the human respiratory tract. Am. Ind. Hyg. Assoc. J. 41:385.

Stahlhofen, W., J. Gebhart, G. Rudolf, and G. Scheuch. 1986. Measurement of lung clearance with pulses of radioactively-labelled aerosols. J. Aerosol Sci. 17:333.

Stauffer, D. 1975. Scaling theory for aerosol deposition in the lungs of different mammals. J. Aerosol Sci. 6:223.

Stewart, N. G., R. N. Crooks, and E. Fisher. 1955. The radiological dose to persons in the UK due to debris from nuclear test expolsions. AERE HP/R1701. Harwell: Atomic Energy Research Establishment.

Stuart, B. O. 1984. Deposition and clearance of inhaled particles. Environ. Health Perspect. 55:369.

Sweeney, T. D., J. D. Brain, A. F. Tryka, and J. J. Godleski. 1983a. Retention of inhaled particles in hamsters with pulmonary fibrosis. Am. Rev. Respir. Dis. 128:138-143.

Sweeney, T. D., W. A. Skornik, J. J. Godleski, and J. D. Brain. 1983b. Collection efficiency is decreased in hamsters with pulmonary fibrosis. Fed. Proc. 42:1349.

Sweeney, T. D., J. D. Brain, W. A. Skornik, and J. J. Godleski. 1985. Chronic bronchitis produced by SO_2 alters the deposition pattern of inhaled aerosol in rats. Am. Rev. Respir. Dis. 131:A195.

Sweeney, T. D., W. A. Skornik, J. J. Godleski, and J. D. Brain. 1986. Collection efficiency of inhaled particles is decreased in hamsters with pulmonary emphysema. Am. Rev. Respir. Dis. 133:147.

Sweeney, T. D., J. D. Brain, S. A. Leavitt, and J. J. Godleski. 1987. Emphysema alters the deposition pattern of inhaled particles in hamsters. Am. J. Pathol. 128:19-28.

Swift, D. L., and D. F. Proctor. 1982. Human respiratory deposition of particles during oro-nasal breathing. Atmos. Environ. 16:2279.

Szalai, A. 1972. The Use of Time: Daily Activities of Urban and Suburban Populations in 12 Countries. Paris: Mouton.

Taplin, G. V., D. P. Tashkin, S. K. Chopra, D. E. Anselmi, D. Elam, B. Calvarese, A. Coulson, R. Detels, and S. N. Rokaw. 1977. Early detection of chronic obstructive pulmonary disease using radionuclide lung-imaging procedure. Chest 71:567-575.

Taulbee, D. P., C. P. Yu, and J. Heyder. 1978. Aerosol transport in the human lung from analysis of single breaths. J. Appl. Physiol. 44:803-812.

Thomson, M. L., and D. Pavia. 1973. Particle penetration and clearance in the human lung. Arch. Environ. Health 29:214-219.

Thurlbeck, W. M. 1982. Postnatal human lung growth. Thorax 37:564-571.

Tu, K. W., and E. O. Knutson. 1984. Total deposition of ultrafine hydrophobic and hygroscopic aerosols in the human respiratory system. Aerosol Sci. Technol. 3:453.

United Nations Scientific Committee on the Effects of Atomic Radiation. 1982. Ionizing Radiation: Sources and Biological Effects. 1982 Report to the General Assembly. New York: United Nations.

Valberg, P. A., J. D. Brain, S. R. LeMott, S. L. Sneddon, and A. Vinegar. 1978. Dependence of aerosol deposition site on breathing pattern. Am. Rev. Respir. Dis. 117:262A.

Valberg, P. A., J. D. Brain, S. K. Sneddon, and S. R. LeMott. 1982. Breathing patterns influence aerosol deposition sites in excised dog lungs. J. Appl. Physiol. 53:824-837.

Van As, A. 1977. Pulmonary airway clearance mechanisms: a reappraisal. Am. Rev. Respir. Dis. 115:721.
Walton, W. H., ed. 1971. Inhaled Particles III. Surrey, England: Unwin Brothers.
Walton, W. H., ed. 1977. Inhaled Particles IV. Elmsford, N.Y.: Pergamon Press.
Walton, W. H., ed. 1982. Inhaled Particles V. Oxford: Pergamon Press.
Wanner, A. 1977. Clinical aspects of mucociliary transport. Am. Rev. Respir. Dis. 116:73-125.
Weibel, E. R., and J. Gil. 1977. Structure-function relationship at the alveolar level. Pp. 1-81 in Lung Biology in Health and Disease, Vol. 3, Bioengineering Aspects of the Lung, J. B. West, ed. New York: Marcel Dekker.
Xu, G. B., and C. P. Yu. 1985. Theoretical lung deposition of hygroscopic NaCl aerosols. Aerosol Sci. Technol. 4:455.
Xu, G. B., and C. P. Yu. 1986. Effects of age on deposition of inhaled aerosols in the human lung. Aerosol Sci. Technol. 5:349.
Yeates, D. B., N. Aspin, H. Levison, M. T. Jones, and A. C. Bryan. 1975. Mucociliary transport rates in man. J. Appl. Physiol. 39:487.
Yeates, D. B., T. R. Gerrity, and C. S. Garrard. 1982. Characteristics of tracheobronchial deposition and clearance in man. Ann. Occup. Hyg. 26:259.
Yu, C. P., and C. K. Diu. 1982. A probabilistic model for intersubject deposition variability of inhaled particles. Aerosol Sci. Technol. 1:355-362.
Yu, C. P., and G. B. Xu. 1986. Predictive models for deposition of diesel exhaust particulates in human and rat lungs. Aerosol Sci. Technol. 5:337-347.
Zeltner, T. B., T. D. Sweeney, W. A. Skornik, and J. D. Brain. 1988. Effects of exercise on the retention of inhaled 0.9 μm particles in hamsters. Am. Rev. Respir. Dis. (Abs.) 137:315.

8

Cells of Origin for Lung Cancer

OVERVIEW

The overall goals of this chapter are to estimate the alpha-energy dose to cells in the respiratory tract that give rise to lung cancers and to compare this dose at a particular level of exposure in uranium miners to that in members of the general population. The key question that the committee seeks to address and resolve in this chapter is which cells are specifically at risk for malignant transformation and the ultimate development of a lung cancer. The committee's dosimetry model incorporates the location of target cell nuclei in the epithelium. If this location could be determined precisely or even narrowed to a particular population of lung cells, then more accurate and less uncertain estimates of comparative doses to critical cell populations in different environments could be made. Thus, the committee reviewed evidence relevant to the question of whether all cells of the respiratory tract can be transformed to produce lung cancer or whether the risk is limited to a particular subpopulation such as the basal cells alone or basal cells and secretory cells. Also relevant is the location in terms of airway branching where the lung cancers originate, whether they arise exclusively from the cells in the central respiratory tract, whether they also can arise from cells of the lung periphery, or whether they can develop anywhere within the respiratory tract.

This chapter reviews current concepts and hypotheses of the cellular origins of cancer in general and lung cancer in particular. It considers the current debate about the proliferative capacity of secretory and basal cells and the possible roles of these cells in the origin of human lung cancers. Also considered are data concerning the sites in the respiratory tract where lung tumors originate, particularly the question of whether they originate exclusively in the central respiratory tract. The evidence on histological types of lung cancers found in uranium miners and in the general population is reviewed for an indication of differences suggesting that different targets in the lung are affected in miners and nonminers. Based on these considerations, the committee sought to identify the cells that are most likely to be the progenitors of lung cancers so that dosimetry studies and modeling can be focused on this population.

To accomplish this goal, this chapter has been organized to review several lines of evidence. The structures of lung and airway epithelia are described, and the organization and function of the lung are considered briefly. The pathologies of major types of lung cancer are reviewed to provide a basis for understanding lung tumor types and their locations within the lung. Ultimately, the goal is to understand the cell type of origin for the tumors and the location of these vulnerable cells within the respiratory tract. Consideration of the cellular origin of lung tumors proceeds along three lines. First, the chapter explores the experimental evidence based on the cells of the lung that have the capacity for proliferation. These data suggest that the basal cells and secretory cells, the proliferative populations of the bronchi, are the probable cells of origin for the tumors. Second, this conclusion is supported by observations that bronchial cells can be propagated in cell culture and can be malignantly transformed. These cell cultures, which are presumed to arise from the proliferative cells within the bronchi, can be transformed by gene transfection and can give rise to a spectrum of cancers that resemble the major varieties of lung cancers. Third, observations of lung cancer in uranium miners show that they do not have a single type or a proportionate distribution of major types of lung cancer discrepant from the general population's. These three types of evidence indicate lung cancer development in the miners is not clearly distinct from that in the general population. Thus, although lung cancer occurs with an unusually high incidence in uranium miners, the process appears to be the same as that which occurs in the general population. For this reason it is presumed that the basic mechanisms of development of lung cancer, for example, the cells of origin and the location of these cells within the respiratory tract, should be the same in the two populations, miners and nonminers. These conclusions can be applied to guide dosimetric models for exposure of the cells at risk to radiation from exposure to radon daughters.

STRUCTURE OF THE LUNG

The lung has the principal function of exchanging gases between the body and the outside air. In its simplest description, the lung contains a conducting tube system composed of bronchi, bronchioles, and alveolar ducts for the flow of gases to and from the gaseous exchange surface in the alveoli.

In the alveoli, gases in the blood can interchange with gases in the alveolar spaces, facilitating the entrance of oxygen into the body and the exit of carbon dioxide. A complex system of blood vessels carries less oxygenated blood coming from the right side of the heart through progressively smaller divisions until the blood reaches the small capillaries of the alveolar wall, where gas exchange occurs. After leaving these alveolar capillaries, the oxygenated blood passes through progressively larger vessels until it is returned to the left side of the heart for circulation throughout the body. Many systems within the lung support this primary function. The mucociliary apparatus includes mucous glands to produce secretions that protect the lung from microbial agents and mobilize inspired particles. Macrophages provide another defense mechanism. The cartilage and smooth muscle surrounding the bronchi provide mechanical structures that prevent the bronchi from collapsing. Other connective tissue cells and products produce a cohesive structure, yet one which preserves the high degree of elasticity needed for expansion and contraction of the lung to move gases within airways and air spaces.

The structure of the airways and, in particular, their epithelial morphology are very relevant for respiratory carcinogenesis. The epithelium of the bronchi is of pseudostratified columnar structure with ciliated cells, mucous goblet cells, and simpler secretory mucous cells (Morgenroth et al., 1982). The bronchial epithelium also contains basal cells and argentaffin neuroendocrine (Kulchitsky) cells. The Kulchitsky cells, despite producing neuroendocrine secretory products, may originate from the same endodermal precursors as the other cells of the bronchi and bronchioles. Submucosal mucous glands are found only around the bronchi. The bronchiolar epithelium is composed of ciliated cells, goblet cells, and Clara cells. Goblet cells progressively decrease and Clara cells increase in proportion from the proximal to the distal bronchioles. Epithelial thickness decreases from the main bronchi to the terminal bronchioles, although the decrease may not be uniform with progressive branchings of airways (Gastineau et al., 1972). Figure 8-1 illustrates the morphology of bronchial and bronchiolar epithelia. The alveolar epithelium is composed primarily of flat type 1 pneumocytes that line most of the alveolar surface and lesser numbers of larger, rounder type 2 pneumocytes.

The branching of the airways has been discussed previously in Chapter 5. Bronchi are identified by having smooth muscle, cartilage, and mucous glands. The mainstem bronchi begin at the branching of the trachea and divide to form the lobar bronchi. After two additional branchings these give rise to the segmental bronchi. After an average of about eight branchings of bronchi, the transition between bronchi and bronchioles is reached. There are numerous (about eight) additional branchings of the bronchioles before the transitions between bronchioles, alveolar ducts, and alveoli.

CELLS OF ORIGIN FOR LUNG CANCER 169

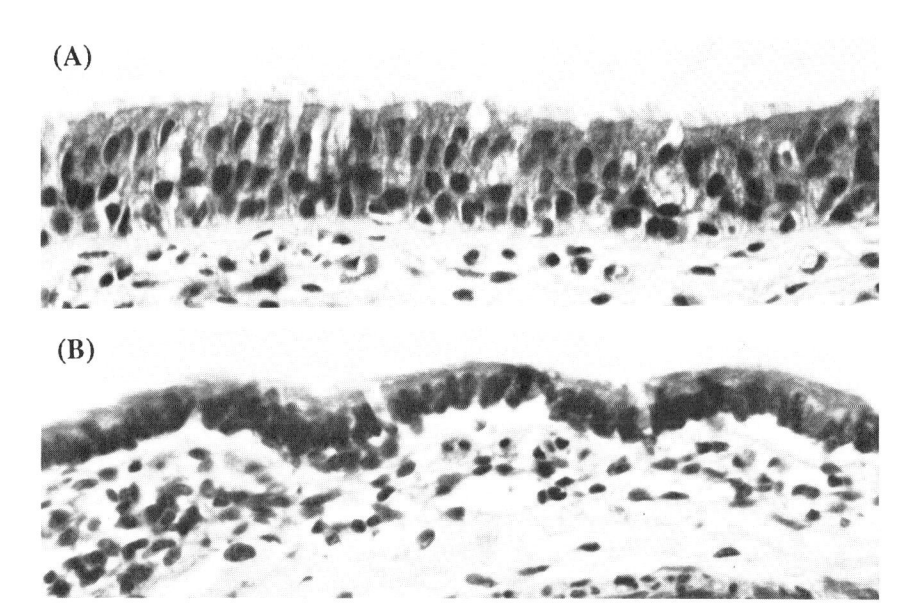

(A)

(B)

FIGURE 8-1 (A) Histology of the bronchial epithelium of a normal human lung. (B) Histology of the bronchiolar epithelium of a normal human lung. The bronchial epithelium (A) is organized as a pseudostratified columnar epithelium with ciliated and secretory mucous goblet cells lining the surface. Basal cells are seen occasionally along the basement membrane. Kulchitsky cells with argentaffin granules cannot be identified in this hematoxylin and eosin-stained slide. The bronchiolar epithelium is thinner and varies in organization from pseudostratified columnar to cuboidal epithelium, depending on how peripheral it is in the branching of bronchioles. The surface is lined by ciliated cells and goblet cells and/or Clara cells, depending on the level of branching of the bronchioles. Magnification = ×625. Hematoxylin and eosin.

PATHOLOGY OF LUNG CANCER

Cancer of the lung is the leading cause of cancer death in the United States (U.S. Department of Health and Human Services [DHHS], 1989). For males it has held this ranking for over two decades. It has recently emerged as the leading cause of cancer deaths among women in the United States. The majority of lung cancers have been attributed to the habit of cigarette smoking. The differing times that lung cancer became the preeminent lethal cancer in males and females in the United States relates to the differences in the demographics of cigarette smoking in cohorts of males and females during the past century. Similar observations have been made in other countries. Currently, the number of cigarette smokers is growing in a number of countries, and rising lung cancer rates have been observed or can be expected in the future in these locations.

Malignant tumors might be expected to develop from any of the cell types present in the lung that preserve the capacity to proliferate. In fact, the vast majority of lung cancers are thought to develop from epithelial cells that line the airways (Spencer, 1977; Dunnill, 1982; Askin and Kaufman, 1985). Although benign and malignant neoplasms may develop from the connective tissues of the lung and tumors of mixed components (mixed tumors) have been reported, these are distinctly uncommon and not typically considered part of the lung cancers referred to here. For the purpose of this review, these are rare biological oddities that will be ignored. The epithelial cells that line the outer surface of the lung and the pleural cavities of the thorax are the source of tumors known as malignant mesotheliomas. Although these tumors are not exceptionally rare, and they do arise from cells that line the lung surface, they also are not typically considered part of the lung cancers referred to here. They are not included in the statistics on bronchogenic lung cancer and are not known to be associated with exposure to radon. These tumors are also excluded in this review.

As noted above, most lung cancers develop from the epithelial cells that line the airways. Lung cancers are also known as bronchogenic carcinomas because of their presumed origin from these airway epithelial lining cells. Lung cancers are a class of tumors that have several typical morphological patterns based on light microscopic examination. The most common morphological types of lung cancers are squamous cell carcinoma, small cell undifferentiated carcinoma, adenocarcinoma, and large cell undifferentiated carcinoma. Bronchioloalveolar cell carcinomas are a less common form of adenocarcinomas which are believed to arise from the cells that line the small peripheral airways (bronchioles) and/or alveolar spaces (Dunnill, 1982). These tumors represent a larger proportion of lung cancers in nonsmokers, in whom adenocarcinomas of all kinds are the most common type of lung tumor (Berg, 1970; DHHS, 1989). Again, as noted above, most types of lung cancers other than bronchogenic cancer are quite rare. For the purpose of this discussion, the four most common types of lung cancers are those listed above (see also Table 8-1). The four leading types of lung cancers have notable differences in their prognoses, response to therapy, and symptoms and signs that lead to their diagnosis. These four major lung cancer types are diagnosed and classified on the basis of histologic features, as determined by conventional light microscopy. The principal characteristics that distingish the four tumor types are illustrated in Figure 8-2.

Squamous cell or "epidermoid" carcinoma (Figure 8-2A) is the most common form of lung cancer found in American men during the past 50 years. The primary lesion in these cancers is typically located in the central or middle zones of the lung, within the first four branchings of the bronchus, and they generally arise in an area of the bronchus with dysplastic squamous epithelium (Auerbach et al., 1957). Histologically, tumors of this type express characteristics of squamous epithelium. The tumor cells typically contain keratin and grow as if forming squamous surfaces, as in the skin. When these tumors are well differentiated, they form aggregates of keratin known as keratin pearls. Squamous cell tumors may be poorly differentiated with a solid sheet of polygonal cells in a mosaic pattern and may show less evidence of keratinization

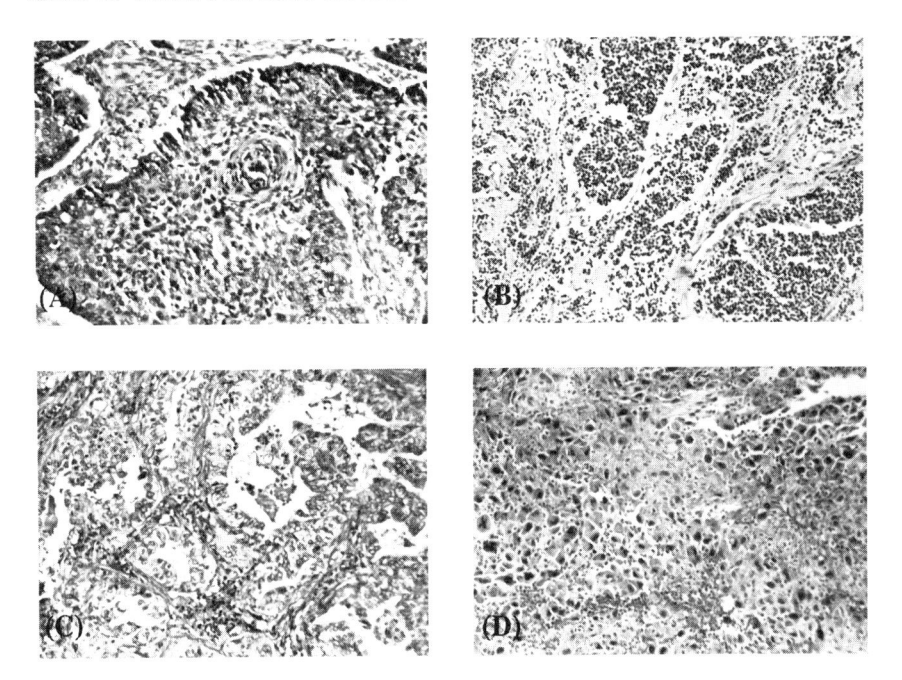

FIGURE 8-2 Photomicrographs of illustrative examples of each of the four major types of bronchogenic carcinoma. (A) Squamous cell carcinoma. (B) Small cell undifferentiated carcinoma. (C) Adenocarcinoma. (D) Large cell undifferentiated. Magnification = ×150. Hematoxylin and eosin stain.

or of intercellular bridges. These poorly differentiated regions may have central areas of necrosis. Cytologically, all of these tumors show various degrees of abnormal and heterogeneous nuclear structure.

Small cell undifferentiated carcinoma (Figure 8-2B), often referred to as oat cell carcinoma, most commonly presents as a central tumor mass. Unlike other lung cancers, the bulk of the tumor is often found in the lymph nodes and connective tissue at the hilum of the lung. Because the cells of these tumors often are comparatively small and quite homogeneous, and because they often fill the hilar lymph nodes, in the early years of this century they were thought to represent lymphomas. More recently, these tumors have been recognized as primary lung cancers. The tumor masses within the lung, which often are less noticeable than the metastatic secondary lesions, have been recognized as primary lesions. These tumors also arise in the epithelium of the bronchi in their first few branchings. Small cell undifferentiated carcinomas may have a variety of morphological patterns, including the classical comma-shaped oat cell variety, as well as small homogeneous round cells, fusiform cells, and larger intermediate cells. The cells of the tumors are usually very homogeneous, show little or no signs of any pattern of differentiation, and often have extensive areas of necrosis within the tumors. Mitotic activity is often very extensive, and rapid growth that outstrips blood supply is thought to explain the frequent finding of necrosis within these tumors. All of the morphological forms of small cell undifferentiated carcinoma have poor prognoses and similar responses to therapy.

There are a variety of types of adenocarcinomas of the lung. Two of the less common types of adenocarcinomas include one rather rare tumor that resembles salivary gland neoplasms and is thought to arise from the cells of bronchial glands. A second type, the bronchioloalveolar carcinoma, is found peripherally in the lung and is thought to arise from the cells that line the terminal airways and/or alveoli. These tumors have a unique histological pattern characterized by growth along the alveolar septa, but without destruction of the underlying lung architecture. These less common tumors are not considered further in this review. The majority of adenocarcinomas (Figure 8-2C) are found centrally in the lung and are thought to arise from the bronchial epithelium. The term adenocarcinoma of the lung is usually applied to this type of neoplasm. Histologically, these adenocarcinomas show a variety of secretory or glandular patterns. They have more orderly glandular patterns in better differentiated tumors and are more solid in the less differentiated tumors. They generally contain mucins that are demonstrable with special stains. Because many cancers, particularly adenocarcinomas, metastasize to the lung, it is important to exclude the possibility that adenocarcinomas in the lung represent metastases from another organ. Adenocarcinomas are also thought to arise in scarred lungs, but the pathogenesis of so-called scar carcinomas is controversial. It is unclear in these cases whether the scar contributes to the formation of the tumor or whether the conditions that produce the tumor also produce a scar. Either or both situations may apply in specific cases (Madri and Carter, 1984).

Large cell undifferentiated carcinoma (Figure 8-2D) is the other major type of human epithelial lung cancer. These tumors are also typically localized in the central region of the lung. The major characteristic that has been used to distinguish this type of tumor is the large size of the cells and the almost complete lack of differentiated histological features when viewed by light microscopy. The tumor cells are typically very pleomorphic and have very prominent nucleoli. By light microscopy these tumors lack the cell bridging or nesting characteristic of squamous cell carcinomas and they lack keratin. Some of these tumors with pleomorphic cells and no evidence of glandular differentiation still showed the presence of mucin upon staining with special stains. Giant cell carcinomas of the lung are also thought to be a related lesion.

There have been a number of classification systems developed to distinguish different types of lung cancers, but the system developed by the World Health Organization is the most widely used (World Health Organization, 1982). Despite the availability of criteria, histopathological diagnosis is a subjective process that is affected by the quantity and quality of the specimen provided

TABLE 8-1 Most Common Types of Lung Cancer

Common Lung Cancer Type	Percentage of Total Lung Cancers
Squamous cell carcinoma	40
Small cell carcinoma	20
Adenocarcinoma	25
Large cell carcinoma	15

SOURCE: Askin and Kaufman (1985).

for study. Differences in classification of lung cancer histopathology between observers and institutions have been well documented (Stanley and Matthews, 1981; Ives et al., 1983). Reliance on clinical reports unsubstantiated by histological diagnosis may introduce substantial misclassification. Because of observer variability, optimal classification of lung cancer cell types should incorporate a standardized review of original histological material by multiple pathologists using defined criteria. However, even such efforts to reduce errors of diagnosis do not ensure comparability between studies. Therefore, it should be noted that evidence on the distribution of tumor types varies among investigators and studies. This reflects, in part, the criteria used to specify tumor type and the techniques used to characterize the tumor. The most common types of pulmonary carcinomas and their approximate proportionate distributions, as observed in large studies of the general population using widely recognized and applied criteria (Carter and Eggleston, 1981; World Health Organization, 1982), are listed in Table 8-1. Bronchioloalveolar cell carcinomas and mesotheliomas, which occur with lower frequencies, and other rare types of lung cancers have been omitted.

That the above estimates of the frequency of occurrence of the most common lung cancers are reasonable can be seen if one considers them in comparison with another recent study (Percy et al., 1983). The overall proportions of squamous cell carcinoma, small cell undifferentiated carcinoma, adenocarcinoma, and large cell undifferentiated carcinoma are, respectively, about 35, 17, 25, and 9% of all lung cancers in the male population of the United States. If one omits the other 14% of tumors in this study, the proportions become 41, 20, 29, and 10%, respectively, which are similar to the distributions of lung cancer types listed above. In yet another study (Stanley and Matthews, 1981), the four major tumor cell types constituted 97% of the tumors, and the proportions were 38, 28, 18, and 13%, respectively. One of the biggest differences between these studies is the relative proportions of adenocarcinomas and large cell carcinomas, which, as noted above, may reflect the thoroughness of evaluation of the cases in the different studies. Table 8-2 compares the percentage distributions of lung cancers between smokers and nonsmokers. For nonsmokers, the most common type of lung cancer is adenocarcinoma, whereas small cell undifferentiated

TABLE 8-2 Distributions of Lung Cancers Between Smokers and Nonsmokers

Tumor Type	Percentage of Lung Cancers	
	Smokers	Nonsmokers
Squamous cell carcinoma	47	14
Small cell carcinoma	17	4
Adenocarcinoma	10	57
Large cell carcinoma	17	8

SOURCE: Data are for U.S. males and are from Berg (1970).

carcinoma is very uncommon (Berg, 1970). Therefore, cigarette smoking is an example of an environmental exposure that appears to shift the distribution of histological types of lung cancers.

The information presented above largely concerns the characteristic of lung cancers as determined by light microscopy. The morphology of lung cancers becomes more complicated as these features are explored in greater depth and with more powerful investigative techniques (see next section). Among the examples of such problems are the unusual types of differentiation seen in lung tumors and the presence of characteristics of more than one pathway of differentiation within the same tumor and even within the same cells (Gatter et al., 1985).

The epithelial lining of the lower respiratory tract is not normally composed of squamous cells. Squamous epithelium arises in this location as a consequence of irritation of the normal epithelium. Cigarette smoke, infections, and environmentally and occupationally related inhalants are likely sources of irritation (Frank, 1982). Acute injuries to the respiratory epithelium can cause the loss of ciliated cells and produce a reparative reaction in which the injured area becomes covered by a thin and undifferentiated epithelial cell lining (see below). If the injury is severe, there can be loss of other cellular constituents with marked thinning of the epithelium to one-half or less of its original thickness. The appearance of the epithelial lining under these conditions can be that of a very irregular, cuboidal regenerative cell population with little differentiation. As irritation becomes repetitive or chronic, the resulting change in the airway lining is normally differentiated metaplastic squamous epithelium. This change can give rise to an epithelium with a thickness as much as twice that of the normal respiratory epithelium. The process leading to abnormal, atypical squamous epithelium is less well understood and occurs over a prolonged period of time (Saccomanno et al., 1965). This proposed sequence has been confirmed by the occasional observation of very early (occult) carcinomas of the lung. In such a patient, malignant cells may be found in the sputum on cytologic examination, and bronchoscopy is likely to show a ragged mucosal surface in one of the subsegmental bronchi with dysplastic squamous epithelium and invasive squamous cell carcinoma contiguously (Auerbach et al., 1979). Squamous carcinoma of the lung, therefore, is presumed not to arise directly from the original respiratory epithelium but rather in areas of injury, repair, and squamous metaplasia and dysplasia that are induced in the bronchial epithelium.

Another issue relevant to dosimetry is the origin of small cell undifferentiated carcinomas and the relationship of these tumors to neuroendocrine cells of the respiratory tract. It is clear that the bronchial epithelium, like many other tissues and organs, contains a small proportion of cells with neuroendocrine secretory granules. These cells, which are most common in fetal life and progressively decrease with advancing age, are readily demonstrated with a Grimelius stain for neurosecretory granules. By electron microscopy, these cells have characteristic membrane-bound cytoplasmic granules that have the appearance of neurosecretory granules. It is believed that these cells can give rise to carcinoid tumors in the lungs and in other tissues. In the lungs, these tumors are usually not malignant. Because these cells have neurosecretory products and some of the same types of peptide hormones are found in small cell undifferentiated carcinomas, it has been suggested that these two types of tumors are related and are even in the same spectrum of tumors (Gould et al., 1978, 1983a,b). In experimental studies in hamsters, it has been reported that exposure to cigarette smoke caused an increase in the number of neuroendocrine cells of the respiratory tract (Tabassian et al., 1989) and that exposure to nitrosamine carcinogens combined with hyperoxia causes the development of neuroendocrine lung tumors (Schuller et al., 1988, 1990). Conversely, others have proposed that the small cell undifferentiated carcinomas arise from transformation of one of the major cell types of the respiratory epithelium (Pfeifer et al., 1989). This issue remains unresolved and requires further study, but one must entertain the possibility that small cell undifferentiated carcinomas may not originate from neuroendocrine cells of the respiratory tract. For example, other lung tumor cell types, particularly squamous cell carcinoma, are seen admixed with small cell carcinoma cells. It is possible that small cell undifferentiated carcinoma represents a more primitive form of these tumors or that these are tumors of a more primitive epithelial cell precursor (e.g., a stem cell). In these tumors of mixed tumor cell types, the clinical course generally reflects the poor prognosis of the small cell undifferentiated cancers.

For large cell undifferentiated carcinomas, the number of these cancers in a clinical series depends on the thoroughness with which cases are investigated and the techniques and criteria chosen to make this distinction (electron microscopy and histochemical staining versus light microscopy alone). When studied extensively, a substantial fraction of these tumors can be reclassified, most as adenocarcinomas and some as squamous cell carcinomas.

This problem of diagnostic specificity is not unique for large cell tumors. When lung cancers of most types are investigated extensively by electron microscopic, histochemical, immunohistochemical, and molecular techniques, the system of classification described above is found to be too simple (Gatter et al., 1985). Many tumors have features of more than one lung cancer cell type. For example, some squamous tumors contain cells that have mucous granules and even neurosecretory granules in addition to keratins (McDowell et al., 1981). Thus, the separation of different types of tumors by light microscopy may be useful for guiding clinical therapy, but does not necessarily fully segregate the cellular products and structures in various lung cancer types. Different types of lung cancers appear to have a far greater degree of overlap than previously considered. This overlap suggests that various types of lung cancers may arise from cells of origin that are in the same family, perhaps a common bronchial precursor or stem cell (Gatter et al., 1985).

It is also important to consider some past observations and some recent trends in lung cancer epidemiology, since these may influence the types of lung cancers that will emerge in the future and thus may be relevant to radon exposure (DHHS, 1989). Although cigarette smoking is recognized as the leading cause of lung cancer, it is clear that several other environmental factors can influence lung cancer incidence by additive or synergistic interactions. Examples are occupational exposures to asbestos or *bis*-chloromethyl ether, which, when combined with cigarette smoking, causes a much higher than expected incidence of lung cancer. With increased use of cigarettes by females, lung cancer rates in females have risen over the past decades. The overall decrease in smoking, particularly among males during the past few years, has begun to be reflected in a decreasing lung cancer rate in young cohorts of males in the United States (Devesa et al., 1989). If this trend continues, there may be a general downturn in the overall incidence of lung cancers in the near future. This effect also could reflect the changed characteristics of cigarettes that are smoked now as compared with the characteristics of those consumed 30 to 50 yrs ago. Consistent with this point is the observation that the proportion of squamous cell carcinomas in a population of smokers is related to the number of cigarettes that they smoke each day (Weiss et al., 1977). Another recent trend is an increasing proportion of adenocarcinomas among lung cancer patients (Wu et al., 1986). It remains to be determined whether this change reflects different diagnostic techniques and criteria implemented in recent years, or whether this is a real change that has resulted from different patterns of cigarette smoking.

Observations of the pathology of human lung cancer lead to many unanswered questions concerning the basic mechanisms of the pathogenesis of lung cancers and the relationships between various types of lung cancers. The questions in turn lead to the pursuit of studies that apply the power of scientific investigation and apply modern experimental techniques toward their solution. It is these points that are addressed in the following section.

BASIC CONCEPTS OF LUNG CANCER DEVELOPMENT

Multistep Process of Lung Cancer Development

A body of clinical and experimental evidence supports the concept that the development of cancer is a multistep process (Armitage, 1985). Data from cancer registries show that the incidence of most cancers rises with age (Doll, 1971). The rising incidence of lung cancer with increasing age is consistent with a multistep process (Cross, 1987). The accumulation of multiple genetic alterations is thought to occur as the result of exposures to exogenous carcinogens such as tobacco smoke, together with a background of genetic changes that are believed to occur spontaneously. Clinical evidence also indicates that most cancers pass through premalignant stages prior to the development of overt clinical disease (Doll, 1971). The most thoroughly studied example of this concept is the evolution of squamous cell carcinoma of the uterine cervix through multiple premalignant and malignant steps. That this is a valid interpretation of the pathogenesis is attested to by the reduced incidence of invasive tumors when clinical intervention removes the early lesions. Similarly, a sequence of premalignant lesions of the bronchial epithelium in cigarette smokers has been shown to precede invasive lung cancer (Auerbach et al., 1961).

Compelling evidence in support of the concept that lung cancer development is a multistep process was derived from prospective cytologic studies in which the development of lung cancer in uranium miners was observed to occur over the course of several years (Saccomanno et al., 1964, 1965, 1974). By studying the cytologies of repetitive sputum samples from uranium miners, cytologic changes were found to proceed through a sequence of progressive steps from squamous metaplasia through various degrees of dysplasia, in situ carcinoma, and invasive carcinoma. Subsequently, comparable sequences of epithelial lesions have been found to precede overt cancers in a number of other tissues (Farber, 1984).

A framework for a more complete explanation of cancer as a multistep process has focused on recent developments in understanding the molecular genetics of human colon cancers (Fearon and Vogelstein, 1990). This analysis combines the results of chromosomal alterations detected in colon cancer cells from spontaneous cases and genetic predisposition syndromes and observations of changes in oncogenes and other genes by molecular genetics techniques. In the case of colon cancers, consistent deletions were found in chromosomes 5q, 17p, and 18q, and genetic loci at these deletion sites were incorporated into a model in which the alteration or loss of these normal genes was related to progressive transitions in the natural history of human colon cancers. One of these chromosomal loci (17p) contains the gene identified as p53, which has been identified as a tumor suppressor gene. When both alleles of this gene are

Normal Cell ⟶ A ⟶ B ⟶ C ⟶ D ⟶ E ⟶ Cancer Cell
<not necessarily in this order>

A,B = loss of two alleles of a tumor supressor gene via deletion or mutation

C = point mutation in Ha-ras or Ki-ras or related gene

D = genetic changes in the regulation of growth factor gene or receptor

E = altered regulation of c-myc or N-myc

FIGURE 8-3 Hypothesis of lung cancer development. Modified from Fearon and Vogelstein (1990).

functioning normally, they impede or prohibit the full neoplastic development of colon cells (and perhaps other cell types as well). When the functioning of the two alleles of this gene are lost by significant mutations of its DNA sequence and/or by partial or complete deletion of the gene, the suppressing effect on tumor development is lost. The identity and functions of the genes at the 5q and 18q loci are less well defined at present, but speculative interpretations have been offered based on available data. In addition to these changes, many colon cancers have been found to have mutated *ras* genes and hypomethylated DNA. These observations have been combined in a hypothetical genetic model of human lung cancer (Figure 8-3). Although this model concerns human colon cancer, it is reasonable to assume that the path to other human cancers, including lung cancer, is quite similar.

Genetic and Chromosomal Alterations in Lung Cancer

Chromosomal alterations are found in a large proportion of lung cancers. In small cell undifferentiated lung cancers, a high proportion of the cells have amplification of one of the *myc* genes, usually either N-*myc* or L-*myc*, and many have a deletion in the p21-p24 region of chromosome 3 (Waters et al., 1988). The human c-*erbA*-beta gene has been localized to the 3p21-3p25 region, and at least one copy of it was deleted in all six of the small cell lung cancer cell lines tested (Dobrovic et al., 1988). This gene is believed to specify a DNA-binding protein that has the features of a hormone receptor. As such, it may function as a tumor suppressor gene for small cell carcinomas and perhaps other lung cancers. In other studies of small cell carcinomas, loss of heterozygosity was found at 3p, 13q, and 17p (Harbour et al., 1988). In view of the fact that the retinoblastoma gene is located at 13q14, it was thought that alterations in the structure of this gene might be detected in small cell carcinomas. When a number of primary tumors and cell lines from small cell carcinomas were tested, a number of structural abnormalities were detected in the retinoblastoma gene in these cells. Thus, the analysis of cytogenetic changes in lung cancers can be used to identify likely sites of recessive tumor suppressor genes that are involved in the malignant transformation of cells.

Weston et al. (1989) examined non-small-cell lung cancers for evidence of consistent DNA sequence deletions. Deletions from chromosomes 3, 11, 13, and 17 were found frequently, but the deletion frequency varied with the locus and tumor type. Deletions were detected by observations of loss of heterozygosity of the tested genetic loci in tumors of patients in comparison with the heterozygous state in uninvolved tissue from the patient. Sequence deletions in chromosome 17p were found in 8 of 9 squamous cell carcinomas but only 2 of 11 adenocarcinomas. Furthermore, in many of the cases of squamous cell carcinoma with 17p deletions, there was an associated chromosome 11 sequence deletion. For other loci, there were similar frequencies of alterations in squamous cell carcinomas and adenocarcinomas. Data for deletions in large cell carcinomas generally resembled those found in adenocarcinomas. These sites of chromosomal alteration are in proximity to the chromosomal location for previously identified tumor suppressor genes (e.g., p53 on 17p). Even though an earlier report had claimed that deletions in chromosome 3 were a consistent change in non-small-cell lung cancer, just as in small cell cancers (Kok et al., 1987), Weston et al. (1989) observed that only about half of the lung tumors showed DNA deletions in chromosome 3. A similar lower rate of allelic loss in the short arm of chromosome 3 was also observed by Rabbits et al. (1989). Therefore, this body of evidence shows that cytogenetic changes are valuable for identifying essential genetic alterations and guiding mechanistic studies of non-small-cell lung cancer.

A long list of alterations in the expression and functioning of oncogenes and growth factors and their receptors has been observed in lung cancers. Overexpression and amplification of genes of the *myc* family are very common in all forms of lung cancer but particularly notable in small cell lung cancers, which typically have overexpression and amplification of the N-*myc* and L-*myc* genes (Waters et al., 1988). In this case the extent of overexpression of these genes is related to the clinical progression of the cancers and is inversely related to prognosis and survival. Nigro et al. (1989) found that mutations in the p53 gene are common in various types of human cancers, including lung cancers. Most of these mutations occur in the portions of the gene that are highly conserved between species and thus likely to code for important functional sites of the p53 protein (gene product).

The c-*raf-1* and c-*myc* oncogenes are commonly overexpressed in small cell carcinomas. To evaluate the role of the abnormal expression of these genes in the induction of lung cancer, these two genes were transfected into Simian virus 40 large T antigen-immortalized cultures of human bronchial epithelial cells (Pfeifer et al., 1989). The transfected cells gave rise to malignantly transformed lung cancer cells with a tumor morphology resembling that of large cell tumors but with evidence of neuroendocrine markers that are suggestive of small cell carcinomas. In fact, Harris and collaborators (Reddel et al., 1988; Pfeifer et al., 1989) have shown that they could malignantly transform human bronchial epithelial cells by transfection with a variety of oncogenes. When the transformed cells were xenotransplanted into nude mice, they gave rise to tumors that reproduced the entire spectrum of human bronchogenic carcinomas. Furthermore, the type of tumor that was generated appeared to relate to the oncogene that had been transfected. From these results it appears that each of the major types of bronchogenic cancers may have a common cell of origin, and also that the morphological type of tumor that results may be determined by the specific kind of genetic change that has occurred during the evolution of the tumor cell from this common precursor cell of origin. However, it is not known whether this common cell of origin is the basal cell, secretory cell, or some other unidentified bronchial cell.

CELLS FROM WHICH LUNG CANCERS DEVELOP

There are two important questions that must be addressed in determining where lung cancers originate. First, which kinds of cells in the human respiratory tract may be the progenitor cells from which lung cancers develop? Second, where in the branching of the respiratory airways do the cancers originate? If these target cells for cancer development, particularly those cancers caused by radon, can be defined, and if their distributions within the lung can be determined, this information could be used in formulating more valid dosimetry models of radon exposure. A refined dosimetry model of this type could specifically focus on the estimated location of the nuclei of these target cells in the airway epithelium (Crawford-Brown, 1987).

The first question is which cells in the human respiratory tract may be the progenitor cells from which lung cancers develop? Which cells are at risk for malignant transformation? Cancer typically occurs in tissues with cells that have a high rate of proliferation or in tissues in which cell proliferation occurs in response to injury. In adult tissues or for cell types in which cell proliferation does not occur, cancer is extremely rare. Thus, chronic irritation or injury was thought to be the etiologic factor for the development of cancer. Subsequently, the role of a variety of specific carcinogenic etiologic agents has come to be recognized, but cell proliferation clearly plays a significant role in the evolution of cancers (Kaufman and Cordeiro-Stone, 1990). The influence of cell proliferation as a contributing factor in the development of cancer presumably results from effects on the mitotic process and on DNA synthesis. Replicating DNA is vulnerable for a variety of reasons. First, replicating DNA is affected to a greater extent by chemical carcinogens than is nonreplicating DNA (Cordeiro-Stone et al., 1982). Second, replication of DNA that contains carcinogen adducts may cause incorporation of incorrect nucleotides at sites of altered or excised bases. Third, some carcinogens may modify nucleotide precursors, and altered precursors may be incorporated into DNA. Fourth, DNA replication itself occurs with a low, but non-zero error rate. Situations that increase cell replication are likely to cause mutations strictly as the result of these errors. There appears to be a critical interrelationship between the repair and replication of DNA as factors in the etiology of cancer (Kakunaga, 1975; Kaufman and Cordeiro-Stone, 1990). If DNA replication proceeds through a damaged region prior to repair, there is a substantial risk of making errors during replication, which may cause a mutation to occur as the result of an alteration of the base sequence of the complementary DNA strand. A mutation does not occur if the repair of the damage precedes replication. Consequently, the relationship in time between the repair and replication of DNA may be a major determinant of the potential for the occurrence of mutations and also, presumably, of carcinogenesis.

It is presumed that the cells of the lung that are sensitive to radon-induced cancer are those that are not terminally differentiated but still have the capability for division and differentiation (Crawford-Brown, 1987). In the normal central respiratory tract, cell proliferation occurs at a very low rate. The turnover time for epithelial cells of the trachea and bronchi of normal mice and rats has been estimated to average about 3 to 7 weeks or even as long as 14 weeks (Shorter, 1970). In response to injury, cell proliferation rates are greatly increased, with a doubling time of less than 48 h. It was thought that cell renewal resulted from increased proliferation of basal cells of the tracheobronchial tract. Basal cells were thought to be the stem cells of the epithelium, and the progeny of the basal cells were thought to differentiate into the secretory, ciliated, and goblet cells that line the respiratory epithelium.

This belief was challenged by studies that examined the response of tracheal epithelium to mechanical injury (McDowell et al., 1979; Kauffman, 1980; Keenan et al., 1982). It was found that injury caused loss of the epithelial cells, and cells from the margins then migrated into the injured area and reepithelialized the surface. Flattened mucous cells and basal cells migrated over the denuded basement membrane and both types of cells proliferated to generate new cells in situ. McDowell et al. (1979) scraped the epithelium from hamster trachea and observed the regenerative process over the next several days. All cells had sloughed from the scraped area by 2 h, leaving a bare basement membrane. Simple flat cells migrated into the denuded area from adjacent epithelium by 6 to 12 h. By 24 h the defect was covered by a one-to-two-cell-thick layer of squamoid cells. At 48 to 72 h the layer of cells was thicker, with squamoid cells and undifferentiated cells. By 96 h the epithelium was restored and looked normal. This process was shown to involve migration and proliferation of mucous-secretory and basal cells and, initially, diminished differentiation of these cells. It was observed that mucous-secretory cells have

```
                      ┌──► Basal cells        ┌──► Ciliated cells    [end stage]
    Basal cells ──────┤                       │
                      └──► Secretory cells ───┴──► Secretory cells
        or
    Basal cells     ─────────►  Basal cells and Ciliated cells
    Secretory cells ─────────►  Secretory cells ─────────► Ciliated cells
```

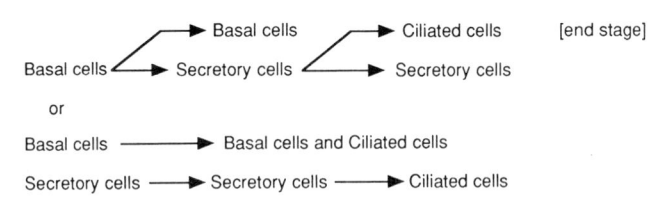

FIGURE 8-4 Pathways of bronchial cell proliferation and differentiation.

a greater proliferative potential than do basal cells during this regenerative process (Keenan et al., 1982). Furthermore, mucous-secretory cells give rise to both squamous and mucous cells. Ciliated cells form from mucous cell precursors via a longer maturation process (Keenan et al., 1982). In these areas following cell proliferation and migration, they often found squamous metaplasia. Therefore, they concluded that squamous metaplasia can arise from mucous cells as well as basal cells. Presumably, both cell types can undergo differentiation to either mucous or squamous cells. This range of differentiation for these cells was thought to explain the occurrence of human lung tumors with combined squamous and glandular differentiation (McDowell et al., 1978).

The idea that basal cells were not the only precursor of the airway epithelium was further strengthened by the results of studies in which isolated rat airway epithelial cells were used to repopulate denuded tracheal grafts. Nettesheim et al. (1990) showed that basal cells from the bronchi and Clara cells from the bronchioles are distinctly different stem cells. On repopulation of denuded tracheas, isolated Clara cells give rise to a cuboidal epithelium with only Clara cells and ciliated cells, such as are found in bronchioles. Isolated tracheal basal cells give rise to higher pseudostratified epithelium with the typical tracheobronchial cell constituents. Secretory cells were able to give rise to the same mixed population as basal cells (Nettesheim et al., 1990). In another study isolated rat basal cells and secretory cells were used separately to repopulate denuded tracheal grafts (Johnson and Hubbs, 1990). In contrast to the results of Nettesheim et al. (1990), basal cells gave rise to an epithelium of only basal and ciliated cells. Secretory cells gave rise to an epithelium of secretory, basal, and ciliated cells. This observation suggests that the secretory cell is the major progenitor for the bronchial epithelium and that basal cells may have a limited differentiation capacity (Johnson and Hubbs, 1990).

Results of these studies demonstrated that the basal and secretory cells throughout the respiratory epithelium have the capacity for cell proliferation. In fact, the vast majority of cell proliferation may occur in the secretory cells. This suggests two pathways for respiratory tract cell proliferation and cell renewal (Figure 8-4).

Thus, not only basal cells but also secretory cells of the tracheobronchial epithelium can proliferate. Therefore, all cells in the respiratory epithelium except ciliated cells appear to have a proliferative capacity. Presumably, the same proliferative capacity is held by both basal and secretory cells in the human respiratory tract. Stimulation of cell proliferation is presumed to have an important role in lung cancer development in humans. It is plausible that injuries from infectious, toxic, or carcinogenic agents may stimulate cell proliferation, and that this proliferation may be critical for malignant transformation. Important genetic damage may occur during tissue repair. Therefore, mucosal areas with sustained increases in cell proliferation, such as focal metaplastic lesions, may be at particular risk. This kind of damage and stimulated cell division in bronchial epithelium is produced by cigarette smoke and may also be produced by exposure to radon daughters.

In humans, various etiological agents cause necrosis and shedding of the tracheobronchial epithelium and lead to acute and chronic forms of bronchitis (Spencer, 1977). As in experimental animals, repair of the denuded basement membrane involves an initial covering of the surface by a simple, squamoid epithelial layer, which is presumably composed of basal cells and undifferentiated secretory cells. This surface lining is often one-half (or less) than the thickness of the normal epithelium. If the injury to the bronchial epithelium is chronic or recurrent, squamous metaplasia often results. Such metaplastic lesions can be twice the thickness of the normal bronchial epithelium. Because cell proliferation may be particularly high, these mucosal lesions may be extremely vulnerable sites for the development of lung cancer.

The cell of origin for small cell undifferentiated carcinoma (oat cell lung cancers) is more obscure and remains a topic of debate. It is uncertain whether these tumors arise from typical airway epithelial cells or whether they are derived from neuroendocrine cells. It has been suggested that small cell and squamous cell cancers have the same cell of origin and that the differences in phenotypes between the two types of tumors result from different genetic changes (see the preceding section). Another possibility is that small cell tumors develop from the common precursor cell when the precursor cells are exposed to carcinogenic agents at exceptionally high rates.

Kreyberg and others thought that small cell carcinomas were undifferentiated squamous cell carcinomas (Dunnill, 1982). Because of cytological similarities between bronchial carcinoids and small cell undifferentiated lung cancers, opinion has evolved to regard them as the benign and malignant ends of a spectrum of neuroendocrine lung tumors (Gould et al., 1978, 1983a,b). Bronchial carcinoid tumors presumably originate from neuroendocrine Kulchitsky cells of the airway epithelium; therefore, it was thought that the small cell tumors also arose from these cells. The presence of neouroendocrine secretory granules in some small cell carcinomas was one part of the evidence suggesting that small cell undifferentiated carcinomas were neuroendocrine tumors. Godwin and Brown (1977) doubted that carcinoid and small cell undifferentiated carcinoma shared a common precursor cell of origin. They noted that carcinoids and small cell carcinomas have very different epidemiologies. Small cell tumors have a marked predominance in males, and they are strongly linked to cigarette smoking and to occupational exposures; the epidemiology of bronchial carcinoids is not consistent. In fact, it has been suggested (Masse and Cross, 1989) that bronchogenic carcinomas of the four major cell types may all arise from the same endodermal progenitor cells. Secretory cells are the most likely progenitor cells for all of the bronchogenic tumor types, although basal cells may also have a role. From this perspective, neurosecretory cells may not be the origin of small cell undifferentiated carcinomas.

Lung cancers often have mixtures of histological patterns (McDowell et al., 1978; Cotran et al., 1989). Most commonly, small cell undifferentiated and squamous cell patterns or squamous cell and adenocarcinoma patterns are found together. Based on these observations, it has been suggested that all histological variants of bronchogenic carcinomas (small cell undifferentiated carcinoma, squamous cell carcinoma, adenocarcinoma, and large cell undifferentiated carcinoma) have a common origin in the endodermal cells that make up the respiratory epithelium. From this perspective, it is possible that all of the histological types of bronchogenic carcinoma represent a phenotypic (and genetic) spectrum of malignant disease derived from a common cell of origin.

The second question is where in the branching of the respiratory airways do the cancers develop? This is another topic that has been the subject of much debate. There are about eight branchings of large airways (bronchi) and about five branchings of bronchioles (James, 1988). Some believe that most bronchogenic carcinomas arise near the hilus of the lung (Ellett and Nelson, 1985; Cotran et al., 1989; Bair, 1990). In fact, it has been suggested that 75% originate from first-, second-, or third-order bronchi, and most of the remainder arise in more distal bronchi (Cotran et al., 1989). If one accepts this position, few lung cancers arise peripherally, and those that do are largely adenocarcinomas. There are several forms of adenocarcinoma, and this diversity adds to the uncertainty about the site of origin for adenocarcinomas. Acinar adenocarcinomas are thought to arise in segmental or subsegmental bronchi (Gibbs and Seal, 1984) and to develop from bronchial surface epithelium or bronchial mucous glands (Dunnill, 1982; Kodama et al., 1982). In contrast, there seems to be a general consensus that the bronchioloalveolar cell form of adenocarcinoma arises from peripheral lung cells, but it is still unclear to what extent alveolar type II cells and bronchiolar Clara cells are the cells of origin.

Spencer's (1977) views concerning the site of origin of lung cancers within the lung are widely cited. He noted that estimates of the proportion of lung tumors that are peripheral in origin vary because of the methods used to determine location (X ray or tissue specimen, or other approaches) and the patient population studied (specimens from surgical patients versus autopsy specimens). His estimate was that 50 to 60% of lung cancers were peripheral, where *peripheral* implied small distal bronchi and bronchioles. By this accounting, 40% of the lung cancers were more proximal than the segmental bronchi. In contrast, in large studies of autopsy cases, Bryson and Spencer (1951) found that only 2.8% were peripheral tumors and Strang and Simpson (1953) found that 8% were peripheral. Walter and Pryce (1955) noted that 50% of resected lung cancers arose from central bronchi and 50% arose from smaller bronchi or bronchioles. Since these cancers were thought to be resectable clinically, they probably represent a selected subpopulation of all lung cancers.

Bronchial basal cell hyperplasia and squamous metaplasia are typically antecedents to dysplastic proliferative lesions and occur in areas near bronchogenic cancers of various histologic types (Dunnill, 1982). Squamous metaplasia, dysplasia, and carcinoma in situ of the bronchus are usually multifocal and commonly occur in the central bronchi prior to the appearance of lung cancers (Spencer, 1977). Metaplasia, dysplasia, and carcinoma in situ are most common at sites of bronchial bifurcations. Auerbach et al. (1957) found widespread carcinoma in situ and early invasive carcinoma in the central respiratory tract of about 90% of patients with invasive carcinomas. Similarly, Valentine (1957) found extensive squamous metaplasia in the bronchi of patients with lung cancer. The occurrence of hyperplastic, metaplastic, and dysplastic lesions as well as localized cancers in major bronchi strongly argues for the importance of the central region of the respiratory tract as the site of origin for most lung cancer development.

To evaluate the risk for development of lung cancer, it is necessary to consider more than the type of cell at risk and its location within the respiratory tract. Factors that influence the risk for lung cancer development in uranium miners exposed to radon at work must be considered, and this must be compared with the effects of radon in the average population at home. In both populations the risk may depend greatly on the rate of cell proliferation in the respiratory tract cells and the effects of exposure to radon at different dose levels and to cigarette smoke. In nonsmokers the rate of turnover of tracheobronchial cells may be very low, but it is increased periodically during tissue repair following respiratory tract infections or occasional exposures to inhaled pollutants. The rates of cell turnover in the respiratory tract of individuals chronically exposed to agents that injure and kill cells are higher. The increase in proliferation rates will depend on the types and amounts of injurious agents to which they are exposed. Tobacco smoke is certainly the leading injurious agent; radon and other inhaled agents can also produce respiratory injuries. Furthermore, various exposures can potentiate each other. In the case of chronic pulmonary injuries, the reparative process may produce alterations in pulmonary structure (e.g., brochiectasis and emphysema) that adversely effect pulmonary function, and this in turn may exacerbate the toxicity of the inciting agent by impeding clearance mechanisms or slowing repair. The extent of exposure to radon, tobacco smoke, and other injurious agents, the chronicity of this exposure, and the presence of chronic changes of pulmonary structure and function can affect cell proliferation rates in respiratory epithelia and perhaps have a major effect on the risk of lung cancer.

In conclusion, bronchogenic cancers arise predominantly in the central respiratory tract from epithelial cells with the capacity for proliferation. Secretory cells have the capacity for proliferation and squamous and mucous cell differentiation. Therefore, these cells, not just basal cells, throughout the central respiratory tract must be included in dosimetry calculations (James, 1988). Dosimetry cannot be focused on basal cells alone, but the dose must be averaged over the whole central bronchial epithelium.

TYPES AND INCIDENCE OF LUNG CANCER IN URANIUM MINERS

Next consider the incidence and types of lung cancer in uranium miners and compare these observations with lung cancer in the general population, in people who are not miners but who are exposed to radon in their homes. As reviewed below, there are no clear differences between the lung cancers in uranium miners and lung cancers that develop in the general population. Therefore, the cell of origin and the location of lung cancers in the general population and in uranium miners are presumed to be the same.

The issue of the percentage distribution of major types of lung cancer in uranium and other underground miners exposed to radon daughters has been reviewed thoroughly in a previous report (NRC, 1988). Samet (1989) reviewed nearly 20 studies of the relationship between exposure to radon and its decay products and the risk of lung cancer. Several studies showed an excess of small cell cancers compared with lung cancers in the general population. Cancers observed in Schneeberg miners were identified as lymphosarcomas, but probably were small cell undifferentiated lung cancers. Tumors found later in autopsies of miners from the adjacent Joachimshal region had a preponderance of small cell undifferentiated carcinomas. An excess of small cell undifferentiated carcinomas of the lung have also been reported among more recent uranium miners from Canada and New Mexico and iron ore miners from Sweden and Great Britain who were exposed to radon daughters (NRC, 1988). Nearly all of these miners were also cigarette smokers.

The most extensive studies of radon daughter exposure and lung cancer were those by Saccomanno and associates on the uranium miners of the Colorado Plateau (Saccomanno et al., 1971, 1974, 1986, 1988, 1989). The studies considered a population of 16,000 uranium miners from 1957 to 1987 and provided extensive data about the types of lung cancer that resulted from radon daughter exposure in comparison to the estimated total radon exposure. There were 383 cases of lung cancers in this population: 356 occurred in smokers, and 25 were nonsmokers (2 cases were unknown). Estimates of radon exposure levels were available for 82.5% of cases. The majority of lung cancers found in this population were small cell undifferentiated carcinomas. Whereas small cell tumors constituted as much as 76% of lung cancers in this population in 1964, it had declined to 22% by the late 1970s. During this time the proportion of squamous cell carcinomas increased to make up the difference. The small cell tumor excess found earlier (Saccomanno et al., 1989) occurred in smoking miners who had been exposed to very high levels of radon (>1,200 working level months [WLM]). Overall, there is very little difference now in the histology of lung cancers in uranium miners compared with that in the general population (Saccomanno et al., 1989). Although three-fourths of the miners were smokers, more than 90% of the lung cancers were in smokers. They interpreted this as evidence of a synergistic interaction between the effects of cigarette smoking and radon exposure.

Although data for nonsmoking uranium miners are sparse, there is no exceptional association between uranium mining and small cell cancers in nonsmoking miners. In fact, the cell type distribution in nonsmoking uranium miners was comparable to the distribution observed in the general population (NRC, 1988). Small cell carcinomas did not predominate in the 21 cases of lung cancer occurring in Navajo uranium miners. Seven of these cancers were small cell carcinomas, eight were squamous cell carcinomas, four were adenocacinomas, and two were large cell carcinomas. While small cell carcinomas was not the predominant cell type in this series, the proportion of subjects with this cell type (33%) was greater than that expected from the distribution of lung cancer histopathology in non smokers. Saccomanno et al. (1989) reviewed the 383 cases of lung cancer to characterize the histological type of the lung cancers in smoking and nonsmoking miners (Table 8-3). The number of tumors in nonsmokers was small, but they interpreted the distribution of tumor types as not dramatically different between smokers and nonsmokers. Slightly less than half of the cases of lung cancer in nonsmoking uranium miners from the Colorado Plateau region were small cell carcinomas, while the remainder were of other cell types. All the cases of lung cancer in the nonsmoking miners were in miners exposed to high aggregate radon doses (>301 WLM) (Saccomanno et al., 1988).

These considerations should guide interpretation of the cancer risks of exposure to radon daughters in uranium miners, who generally also smoke cigarettes, as the risk in the overall population who do not mine uranium but who may be exposed to radon daughters at much lower levels is contemplated. In this regard, it is clear that the bulk of U.S. uranium miners who developed lung cancers in previous years were exposed at vastly higher rates than are typical for the general environment or even homes with the highest ambient radon levels. The rate of radon daughter exposure in miners has decreased progressively over a number of years. Most of the U.S. uranium miners were also smokers.

TABLE 8-3 Percentage Distribution of Major Lung Cancer Types Among Miners

	Uranium Miners		
Tumor Type	U.S. Males	Smokers	Nonsmokers
Squamous cell carcinoma	35	36	40
Small cell carcinoma	17	32	24
Adenocarcinoma	25	13	12
Large cell carcinoma	9	12	20

SOURCE: Cothern (1990).

In summary, the proportions of lung cancers of different cell types that are induced by cigarette smoking or by uranium mining or the combination are essentially the same when viewed over a lifetime (Cross, 1987; Cothern, 1990). An excess of small cell undifferentiated carcinomas was seen in U.S. uranium miners, particularly those that smoked cigarettes, in early years of epidemiological observation. Presumably, small cell carcinoma occurs earlier as a result of smoking and exposure to radon daughters. While small cell carcinomas were the first to appear, all forms of lung cancer were increased by exposure to radon daughters (Cross, 1987). The tumors that appeared after a longer latency were predominantly squamous cell tumors. As time has gone by, the proportions of the various types of lung cancers among uranium miners has come to resemble the distribution of lung cancers observed in the general population that largely reflect the consequences of cigarette smoking. Since there are no extreme differences between the lung cancers that develop in uranium miners compared with the lung cancers that develop in the general population, it is presumed that the cells of origin and the locations of origin of cancers within the lungs of the general population and uranium miners are the same. Adenocarcinomas in nonsmokers may be a possible exception as it appears to be less frequent for nonsmoking miners than for nonsmokers in the general population (c.f. Tables 8-2 and 8-3). Future data from nonsmoking miners who die at older ages may provide further information.

DISCUSSION

Lung cancer has been described here as a disease that develops as the result of a process that involves multiple steps and that usually proceeds over many years. This interpretation is derived from observations made over many years in high-risk populations, particularly uranium miners, most of whom smoked cigarettes. This view of lung cancer as a multistep process has led to efforts to define the multiple individual steps. This has met with much success in the past few years as the result of studies of tumor cell cytogenetics and genetic changes in oncogenes and other genetic loci. As the consistencies increase in the catalog of changes found in lung cancers, the pattern of critical genetic changes in these tumors begins to emerge. Concurrent with these advances, there has been an advance in understanding of the growth and differentiation potential of cells of the respiratory tract. It is recognized that basal cells are not the only proliferative cells in the central respiratory tract and that the cells with mucous differentiation have the capacity to proliferate and, as such, may be cells at risk for malignant transformation.

With regard to the identification of the critical target cell population in the lungs that may be the progenitor for human lung cancers, there are several points that can be made. The vast majority of the lung cancers that are seen in smokers and uranium miners develop in the central respiratory tract. They have been estimated to arise largely in the first three to five branchings of the bronchi; to add a margin of safety to exposure estimates, it might be useful to consider exposures that occur in the bronchi through eight branchings. This appears to exclude the lung periphery, which is thought to be the site of origin of bronchioloalveolar cell tumors; these tumors are relatively uncommon in the U.S. population. Whereas peripheral lung adenocarcinomas and even small cell carcinomas may arise in the peripheral lung of rodents, these appear to be very uncommon in humans, presumably based on patterns of exposure and the cell populations in the human respiratory tract at that level. Since adenocarcinomas are, if anything, underrepresented among the lung cancers in uranium miners, it is likely that effects of radon on bronchial gland cells are minimal. The observation that the secretory cells of the central airways can proliferate changes the way that the central respiratory tract must be viewed with regard to the cells at risk for transformation. It appears that the majority of the cells in the central bronchial epithelium, and not just the basal cells, could undergo transformation. Although the number of lung cancers in nonsmoking uranium miners is small, there are no strong differences between the proportions of lung cancers of different histologic types in smoking miners, nonsmoking miners, and the general population. The uranium miners have had a small relative increase in small cell lung tumors and a relative paucity of adenocarcinomas. Consequently, it is unlikely from the available data that the tumors induced in the lungs by low-level ambient exposures to radon differ greatly in type or location from those found in miners. This indicates that exposure modeling should focus on the basal and secretory cells of the bronchial epithelium in the first five (or eight) branchings of the bronchi as the likely target cells for lung cancer development.

REFERENCES

Armitage, P. 1985. Multistage models of carcinogenesis. Environ. Health Perspect. 63:195-201.
Askin, F. B., and D. G. Kaufman. 1985. Histomorphology of human lung cancer. Pp. 17-21 in Carcinogenesis, A Comprehensive Survey. Vol. 8, Cancer of the Respiratory Tract: Predisposing Factors, M. J. Mass, D. G. Kaufman, J. M. Siegfried, V. E. Steele, and S. Nesnow, eds. New York: Raven Press.
Auerbach, O., J. B. Gere, J. M. Pawlowski, G. E. Muehsem, H. J. Smolin, and A. P. Stout. 1957. Carcinoma-in-situ and early invasive carcinoma occurring in the tracheobronchial trees in cases of bronchial carcinoma. J. Thoracic Surg. 34:298-309.
Auerbach, O., A. P. Stout, E. C. Hammond, and L. Garfinkel. 1961. Changes in bronchial epithelium in relation to cigarette smoking and in relation to lung cancer. N. Engl. J. Med. 265:253-267.
Auerbach, O., E. C. Hammond, and L. Garfinkel. 1979. Changes in bronchial epithelium in relationship to cigarette smoking, 1955-1960 vs. 1970-1977. N. Engl. J. Med. 300:381-385.
Bair, W. J. 1990. Overview of ICRP Respiratory Tract Model. Radiat. Prot. Dosim., in press.
Berg, J. 1970. Epidemiology of the different histologic types of lung cancer. Pp. 93-104 in Morphology of Experimental Respiratory Carcinogenesis, P. Nettesheim, M. G. Hanna, and J. W. Deatherage, eds. Oak Ridge, Tenn.: U.S. Atomic Energy Commission.
Bryson, C. C., and H. Spencer. 1951. Carcinoma of the bronchus: A clinical and pathological survey of 866 cases. Qu. J. Med. 20:173-188.
Carter, D., and J. C. Eggleston. 1981. Tumors of the Lower Respiratory Tract. Atlas of Tumor Pathology. Second Series. Fascicle 17. Washington, D.C.: Armed Forces Institute of Pathology.
Cordeiro-Stone, M., M. D. Topal, and D. G. Kaufman. 1982. DNA in proximity to the site of replication is more alkylated than other nuclear DNA in s Phase 10T1/2 cells treated with N-methyl-N-nitrosourea. Carcinogenesis 3:1119-1127.
Cothern, C. R. 1990. Indoor air radon. Rev. Environ. Contam. Toxicol. 111:1-60.
Cotran, R. S., V. Kumar, and S. L Robbins. 1989. Pp. 797-810 in Pathologic Basis of Disease. Philadelphia: W. B. Saunders Co.
Crawford-Brown, D. J. 1987. Dosimetry. Pp. 173-213 in Environmental Radon, C. R. Cothern and J. E. Smith, eds. New York: Plenum Press.
Cross, F. T. 1987. Health effects. Pp. 215-248 in Environmental Radon, C. R. Cothern and J. E. Smith, eds. New York: Plenum Press.
Devesa, S. S., W. J. Blot, and J. F. Fraumeni, Jr. 1989. Declining lung cancer rates among young men and women in the United States: A cohort analysis. J. Natl. Cancer Inst. 81:1568-1571.
Dobrovic, A., B. Houle, A. Belouchi, and W. E. C. Bradley. 1988. *erb-A*-related sequence coding for DNA-binding hormone receptor localized to chromosome 3p21-3p25 and deleted in small cell lung carcinoma. Cancer Res. 48:682-685.
Doll, R. 1971. The age distribution of cancer: Implications for models of carcinogenesis. J. R. Soc. Med. 134:133-166.
Dunnill, M. S. 1982. Pp. 293-334 in Pulmonary Pathology. New York: Churchill Livingstone.

Ellett, W. H., and N. S. Nelson. 1985. Epidemiology and risk assessment: Testing models for radon-induced lung cancer. Pp. 79-107 in Indoor Air and Human Health, R. B. Gammage and S. V. Kaye, eds. Chelsea, Mich.: Lewis Publishers, Inc.

Farber, E. 1984. Chemical carcinogenesis: A current biological perspective. Carcinogenesis 5:1-5.

Fearon, E. R., and B. Vogelstein. 1990. A genetic model for colorectal tumorigenesis. Cell 61:759-767.

Frank, A. L. 1982. The epidemiology and etiology of lung cancer. Clin. Chest Med. 3:219-228.

Gastineau, R. M., P. J. Walsh, and N. Underwood. 1972. Thickness of the bronchial epithelium with relation to exposure to radon. Health Phys. 23:857-860.

Gatter, K. C., M. S. Dunnill, K. A. F. Pulford, A. Heryet, and D. Y. Mason. 1985. Human lung tumours: A correlation of antigenic profile with histological type. Histopathology 9:805-823.

Gibbs, A. R., and R. M. E. Seal. 1984. The histological varieties of bronchial carcinoma. Pp. 129-145 in Bronchial Carcinoma, M. Bates, ed. Berlin: Springer-Verlag.

Godwin J. D., and C. C. Brown. 1977. Comparative epidemiology of carcinoid and oat-cell tumours of the lung. Cancer 40:1671-1673.

Gould, V. E., A. D. Yannopoulos, S. C. Sommers, and J. A. Terzakis. 1978. Neuroendocrine cells in dysplastic bronchi. Am. J. Pathol. 90:49-56.

Gould, V. E., R. I. Linnoila, V. A. Memoli, and W. H. Warren. 1983a. Neuroendocrine cells and neuroendocrine neoplasms of the lung. Pathol. Annu. 18:287-330.

Gould, V. E., R. I. Linnoila, V. A. Memoli, and W. H. Warren. 1983b. Neuroendocrine components of the bronchopulmonary tract: Hyperplasia, dysplasia and neoplasms. Lab. Invest. 49:519-537.

Harbour, J. W., et al. 1988. Abnormalities in structure and expression of the human retinoblastoma gene in SCLC. Science 241:353-357.

Ives, J. C., P. A. Buffler, and S. D. Greenberg. 1983. Environmental associations and histopathologic patterns of carcinoma of the lung: The challenge and dilemma in epidemiologic studies. Am. Rev. Respir. Dis. 128:195-209.

James, A. C. 1988. Lung dosimetry. Pp. 259-309 in Radon and Its Decay Products in Indoor Air, W. W. Nazaroff and A. V. Nero, eds. New York: John Wiley & Sons.

Johnson, N. F., and A. F. Hubbs. 1990. Epithelial progenitor cells in the rat trachea. Am. J. Resp. Cell Molec. Biol., in press.

Kakunaga, T. 1975. The role of cell division in the malignant transformation of mouse cells treated with 3-methylcholanthrene. Cancer Res. 35:1637-1642.

Kauffman, S. L. 1980. Cell proliferation in the mammalian lung. Int. Rev. Exp. Pathol. 22:131-191.

Kaufman, D. G., and M. Cordeiro-Stone. 1990. The roles of cell proliferation and gene replication in neoplastic transformation. Pp. 143-151 in Growth Regulation and Carcinogenesis, W. R. Paukovits, ed. Boca Raton, Fla.: CRC Press.

Keenan, K. P., J. W. Combs, and E. M. McDowell. 1982. Regeneration of hamster tracheal epithelium after mechanical injury. Cell Pathol. 41:231-252.

Kodama, T., Y. Shimosato, and T. Kameya. 1982. Histology and ultrastructure of bronchogenic and bronchial gland adenocarcinoma (including adenoid cystic and mucoepidermoid carcinomas) in relation to histogenesis. Pp. 147-166 in Morphogenesis of Lung Cancer, Y. Shimasoto, M. R. Melamed, and P. Nettesheim, eds. Boca Raton, Fla.: CRC Press.

Kok, K., et al. 1987. Deletions of a DNA sequence at the chromosomal region 3p21 in all major types of lung cancer. Nature 330:578-581.

Madri, J. C., and D. Carter. 1984. Scar cancers of the lung: Origin and significance. Hum. Pathol. 15:625-631.

Masse, R., and F. T. Cross. 1989. Risk considerations related to lung modeling. Health Phys. 57(Suppl. 1):283-289.

McDowell, E. M., P. J. Becci, L. A. Barrett, and B. F. Trump. 1978. Morphogenesis and classification of lung cancer. Pp. 445-519 in Pathogenesis and Therapy of Lung Cancer, C. C. Harris, ed. New York: Marcel Dekker.

McDowell, E. M., P. J. Becci, W. Schurch, and B. F. Trump. 1979. The respiratory epithelium. VII. Epidermoid metaplasia of hamster tracheal epithelium during regeneration following mechanical injury. J. Natl. Cancer Inst. 62:995-1008.

McDowell, E. M., T. S. Wilson, and B. F. Trump. 1981. Atypical endocrine tumors of the lung. Arch. Pathol. Lab. Med. 105:20-28.

Morgenroth, K., M. T. Newhouse, and D. Nolte. 1982. Pp. 9-96 in Atlas of Pulmonary Pathology. London: Butterworth Scientific.

National Research Council (NRC). 1988. Pp. 445-488 in Health Risks of Radon and Other Internally Deposited Alpha-Emitters. Washington, D.C.: National Academy Press.

Nettesheim, P., et al. 1990. The role of clara cells and basal cells as epithelial stem cells of the conducting airways. Pp. 99-111 in Biology, Toxicology and Carcinogenesis of Respiratory Epithelium, D. G. Thomassen and P. Nettesheim, eds. Washington, D.C.: Hemisphere Publishing Corp.

Nigro, J. M., et al. 1989. Mutations in the p53 gene occur in diverse human tumour types. Nature 342:705-708.

Percy, C., J. W. Horm, and T. E. Goffman. 1983. Trends in histologic types of lung cancer, SEER, 1973-1981. Pp. 153-159 in Lung Cancer: Causes and Prevention, M. Mizell and P. Correa, eds. Deerfield Beach, Fla.: Verlag Chemie International.

Pfeifer, A. M. A., G. E. Mark, L. Malan-Shibley, S. Graziano, P. Amstad, and C. C. Harris. 1989. Cooperation of c-raf-1 and c-myc protooncogenes in the neoplastic transformation of simian virus 40 large tumor antigen-immortalized human bronchial epithelial cells. Proc. Natl. Acad. Sci. USA 86:10075-10079.

Rabbits, P., et al. 1989. Frequency and extent of allelic loss in the short arm of chromosome 3 in nonsmall-cell lung cancer. Genes, Chromsomes, and Cancer 1:95-105.

Reddel, R., et al. 1988. Human bronchial epithelial cells neoplastically transformed by v-Ki-ras: Altered response to inducers of terminal squamous differentiation. Oncogene Res. 3:401-408.

Saccomanno, G., V. E. Archer, R. P. Saunders, L. A. James, and P. A. Bechler. 1964. Lung cancer of uranium miners on the Colorado Plateau. Health Phys. 10:1195-1201.

Saccomanno, G., R. P. Saunders, V. E. Archer, O. Auerbach, M. Kuschner, and P. A. Bechler. 1965. Cancer of the lung: The cytology of sputum prior to the development of carcinoma. Acta Cytol. 9:413-423.

Saccomanno, G., V. Archer, O. Auerbach, M. Kuschner, R. P. Saunders, and M. G. Klein 1971. Histologic types of lung cancer among uranium miners. Cancer 27:515-523.

Saccomanno, G., V. E. Archer, O. Auerbach, R. P. Saunders, and L. M. Brennan. 1974. Development of carcinoma of the lung as reflected in exfoliated cells. Cancer 33:256-270.

Saccomanno, G., C. Yale, W. Dixon, O. Auerbach, and G. C. Huth. 1986. An epidemiological analysis of the relationship between exposure to Rn progeny,

smoking and bronchogenic carcinoma in the U-mining population of the Colorado Plateau—1960-1980. Health Phys. 50:605-618.
Saccomanno, G., G. C. Huth, O. Auerbach, and M. Kuschner. 1988. Relationship of radioactive radon daughters and cigarette smoking in the genesis of lung cancer in uranium miners. Cancer 62:1402-1408.
Saccomanno, G., G. C. Huth, O. Auerbach, and M. Kuschner. 1989. The histology of neoplasia in uranium miners with smoking and radon exposure evaluation. Pp. 53-62 in Proceedings of the 24th Annual Meeting, N. H. Harley, ed. Bethesda, Md.: National Council on Radiation Protection and Measurements.
Samet, J. M. 1989. Radon and lung cancer. J. Natl. Cancer Inst. 81:745-757.
Schuller, H. M., K. L. Becker, and H. P. Witschi. 1988. An animal model for neuroendocrine lung cancer. Carcinogenesis 9:293-296.
Schuller, H. M., H. P. Witschi, et al. 1990. Pathobiology of lung tumors induced in hamsters by 4-(methylnitros-amino)-1-(3-pyridyl)-1-butanone and the modulating effect of hyperoxia. Cancer Res. 50:1960-1965.
Shorter, R. G. 1970. Cell kinetics of respiratory tissues, both normal and stimulated. Pp. 45-61 in Morphology of Experimental Respiratory Carcinogenesis, P. Nettesheim, M. G. Hanna, and J. W. Deatherage, eds. Oak Ridge, Tenn.: U.S. Atomic Energy Commission.
Spencer, H. 1977. Pp. 115-149, 773-860 in Pathology of the Lung. Oxford: Pergamon Press.
Stanley, K. E., and M. J. Matthews. 1981. Analysis of a pathology review of patients with lung tumors. J. Natl. Cancer Inst. 66:989-992.
Strang, C. and J. A. Simpson. 1953. Carcinomatous abscess of lung. Thorax 8:11-28.
Tabassian, A. R., E. S. Nylen, R. I. Linnoila, R. H. Snider, M. M. Cassidy, and K. L. Becker. 1989. Stimulation of hamster pulmonary neuroendocrine cells and associated peptides by repeated exposure to cigarette smoke. Am. Rev. Resp. Dis. 140:436-440.
U.S. Department of Health and Human Services (DHHS). 1989. A Report of the Surgeon General. Reducing the Health Consequences of Smoking, 25 Years of Progress. Rockville, Md.: U.S. Department of Health and Human Services, Public Health Service, Centers for Disease Control, Office of Smoking and Health.
Valentine, E. H. 1957. Squamous metaplasia of the bronchus. Cancer 10:272-279.
Walter, J. B., and D. M. Pryce. 1955. The site of origin of lung cancer and its relation to histologic type. Thorax 10:117-126.
Waters, J. J., J. M. Ibson, P. R. Twentyman, N. M. Bleehan, and P. H. Rabbits. 1988. Cytologic abnormalities in human small cell lung carcinoma: Cell lines characterized for *myc* gene amplification. Cancer Genet. Cytogenet. 30:213-223.
Weiss, W., S. Altan, M. Rosenzweig, and M. A. Weiss. 1977. Lung cancer type in relation to cigarette dosage. Cancer 39:2568-2572.
Weston, A., J. C. Willey, R. Modali, H. Sugimura, E. M. McDowell, J. Resau, B. Light, A. Haugen, D. L. Mann, B. F. Trump, and C. C. Harris. 1989. Differential DNA sequence deletions from chromosomes 3, 11, 13, and 17 in squamous-cell carcinoma, large-cell carcinoma, and adenocarcinoma of the human lung. Proc. Natl. Acad. Sci. USA 86:5099-5103.
World Health Organization. 1982. International Histological Classification of Tumors, No. 1: Histological Typing of Lung Tumors, 2nd ed. Am. J. Clin. Pathol. 77:123-136.
Wu, A. H., B. E. Henderson, D. C. Thomas, and T. M. Mack. 1986. Secular trends in histologic types of lung cancer. J. Natl. Cancer Inst. 77:53-56.

9

The Committee's Dosimetric Model for Radon and Thoron Progeny

INTRODUCTION

The purpose of this chapter is to describe the underlying assumptions, the specific experimental data, and the formulation of the theoretical dosimetry model that the committee has used to carry out its task of comparing the dose per unit exposure received by underground miners with those received by subjects exposed to radon progeny at home. The modeling procedures that are reviewed and adapted here by the committee were developed as part of an overall review of models and experimental data for radon progeny dosimetry that was sponsored by the U.S. Department of Energy (James, 1990).

DOSIMETRIC ASSUMPTIONS AND MODEL

Both secretory and basal cells in the bronchial epithelium and, to a lesser extent, secretory cells in the bronchioles were identified in Chapter 8 as the principal targets for induction of bronchogenic cancer. The location of these cells within the ciliated epithelium is illustrated schematically in Figure 9-1, which shows a diagrammatic section through the wall of a bronchus. When radon progeny decay, the alpha particles that they emit lose all of their energy within very short distances in the fluid lining the bronchial or bronchiolar airway surfaces and in the epithelial tissue that contains the target cells. The range of the 6-MeV alpha particle from polonium-218 (^{218}Po) in fluid or tissue is 48 μm, and that for the 7.7-MeV alpha particle of ^{214}Po is 71 μm. The location of the target cells in relation to the points at which radon progeny decay (the source

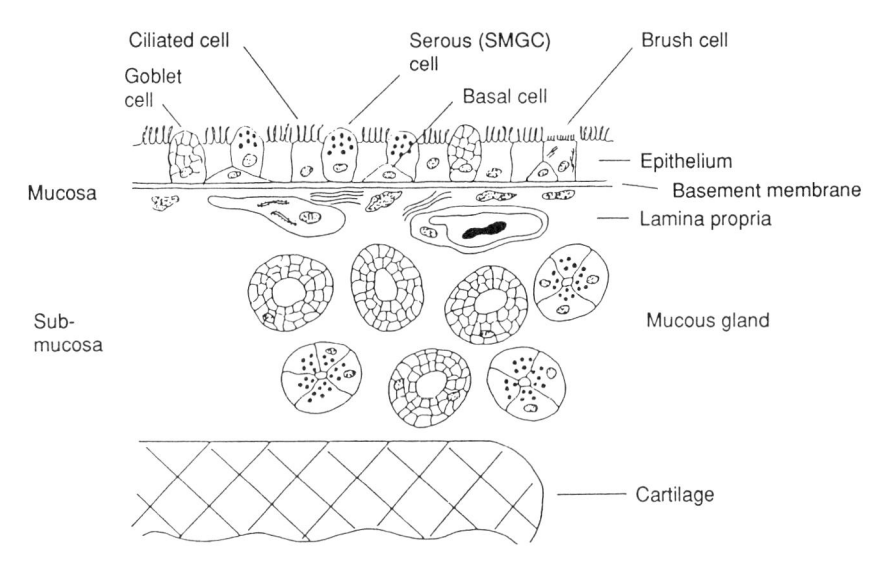

FIGURE 9-1 Diagrammatic section through the wall of a bronchus showing the types of cells found in the epithelium.

of alpha particles) is therefore a critical factor in determining dose. To evaluate the dose received by the respective populations of secretory or basal target cells from each decay of ^{218}Po or ^{214}Po, it is necessary to represent the positions of the source and targets by a geometrical model. The geometry of each bronchial or bronchiolar airway is approximated by a cylindrical tube (Figure 9-2). In the model, the inner surface of the tube is considered to be lined by a thin layer, or sheath, of fluid representing mucus. This inner sheath of mucus is separated from the underlying epithelium by a band of hair-like cilia, which are responsible for clearing the mucus in the direction of the trachea. The cilia are bathed in an aqueous fluid that forms a second, thin layer of shielding material. Both fluid layers have the protective effect of absorbing some of the energy from radon progeny alpha particles. The geometrical model of the airway wall is used to calculate the radiation dose at all points in the underlying epithelium where sensitive targets are found, and particularly the doses received by the nuclear DNA of sensitive cells from alpha-particle decays of the radon progeny source wherever this may be located. Figure 9-2 illustrates two possible locations of the source, within the sheath of mucous "gel" overlying the cilia and within the epithelium itself, if the progeny move into the epithelium.

In a dosimetric model, the distribution of target cell nuclei in the bronchial epithelium can be approximated by the idealized structure shown in Figure 9-3. A recent, detailed study has shown that the thickness of histologically normal human bronchial epithelium is typically about 58 μm in the larger bronchial airways and 50 μm in the most distal bronchi (Mercer et al., in press). Mercer and colleagues found that secretory cell nuclei are distributed approximately uniformly between about 10 and 40 µm below the epithelial surface and that basal cell nuclei are located between about 35 and 50 µm below the surface, as illustrated schematically in Figure 9-3. The thickness of the mucous gel overlying the cilia is difficult to determine in histological preparations. Using the technique of fixation by vascular perfusion, Mercer et al. (1989)estimated that the gel is typically only 2-µm thick in the bronchi. It has been assumed previously that bronchial mucus is substantially thicker, for example, Harley and Pasternack (1982) and National Council on Radiation Protection and Measurements (NCRP, 1984) assume a value of 15 µm for the overall thickness of protective mucus (which, in their case, includes the 6-µm-thick fluid layer bathing the cilia). In view of the uncertainty in the actual value, the committee assumed for the purpose of modeling bronchial doses that the normal thickness of the mucous gel has an intermediate value of 5 µm (Figure 9-3). The thickness of mucus in smokers, and particularly in subjects with chronic bronchitis, may be substantially greater (Chapter 7).

The calculated variation of dose with depth below the surface of the normal bronchial epithelium for the decay of one alpha particle of ^{218}Po or ^{214}Po per cm^2 of airway surface is shown in Figure 9-4. The calculation is based on the mathematical technique described by Harley (1971), except that it uses Armstrong and Chandler's (1973) theoretical stopping power of tissue as a function of alpha-particle energy (Nuclear Energy Agency Group of Experts [NEA], 1983). Both calculations take into account the additional dose contributed by any alpha particles that cross the airway lumen from the opposite wall. An airway caliber of 5 mm in diameter is assumed to typify a bronchus. In adults, the actual airway caliber varies from about 1 cm in

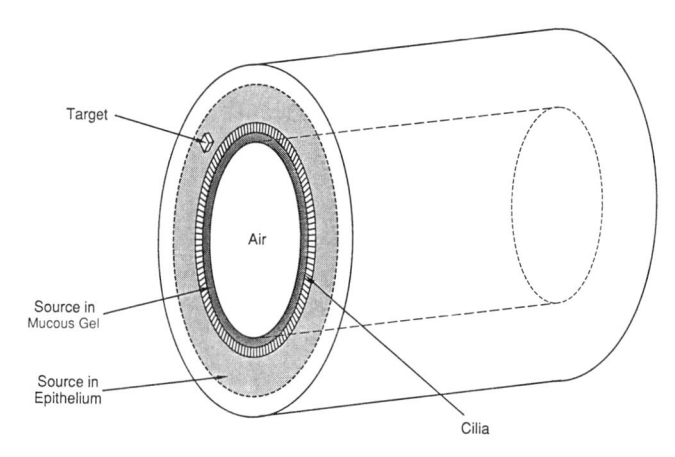

FIGURE 9-2 Cylindrical model of a bronchial or bronchiolar airway used to calculate doses received by target cell nuclei from alpha-particle decays of radon progeny located in mucus or in the epithelium.

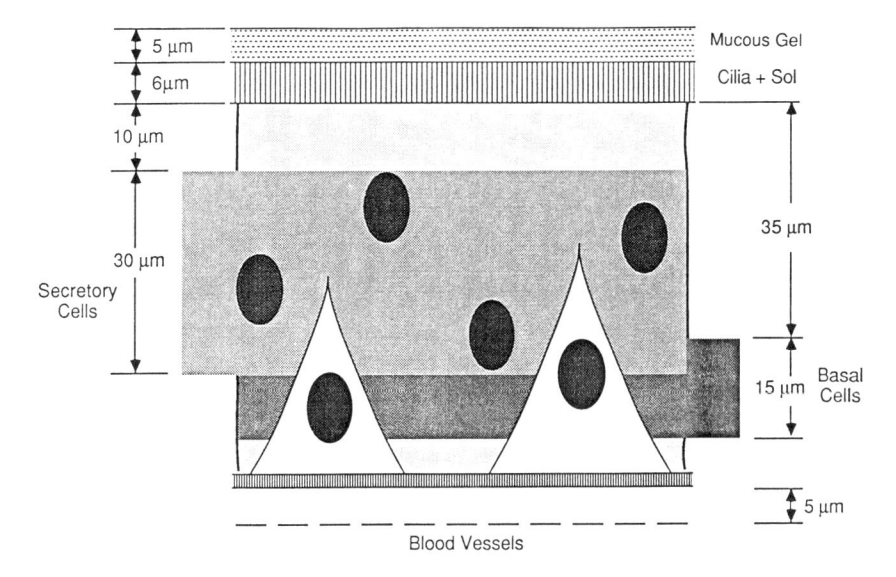

FIGURE 9-3 Model of the location of secretory and basal cell nuclei in bronchial epithelium and of the structure of the mucous layers at the epithelial surface.

the main bronchi (referred to as the first bronchial generation) to about 2 mm in the eighth generation, which the committee took to represent the last and smallest bronchi. Since the range of radon progeny alpha particles in air is on the order of 5 cm, however, it is found that the actual airway caliber has little effect on the dose received by epithelial target cells from a given number of alpha-particle decays per unit surface area.

Figure 9-4 shows two curves, labeled activity in mucus, for each of ^{218}Po and ^{214}Po. In each case, the lower curve represents activity retained in the 5-μm-thick mucous gel overlying the cilia (see Figure 9-3). The corresponding upper curves represent the higher doses calculated for activity mixed in the 6-μm-thick aqueous layer that bathes the cilia. This amount of variation in the location of the mucous source has a relatively minor impact on calculated doses. In contrast, however, the same number of alpha-particle decays from radon progeny located in the epithelium gives rise to a relatively constant dose throughout the tissue. Figure 9-4 also indicates the range of depths at which secretory or basal cell nuclei are assumed to occur. When the depth-dose curves are averaged over these ranges of target depths, it is found that the average dose received by secretory cell nuclei is relatively independent of the location of radon progeny alpha-particle decays, but the dose received by basal cell nuclei is significantly higher if radon progeny decay in the epithelium rather than in mucus. The degree to which radon progeny are taken up by the epithelium is an uncertain factor in the dosimetry model. Its overall impact on the calculated

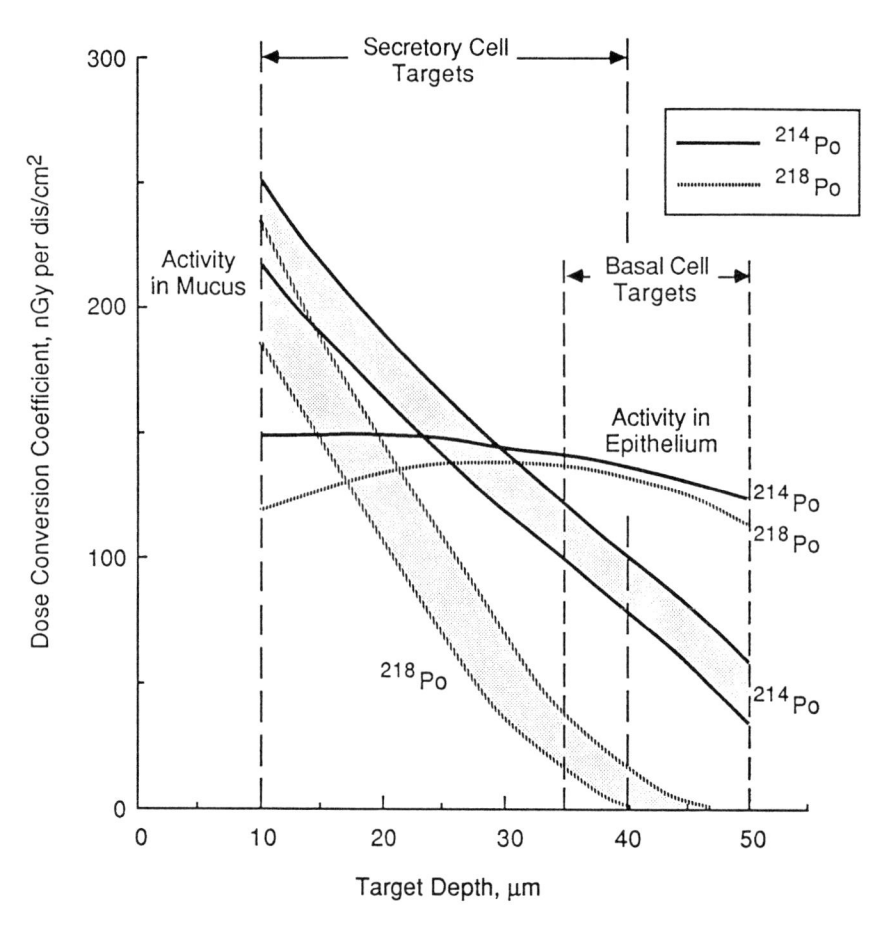

FIGURE 9-4 Variation of dose with depth below the surface of bronchial epithelium for alpha-particle decays of radon progeny in mucus or in the epithelium itself. The ranges of depth at which secretory or basal cell nuclei occur are shown for comparison.

conversion coefficient between exposure and critical dose is examined later in this chapter.

Figure 9-5 illustrates the model of target cell nuclei and mucus assumed by the committee to represent the epithelial lining of the bronchioles. These airways are devoid of basal cells. The sensitive targets are assumed to be the nuclei of secretory cells, which are located between 4 and 12 μm below the epithelial surface (Mercer et al., in press). Both the cilia (the mucous sol layer) and the overlying mucous gel are assumed to be thinner than those in the bronchi (4-μm high and 2-μm thick, respectively). The depth-dose curves calculated for secretory cell nuclei in the bronchiolar epithelium are shown in

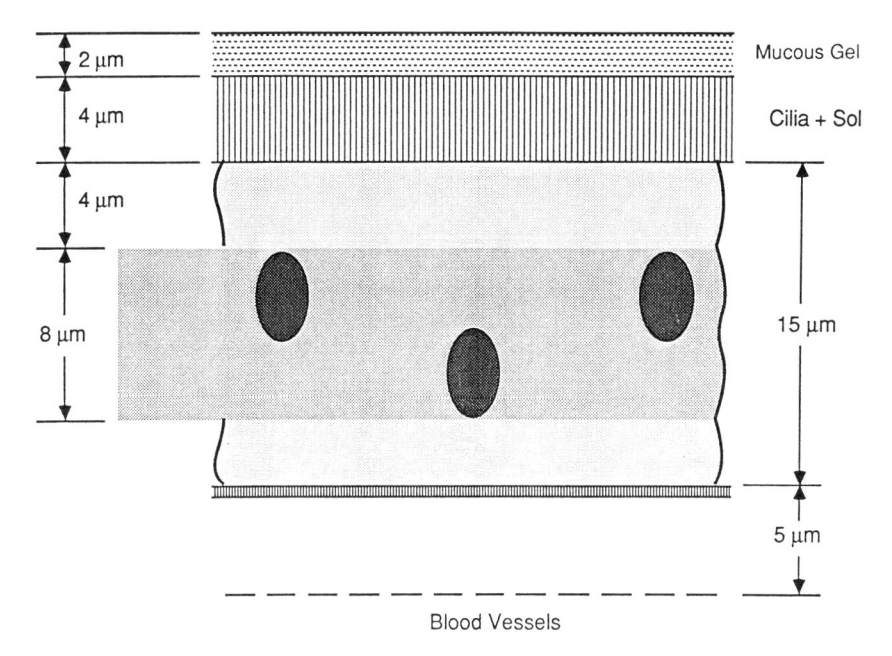

FIGURE 9-5 Model of the location of secretory cell nuclei in bronchiolar epithelium and of the structure of bronchiolar mucus.

Figure 9-6. In this case, it is found that the average dose received by target cell nuclei does not vary significantly with the location of the radon progeny alpha-particle decays, whether or not these occur in mucus (of the assumed normal thickness) or in the epithelium itself.

The values of target cell dose calculated respectively for bronchial and bronchiolar epithelia from the alpha decay of ^{218}Po or ^{214}Po per cm^2 of airway wall are given in Table 9-1. It has been found that the epithelial thickness is the same in infants and children as that in the adult (Gehr, 1987). The tabulated dose conversion coefficients are therefore assumed to apply to all subjects.

The doses received by these various target cell populations for a given subject, under given conditions of exposure, are evaluated by first modeling the number of ^{218}Po and ^{214}Po alpha decays that occur in each airway generation, using the procedures described below. In order to apply the dose conversion coefficients given in Table 9-1, it is necessary to specify the surface areas of the respective airways in each subject. To do this, a model of the variation of airway caliber and length throughout the bronchial tree must be assumed. Several, more or less complete, models of bronchial and bronchiolar airway dimensions in the adult male have been used previously for radon progeny dosimetry (Weibel, 1963; Yeh and Schum, 1980; Phalen et al., 1985). The airway models of these investigators were derived by measuring casts of human lungs that were made

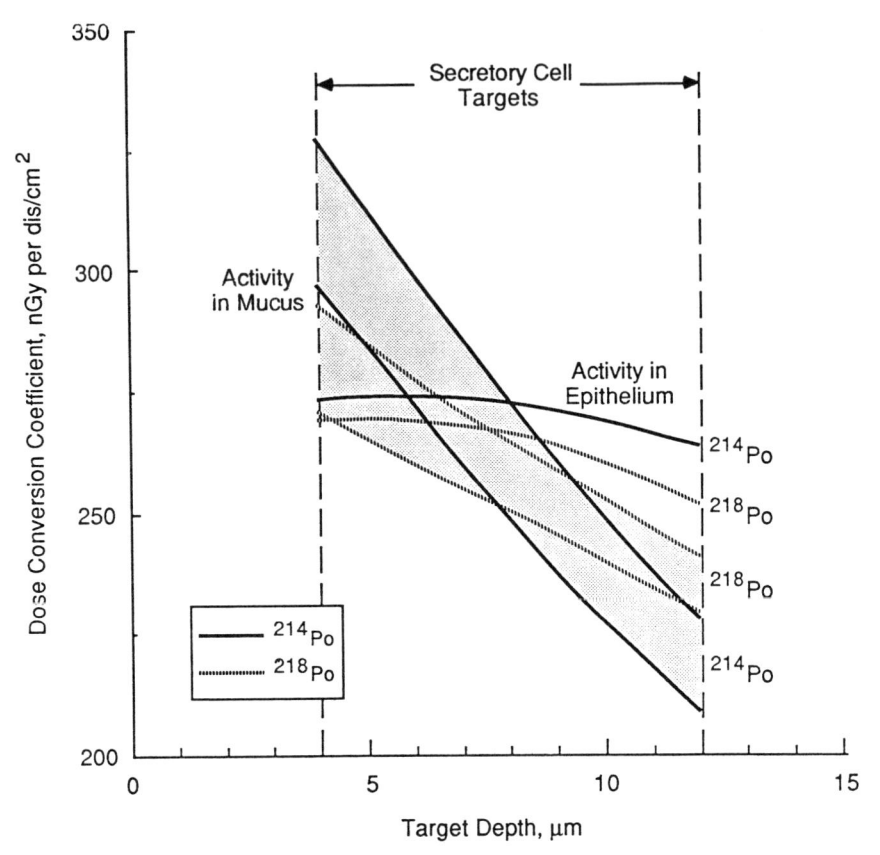

FIGURE 9-6 Variation of dose with depth below the epithelial surface in the bronchioles. This is shown over the range of depths at which secretory cell nuclei are assumed to occur.

at different degrees of inflation in each case. However, even when the reported airway dimensions are scaled to a common lung volume, the different models still exhibit significantly different dimensions (Yu and Diu, 1982). For the purpose of dosimetry, the committee assumes that a "representative" model for the adult male is obtained by averaging the airway dimensions reported in the three studies cited above, after these have been scaled to the normal functional residual capacity (FRC) of 3,000 ml (James, 1988).

The model of airway dimensions is needed for two purposes: first, to calculate the fractions of inhaled radon progeny activity that are deposited in each airway generation throughout the bronchial tree (and also in the alveolated respiratory airways) and then to calculate the surface densities of alpha-particle decays throughout the bronchi and bronchioles, in order to evaluate target cell doses. Previous models of radon progeny dosimetry for adult females (who

TABLE 9-1 Calculated Mean Doses to Bronchial and Bronchiolar Target Cell Nuclei (in nGy) from 1 α-Decay of ^{218}Po or ^{214}Po per cm^2 of Airway Surface

Location of Radon Progeny	Mucous Gel ^{218}Po	^{214}Po	Epithelium ^{218}Po	^{214}Po
Target Cell Nuclei in Bronchi:				
Secretory cells	78	142	131	264
Basal cells	3	68	126	132
Target Cell Nuclei in Bronchioles:				
Secretory cells	251	250	264	271

typically have lungs smaller than those of adult males) and children have assumed uniform scaling of airway dimensions with lung volume (Harley and Pasternack, 1982; Hofmann, 1982; NEA, 1983; NCRP, 1984) or, alternatively (James, 1988), have adopted the scaling of bronchial and bronchiolar dimensions with body height reported by Phalen et al. (1985) from measurements of airway casts that were taken from children and infants. However, the data of Phalen et al. surprisingly indicate that the small bronchioles in young children and infants are not substantially different in size from those in adults. These data yield unrealistically high values of the physiological dead space in young subjects. To give values of dead space that are comparable with physiological data (Table 9-2), the committee has adopted the scaling procedure introduced by Yu and Xu (1987) and also used by Egan et al. (1989). The committee assumes that all airway dimensions in the fully grown lungs of adults (male and female) are scaled according to the one-third power of the subject's FRC.

The factors reported by Phalen et al. (1985) to scale bronchial airway dimensions for young subjects in terms of their height are based on reasonably complete samplings of all airways. These data can therefore be used to evaluate the size of the trachea and bronchi in growing subjects relative to the standard values adopted for the adult male. However, in the absence of comparable data for the bronchioles, their dimensions must be inferred. It is reasonable to adopt Weibel's (1963) inference that, on average, the diameter and length of bronchioles decrease exponentially in successive generations, until they match the dimensions of the first respiratory bronchioles. The size of the latter is determined by considering how the lung grows from birth to maturity. Once the lung is fully developed, the respiratory airways are scaled replicas of those in the adult. Recent data indicate that this development process is complete by age 2, when the growing lung has a full complement of airways and most of

TABLE 9-2 Comparison of Tracheobronchiolar Dead Space Predicted Using the Committee's Model of Airway Dimensions Vis-à-Vis Physiological Data

	Anatomical Dead Space (cm³)						
	Physiological Data				Model		
Subject Age (yr)	Cook et al. (1955)	Hart et al. (1963)	Wood et al. (1971)	Mean V_{TOT}	V_{TOT}	V_{TB}	V_{ET}
Newborn	8.5	—	—	8.5	—	—	1.5
1/12	11	—	—	11	13	10.7	1.9
1	—	—	—	—	25	15.9	4.6
5	—	51	—	51	47	33	14
10	—	88	81	85	85	59	26
Woman	—	126	127	127	118	78	40
Man	—	151	160	155	147	97	50

the final number of alveoli (Gehr, 1987; Zeltner et al., 1987). The respiratory airways are therefore scaled according to the growth in lung volume (FRC).

In order to estimate typical dimensions of respiratory airways in the developmental stage from birth to age 2 yr, it is reasonable to assume that the number of airways is virtually complete at birth, but that alveolization of these immature airways is not. The data of Zeltner et al. (1987) indicate that the number of alveoli at age 1 mo is about 20% of the typical adult value, and that at age 1 yr it is about 80%. Hansen et al. (1975) provided a complete description of the branching structure and dimensions of the respiratory acinus in an adult male. This model has been adopted by the committee to derive respiratory airway dimensions for the lungs of children and infants by scaling for lung volume and number of alveoli, respectively (Egan et al., private communication). The derived dimensions of respiratory bronchioles are then used to infer the size of the most distal generation of bronchioles in young subjects. Figures 9-7 and 9-8 show, respectively, the values of airway diameter and length that are given by the above scaling procedures for adult females, children aged 10 or 5 yrs, and infants aged 1 yr or 1 mo. The reference values for the adult male are also shown, for comparison. Table 9-3 gives the corresponding estimates of the total surface area of the airways in each generation of the bronchial and bronchiolar model for each subject. These values are used with the coefficients given in Table 9-1, to convert the calculated number of radon progeny alpha-particle decays in each airway generation into radiation doses absorbed by target cell nuclei.

CLEARANCE MODEL

The process of breathing continuously deposits radon progeny at all levels in the bronchial tree. The process of mucociliary clearance, which acts in all

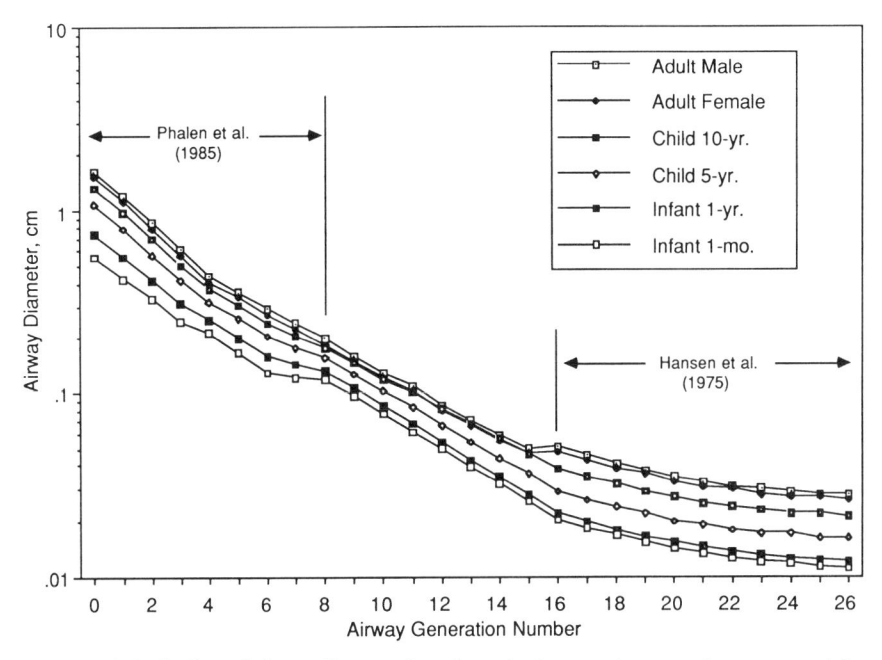

FIGURE 9-7 Scaling of airway diameter throughout the lung respiratory pathway assumed for the calculation of radon progeny deposition and dose in each generation for adults, children, and infants. The trachea is labeled generation 0. Airway generations 1-8 represent the bronchi. These are scaled with body height according to data from Phalen et al. (1985). Generations 9-15 represent the bronchioles. Generations 16-26 represent the respiratory airways whose dimensions are based on the model of Hansen et al. (1975).

ciliated airways, continually redistributes these progeny in a proximal direction (toward the trachea and larynx). This tends to concentrate the alpha particle decays on smaller areas of airway surface (as shown in Table 9-3), since the number of airways is halved each time subsidiary branches converge into a common parent. The combination of these processes with that of radioactive decay determines the number of alpha decays that occur in each airway generation, the surface density of the decays, and thus the dose received by target cells. This complex movement and decay of radon progeny within the bronchial tree is represented mathematically by a so-called clearance model.

Figure 9-9 shows the model used by the committee to calculate the number of radon progeny decays that occur in each airway generation. The model represents the average time taken by a theoretical, discrete packet of mucous gel to traverse a given airway generation by the parameter $1/\lambda_i$, where λ_i is a constant that represents the rate of mucous transport in the ith generation. If k represents each of the short-lived radon progeny in turn (^{218}Po through ^{214}Pb

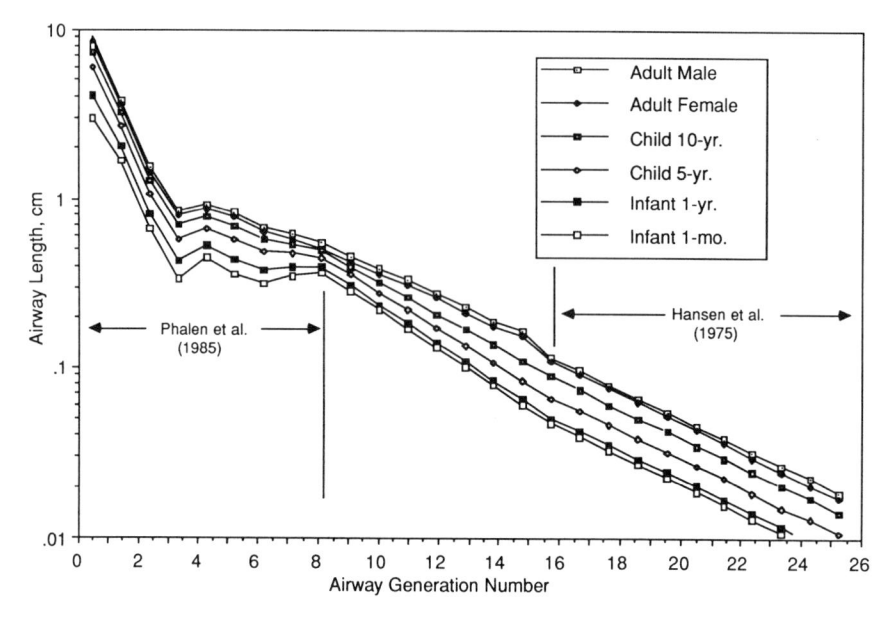

FIGURE 9-8 Scaling of airway length as described for Figure 9-7.

TABLE 9-3 Total Surface Area of the Airways in Each Generation of the Committee's Model of the Bronchial Tree for Adults, Children, and Infants

		Surface Area (cm^2)					
		Adult		Child		Infant	
Airway	Gen. No.	Male	Female	10 yr	5 yr	1 yr	1 mo
Trachea	0	47	41	30	20	9.3	5.1
Bronchi	1	29	25	19	13	7.1	4.4
	2	16	14	11	7	4.1	2.6
	3	13	11	9	6	3.2	2.0
	4	20	17	14	10	6.5	4.6
	5	29	25	20	14	8.5	5.7
	6	39	33	27	19	11.7	8.0
	7	58	50	43	33	22	17
	8	85	74	68	55	41	34
Bronchioles	9	113	98	90	70	51	42
	10	154	134	120	89	63	52
	11	227	196	161	113	77	64
	12	298	258	212	142	95	80
	13	402	348	284	181	116	99
	14	545	476	378	231	145	122
	15	824	721	508	297	179	152

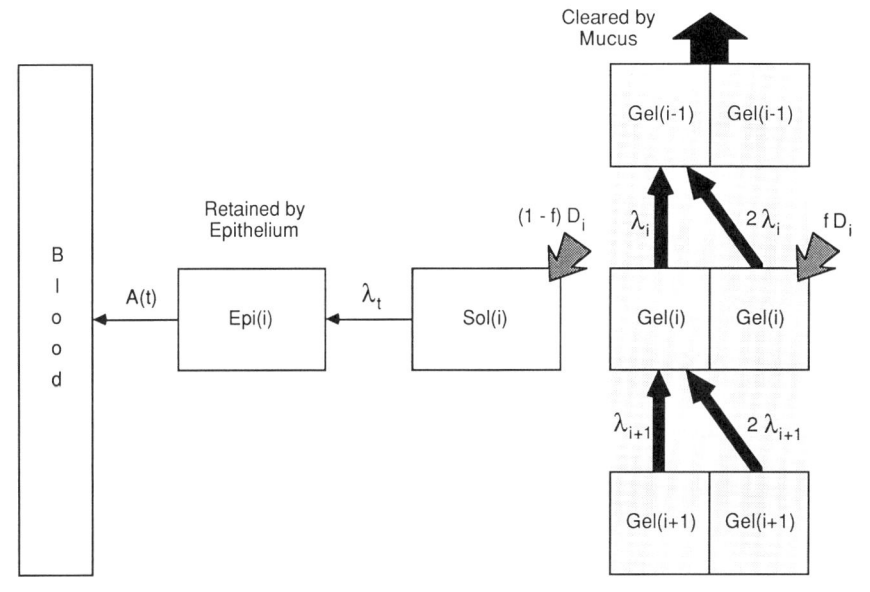

FIGURE 9-9 Model of mucous clearance through the *i*th airway generation. The model also has a pathway along which radon progeny may be transferred through the mucous sol layer to be retained temporarily in epithelial tissue, before radioactive decay or absorption into the blood.

through ^{214}Bi), the number of atoms, n_k, of the kth progeny that decays in generation i is given by the sum of three components

$$n_k = n_{k,1} + n_{k,2} + n_{k,3}$$

where

$$n_{k,1} = A_{k,i}\lambda_k/(\lambda_k + \lambda_i) \tag{9-1}$$

$$n_{k,2} = A'_{k,i}\lambda_k/(\lambda_k + 2\lambda_i) \tag{9-2}$$

$$n_{k,3} = D_{k,i}\lambda_k/(\lambda_k + 2\lambda_i) \tag{9-3}$$

and A_k is the activity of the kth progeny that is cleared into generation i from generation $i + 1$, A'_k is the activity of the kth progeny that is produced in generation i by decay of its parent for $k > 1$, and D_k is the activity of the kth progeny that is deposited from inhaled and exhaled air.

In these equations, the effective rate of mucociliary clearance, $2\lambda_i$, is taken to be twice the value that applies to activity entering generation i from below, since, on average, progeny produced by decay of the parent and by direct deposition have only half the distance to travel (Jacobi and Eisfeld, 1980; NEA, 1983).

TABLE 9-4 Cuddihy and Yeh's (1988) Estimates of the Time Taken by Mucus to Traverse Each Airway Generation Throughout the Bronchial Tree

Airway	Gen. No.	Mucous Transit Time (min)	
Bronchi	1	11	
	2	9	
	3	7	
	4	10	Total
	5	11	~100 min
	6	13	
	7	16	
	8	22	
Bronchioles	9	22	
	10	28	
	11	45	
	12	91	Total
	13	143	~40 h
	14	417	
	15	1,667	

The committee adopted values of λ_i derived recently by Cuddihy and Yeh (1988) from a review of the available human data. These investigators matched the time taken by volunteer subjects to clear various fractions of insoluble radioactive particles with theoretical estimates of the amount of activity deposited at various depths in the bronchial tree. The estimated times taken to traverse each airway generation, $1/\lambda_i$, are given in Table 9-4. In the absence of contrary data, the committee assumed that these values also typify mucous clearance times in children and infants.

The clearance model shown in Figure 9-9 also enables the effect on doses to target cells of any transfer of deposited radon progeny from the surface fluid to the underlying epithelium to be evaluated (NEA, 1983). This is represented by partitioning the activity D_i deposited in generation i into a fraction, fD_i, that is retained and cleared by the mucous gel and a complementary fraction, $(1 - f)D_i$, that is transferred through the sol layer to be retained temporarily in the epithelium. It can be assumed that the rate of transfer through the sol layer is rapid in comparison with the rates at which the progeny decay. However, once the progeny have been taken up in tissue, their biological retention time is likely to be significant. A half-time of approximately 10 h has been observed for the absorption of ^{210}Pb ions from human lung into the blood (Booker et al., 1969; Hursh and Mercer, 1970), and a similar value was measured in dogs (Bianco et al., 1974).

James et al. (1977) observed in rabbits that insoluble particles are distributed more proximally in the bronchial tree than are ^{212}Pb ions introduced simultaneously in a small volume of fluid. Greenhalgh et al. (1982) found that the fraction of ^{212}Pb ions retained in the nasal airways of rats was always greater than that of insoluble particles. They reported that, on average, only 57% of ^{212}Pb ions that were attached to insoluble particles on instillation into the noses of the rats were cleared by mucus over an observation period of about 1 h, whereas 75% of particles were cleared. Lead ions are clearly more mobile on the epithelial surface than insoluble particles are. Greenhalgh et al. (1982) interpreted their observed excess of retained ions as evidence in support of the uptake of radon progeny by bronchial epithelium in rats that was reported earlier by Kirichenko et al. (1970). On the basis of these limited experimental data, it is difficult to determine the fractions of deposited radon progeny that are taken up by the epithelium and the factors that influence uptake. However, the impact of this uncertain aspect of radon progeny behavior on doses to target cells is examined in Chapter 3.

DEPOSITION MODEL

The committee used the theoretical model of aerosol transport and deposition in the lung that was developed by Egan and Nixon (1985) and updated by Egan et al. (1989) to calculate the fractions of inhaled radon progeny that are deposited in each airway generation. This model applies the same mathematical approach as that used by Taulbee and Yu (1975).

The lungs of various subjects (adults, children, and infants) are represented by the regular system of branching airways that was discussed earlier in Chapter 5. In this model, each airway branch within one generation is taken to have identical dimensions, which are a characteristic of the subject. Each point within the lung is then characterized by its axial distance, x, from the origin of the trachea. The properties of the airway generation at that level in the lung are associated with each value of x, for example, the number of branches, the branch diameter and length, and the number of alveoli. Within these model airways, the variation of aerosol concentration, $c(x,t)$, is governed by convection at velocity $u(x,t)$ through the airways, diffusion along the airways [represented by an effective diffusivity $D(x,t)$], and deposition onto the various lung surfaces, which is characterized by the deposition rate per unit length of airway, $L(x,t)$. As described by Egan and Nixon (1985), the combined effect of these processes is represented by the following equation:

$$\frac{\delta}{\delta t}(A_T c) = -\frac{\delta}{\delta x}(A_A u c) + \frac{\delta}{\delta x}(A_A D \frac{\delta c}{\delta x}) - L \qquad (9\text{-}4)$$

In this formalism, $A_A(x,t)$, which is represented in Equation 9-4 by the simpler term A_A, is the cross-sectional area for aerosol transport. This is summed over all airways at distance x from the entrance of the trachea. In the same manner, $A_T(x,t)$ is a cross-sectional area function that allows for the alveolar volume associated with the respiratory airways, where $(A_T - A_A)\,l$ is the additional alveolar volume associated with an airway generation of length, l. All of these variables are functions of time, t, and distance, x.

Equation 9-4 is solved numerically for $c(x,t)$ over several breaths, until ventilation and deposition rates reach equilibrium. The breathing process is simulated by allowing the cross-sectional areas of the alveolated airways to expand and contract about their mean values. Thus, for both A_A and A_T,

$$A(x,t) = A(x)f(t) \tag{9-5}$$

where $A(x)$ is the mean cross-sectional area, and $f(t)$ is a function of time. The function $f(t)$ is chosen to represent the subject's breathing pattern.

The flow fields that are produced in the lung during breathing are complicated, and they induce irreversible mixing between the tidal and reserve air within the lung. Characteristics of this mixing process have been investigated experimentally by Scherer et al. (1975). These investigators represented the dispersion of tidal flow that they measured in a hollow airway cast by an axial diffusion coefficient. The diffusion coefficient was found to be higher on inhalation than on exhalation. This process has the effect of "washing in" radon progeny that are attached to small ambient particles (the so-called accumulation aerosol), to increase their rate of deposition within the lung over successive breaths (Egan and Nixon, 1987). On the basis of the analysis of Scherer and colleagues, the committee's deposition model represents the effective diffusivity of particles within the lung [denoted earlier as $D(x,t)$ in Equation 9-4] by:

$$D(x,t) = D_B + 1.08u(x,t)d \quad \text{(for inspiration)}$$

$$D(x,t) = D_B + 0.37u(x,t)d \quad \text{(for expiration)}$$

where D_B is the Brownian diffusion coefficient (which depends on particle size), $u(x,t)$ is the convective flow velocity, and d is the diameter of an individual airway through which the flow is being considered (Pack et al., 1977; Egan and Nixon, 1985).

The deposition term $L(x,t)$, or L in Equation 9-5, represents the rate at which particles are removed from the airstream per unit length of airway because of the combined effects of impaction, gravitational settling, and Brownian diffusion. The contribution to $L(x,t)$ made by inertial impaction is modeled directly from the experimental data obtained with hollow bronchial casts of human lung (Gurman et al., 1984). Figure 9-10 shows that the efficiencies, η_I, with which particles are deposited by impaction in each bronchus of a particular airway generation, i, can be approximated by an expression of the form

$$\eta_I = \alpha Stk_i^{\beta} \tag{9-6}$$

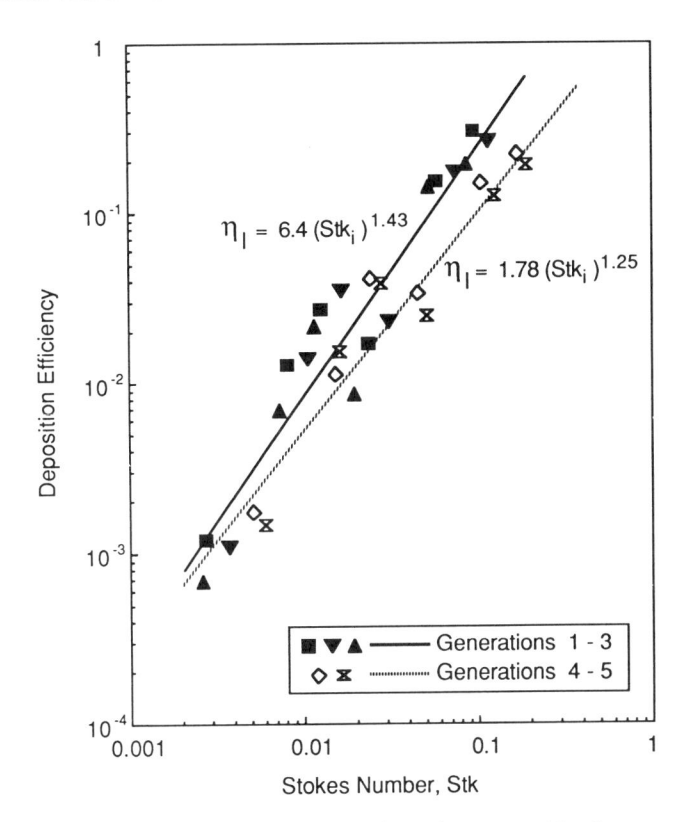

FIGURE 9-10 Efficiency of particle deposition by impaction measured by Gurman et al. (1984) in a hollow cast of a human trachea and five generations of bronchial airways. The symbols show the mean deposition efficiencies measured in each airway generation (1 through 5) as a function of stokes number. The stokes number was varied experimentally by changing both the particle size and the airflow rate through the cast.

where Stk_i is the stokes number of the flow in that airway generation, given by $Stk_i = \rho d_p^2 \, u_i / 18 \, \mu d_i$, where ρ is the particle density, d_p is the particle diameter, μ is the fluid viscosity, and u_i and d_i are the mean flow velocity and airway diameter, respectively, in generation i. The values of the constants α and β are found to be significantly higher in generations 1 through 3 than they are thereafter (these fitted values are shown in Figure 9-10). The experimental data were obtained under a sinusoidally varying flow that approximates the variation of flow rate in vivo over a normal inhalation cycle (Gurman et al., 1980).

The bulk of deposition by gravitational sedimentation and Brownian diffusion occurs in the smaller airways, where the air velocity is low and the flow field is less complicated than that in the bronchi. It is therefore reasonable to apply theoretical treatments to represent these processes. Pich's (1972) theory was found by Heyder and Gebhart (1977) to predict the observed deposition of particles by sedimentation in inclined circular tubes. The committee's deposition model (Egan et al., 1989) uses the theoretical expression derived by Ingham (1975) to evaluate deposition by Brownian diffusion.

The probability that an aerosol particle will be deposited on the wall of a given airway is determined by considering the interaction of the thermodynamic process of Brownian diffusion with the aerodynamic processes of gravitational sedimentation and impaction. The aerodynamic mechanisms can be treated as mutually exclusive. As discussed by Heyder et al. (1985), deposition (DE) in an aerosol filter can then be described by:

$$DE = 1 - (1 - \eta_I)(1 - \eta_G)$$
$$= \eta_I + \eta_G - \eta_I \eta_G \tag{9-7}$$

where η_I and η_G are the mean probabilities that a particle undergoing impaction or gravitational sedimentation is captured in the filter. However, in airways where aerodynamic deposition is predominantly due to sedimentation, this process is competing for the same particles as thermodynamic Brownian diffusion, so that the aerodynamic and thermodynamic processes cannot be considered to be independent of each other. Heyder et al. (1985) studied experimentally the interaction of diffusional and gravitational particle transport in various types of aerosol filters. They found empirically that the interaction can be described by:

$$DE = \eta_D + \eta_G - (\eta_D \cdot \eta_G)/(\eta_D + \eta_G) \tag{9-8}$$

Therefore, Equation 9-9, which has the same form as Equation 9-8, is appropriate to evaluate the effect of simultaneous deposition by thermodynamic (th) and aerodynamic (ae) processes.

$$DE = \eta_{th} + \eta_{ae} - (\eta_{th} \cdot \eta_{ae})/(\eta_{th} + \eta_{ae}) \tag{9-9}$$

An expression of the same form as Equation 9-7 is inappropriate, but this is commonly applied in aerosol deposition models. The proper treatment of thermodynamic and aerodynamic processes as competing processes (represented by Equation 9-9 gives significantly lower estimates of their combined efficiency.

There is some uncertainty in the use of a theoretical expression to evaluate the thermodynamic deposition efficiency in the complex flow fields that occur in the upper bronchial airways. Recent experimental studies of particle deposition by diffusion in hollow casts of bronchial generations 1 through 6 (taken from human lungs) have shown that Ingham's (1975) theory tends to underestimate deposition in these airways (Cohen, 1987; Cohen et al., 1990). The factor by which Ingham's expression is found to under-predict values measured by Cohen et al. (1990) is shown in Figure 9-11. In this figure the correction

$$F = 1 + 80 \exp\{-[\log(80 + 10/d_p^{0.6})]^2\}$$

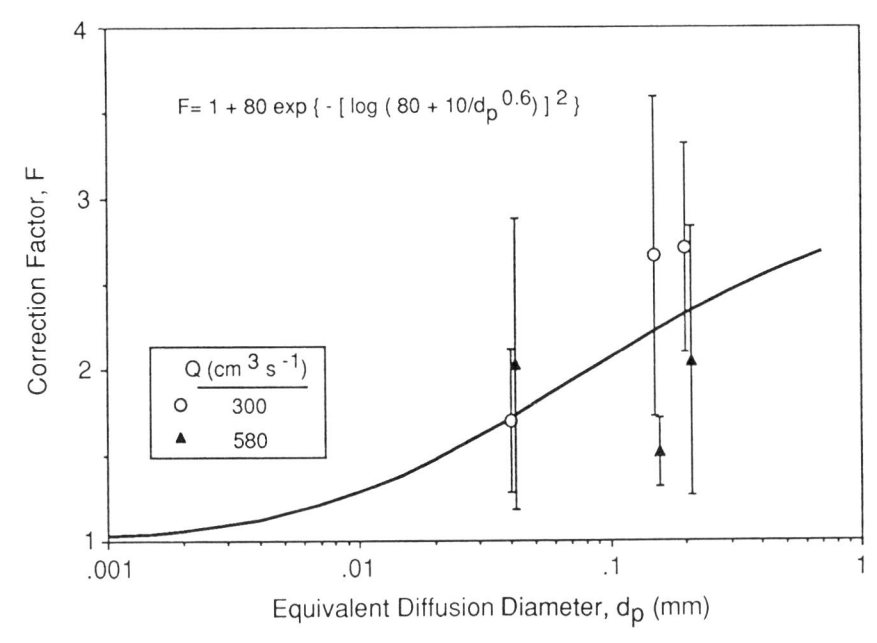

FIGURE 9-11 Factors by which the deposition efficiencies of sub-micron-sized particles measured by Cohen et al. (1990) for cyclic flow in hollow bronchial casts exceed the values calculated using the Ingham (1975) expression for diffusion in constant laminar flow. The symbols represent the values of the correction factor, F, averaged over airway generations 1 through 6, for each combination of mean tracheal flow rate and particle size. The error bars show ±1 standard deviation of individual values of F observed in different airway generations (there was no apparent correlation of F with airway generation number). The curve shows the variation of F with particle diffusion diameter, d_p, that is derived from these data.

factors observed in airways sampled from the six generations included in the bronchial casts are averaged over all generations, since there was found to be no distinct variation with generation number. The error bars shown in the Figure 9-11 reflect a combination of variability in deposition efficiency between the sampled airways and measurement uncertainties. These studies were carried out under conditions of cyclic inspiratory flow, to simulate the in vitro variation of airflow, and at two values of the average inspiratory flow rate. The so-called correction factor, F, is found to be insensitive to the flow rate (denoted in Figure 9-11 by Q, in cm^3/s). As also shown in Figure 9-11, data were obtained for particles of different sizes (of about 0.04, 0.15, and 0.2 μm in diameter. Despite this limited range of particle sizes, Cohen et al. (1990) observed a clear trend in their data which indicates that the enhancement of thermodynamic deposition efficiency decreases for smaller particles. Cohen et al., (1990) reported that the measured deposition in each airway can be approximated by an expression of the form

$$\log(DE) = \gamma + \delta \log(\triangle), \qquad (9\text{-}10)$$

where $\triangle = \pi LD/(4Q)$, L is airway length, D is the particle diffusion coefficient, and Q is the flow rate through the airway. For particles in the size range of unattached radon progeny (of approximately 0.001 μm in diameter), it can be shown that Equation 9-10 converges to predict a slightly lower value of deposition efficiency than that predicted by the Ingham (1975) expression. The variation of the correction factor for thermodynamic deposition efficiency with particle size that is indicated by these experimental data is found empirically to be represented by:

$$F = 1 + 80 \exp\{-[\log(80 + 10/d_p^{0.6})]^2\} \qquad (9\text{-}11)$$

This empirical function is compared in Figure 9-11 with the data of Cohen et al. (1990) from which it is derived.

The committee applied Equation 9-11 to correct values of bronchial deposition predicted by the theoretical model of Egan et al. (1989) for the enhancement effects observed by Cohen and coworkers (1990). It was found that for particles larger than those studied experimentally and at the higher flow rates that occur in exercising subjects, the effect of enhanced thermodynamic deposition is diminished by increased impaction. Figure 9-12 shows how the resulting correction factor for the combined deposition in each bronchial airway calculated by the model of Egan et al. (1989) varies with particle size and flow rate. The committee used the net correction factors shown in Figure 9-12 to evaluate bronchial deposition as a function of radon progeny aerosol size. The effect of this correction for enhanced thermodynamic deposition on calculated doses from inhaled radon progeny is examined in Chapter 4.

Figure 9-13 shows the variation in deposition efficiency with particle size that is predicted for each airway generation in the model of the adult female lung over the size range of concern for deposition of radon progeny. Figure 9-13 relates the number of particles deposited in each airway to the number that enter the trachea on inhalation. Similar patterns of deposition have been calculated for adult males and young subjects. The effect of prefiltration of particles in the nose or mouth, which is important for very small micron-sized particles, is not included in this calculation. It is considered below.

Figure 9-13 shows that the deposition efficiency of particles with diameters of 0.001 μm is calculated to be uniformly high throughout the bronchi (generations 1 to 8). By the time the inspired air reaches the bronchioles (generations 9 to 15), the number of airborne, 0.001-μm particles available for deposition is low. Therefore, deposition is found to decrease rapidly in succeeding generations of the bronchiolar airways. This very small particle diameter of 0.001 μm is within the typical size range of "unattached" radon progeny (Chapter 2). If the particle size is increased to 0.02 μm, the deposition efficiency is found

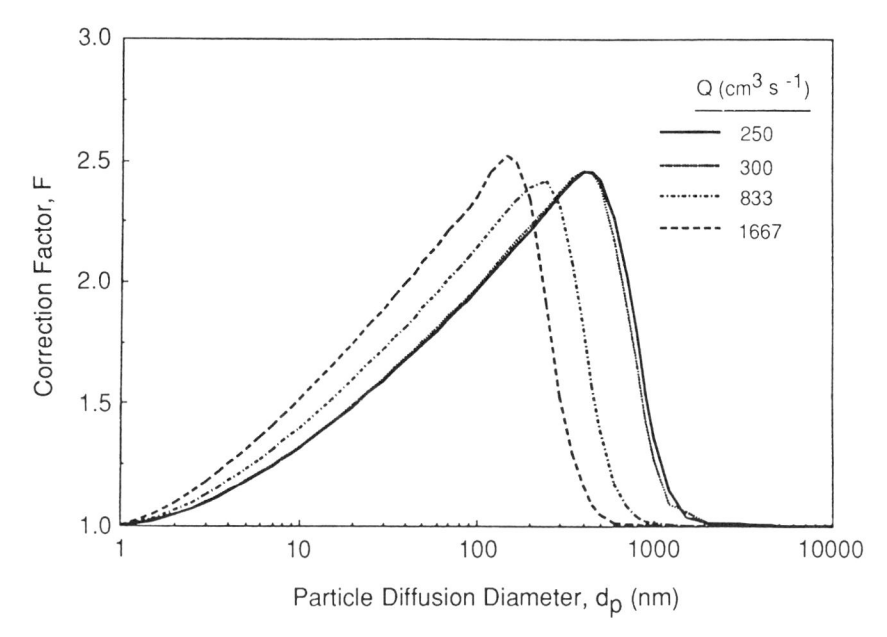

FIGURE 9-12 Factor to correct bronchial deposition calculated by Egan et al. (1989) for the thermodynamic deposition efficiencies measured by Cohen et al. (1990). The correction factor, F, is reduced for larger particles and higher flow rates because of competitive deposition by impaction.

to be about an order of magnitude lower in the bronchi and to reach a broad peak in the respiratory airways (generations 16 to 26). Particles of this size are typical of the radon progeny "growth" mode produced by human activities such as cooking or vacuum cleaning (Chapter 2). It is assumed that the size distributions of both the unattached and growth modes of radon progeny are not affected by the humid environment of the respiratory tract.

The bronchial deposition efficiencies of particles in the size range of attached radon progeny are found to be about two orders of magnitude lower than those of "unattached" progeny. This is shown below to lead to disproportionately large contributions to the dose from exposure to small fractions of radon progeny in the unattached state. The attached or so-called accumulation mode of the radon progeny aerosol has a median size in ambient air that ranges from about 0.15 to 0.25 μm diameter (Chapter 6). However, the carrier aerosol particles are considered to be partly hygroscopic and to grow in the respiratory tract to about double their ambient size (Chapter 6; Sinclair et al., 1974). Figure 9-13 compares the profiles of deposition throughout the lung that are calculated for particles that attain equilibrium sizes of 0.3 and 0.5 μm within the respiratory tract. It is seen that deposition is expected to be relatively independent of particle size over this limited range, at least in resting subjects.

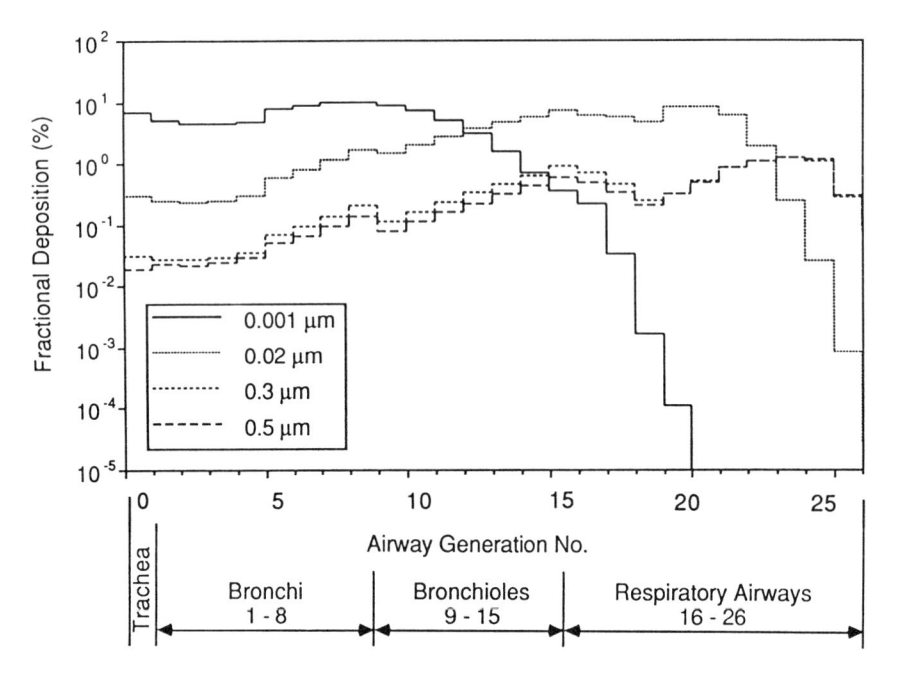

214   COMPARATIVE DOSIMETRY OF RADON IN MINES AND HOMES

FIGURE 9-13 Fraction of particles entering the trachea that is calculated to deposit in each airway generation in the lung of an adult female subject. Histograms are shown for particles of 0.001, 0.02, 0.3 and 0.5 μm diameter within the respiratory tract. The calculation relates to a resting subject (ventilation rate assumed to be 0.39 m³/h).

Figure 9-14 illustrates how the deposition profiles of unattached and attached progeny are expected to be influenced by a subject's breathing rate. An increase in the breathing rate (in this case, from rest to light work) is found to decrease the efficiencies with which inhaled progeny are deposited in the bronchi. In the example considered here (the adult female), as the ventilation rate increases from 0.39 to 1.26 m³/h, the average deposition efficiencies of airway generations 1 through 8 are expected to decrease by 26% for 0.001-μm particles and by 50% for 0.3-μm particles. However, these reduced deposition efficiencies are more than offset by the increased amount of activity inhaled at the higher ventilation rate. The inhaled activity is increased by the factor 1.26/0.39 = 3.23. The net effects of variations in breathing rate and radon progeny aerosol size distributions on doses received by different subjects are examined below.

Another significant factor that must be accounted for in calculating the deposition profile of radon progeny within the respiratory tract is the typical variability or dispersion in size of the aerosol particles. Unattached radon progeny have a relatively narrow distribution of particle size, whereas the

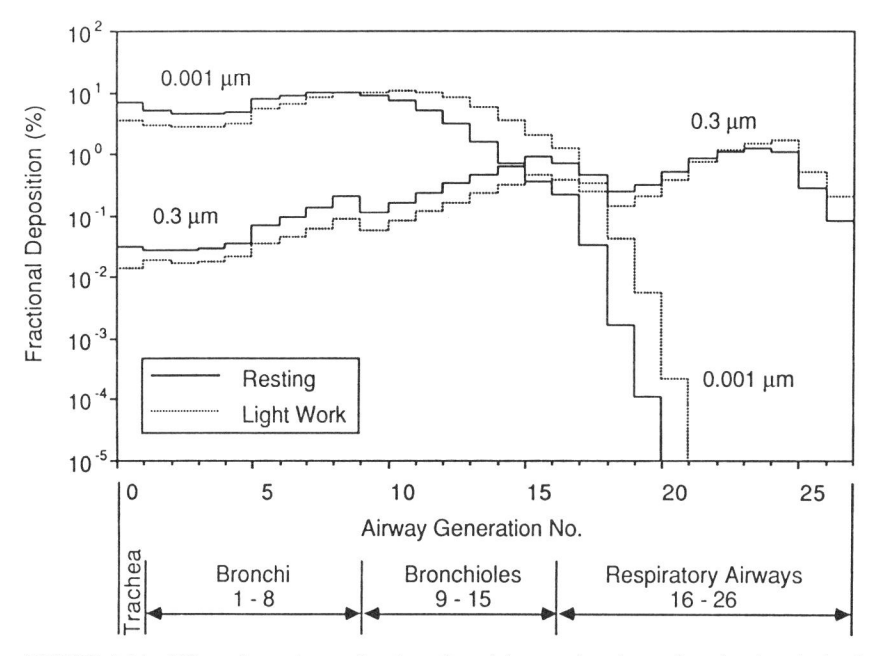

FIGURE 9-14 Effect of exercise on fraction of particles entering the trachea that is calculated to deposit in each airway generation. Histograms are shown for 0.001- and 0.3-μm-diameter particles, for an adult female at two levels of physical activity (resting, 0.39 m³/h, and light work, 1.26 m³/h).

activity-size distribution of progeny attached to ambient aerosol particles is typically broad (Chapter 2). Each component or mode of the radon progeny aerosol is assumed to be represented by a log normal distribution of activity with particle size, in which the geometric standard deviation (denoted by σ_g) is related to the activity median diameter (denoted by AMD) by

$$\sigma_g = 1 + 1.5[1 - (100 \text{ AMD}^{1.5} + 1)^{-1}] \qquad (9\text{-}12)$$

This function is illustrated in Figure 9-15.

NASAL AND ORAL FILTRATION

In evaluating dose to the lung from exposure to radon progeny, it has been customary to assume that the nose filters out 50 to 60% of the unattached progeny from the inhaled air (Jacobi and Eisfeld, 1980; Harley and Pasternack, 1982; NEA, 1983; James, 1984, 1988; NCRP, 1984). Likewise, it has generally been assumed for attached progeny that the filtration efficiency of the nose is either zero or, at most, a few percent. These assumptions were based on the experimental data obtained by George and Breslin (1969) in several

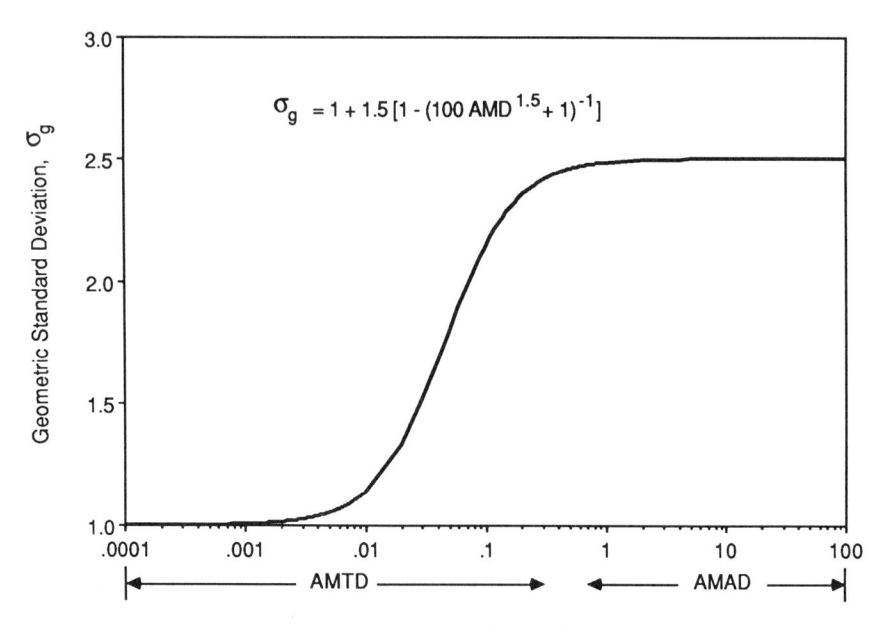

FIGURE 9-15 Variation of geometric standard deviation of particle size with activity median diameter of radon progeny aerosols that is assumed in calculations of deposition in the respiratory tract.

volunteer subjects. These investigators drew freshly formed ^{218}Po in through each subject's nose and out through a mouthpiece at various flow rates. They measured the fractional penetration of the unattached ^{218}Po activity through the nasal passages and oral cavity. Penetration of radon progeny attached to condensation nuclei was measured in the same way. Their experimental data are reproduced in Figure 9-16.

More recently, hollow casts of the human nasal and oral passages have been used to study the mechanisms of particle deposition, in particular, the influence of flow rate and particle diffusion coefficient on the deposition of particles in the size range of 0.2 to 0.005 μm in diameter (Cheng et al., 1988, 1990). For these particles, the deposition efficiency, E, of both the nasal and oral passages was found to be represented by an empirical expression of the form

$$E = 1 - \exp(-kQ^{-1/8}D^{2/3}) \qquad (9\text{-}13)$$

where k is a constant, Q is the flow rate (in liters/min), and D is the particle diffusion coefficient (in cm^2/s). Using postmortem casts, Cheng et al. (1988, 1990) found similar deposition efficiencies for the nasal and the oral pathways through to the trachea and, thus, similar values of the constant k in Equation 9-13.

DOSIMETRIC MODEL FOR RADON AND THORON PROGENY

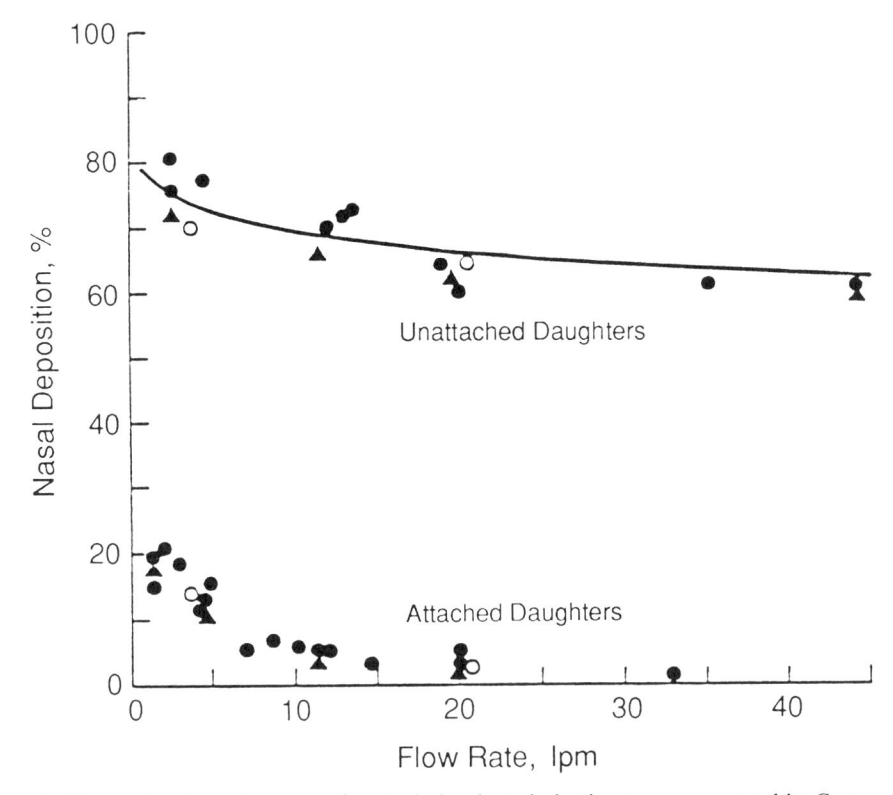

FIGURE 9-16 Nasal deposition of unattached and attached radon progeny measured by George and Breslin (1969) in several human subjects.

On extending this work to include two casts made with magnetic resonance imaging (MRI) data of the nasal and oral passages obtained in vivo and also to include smaller particles (unattached ^{212}Pb with a measured diffusion coefficient of 0.02 ± 0.004 cm^2/s, corresponding to a 0.0018-μm particle diameter), Cheng et al. (1989) found that it was necessary to modify the exponent of the particle diffusion coefficient in Equation 9-13 from 2/3 to 1/2 to fit the measured deposition efficiencies. Thus, their revised expression is:

$$E = 1 - \exp(-k'Q^{-1/8}D^{1/2}). \quad (9\text{-}14)$$

The data of Cheng et al. (1989) for inspiration of particles through the nasal passages are shown in Figure 9-17, together with the fitted efficiency function given by $k' = 13.2$. Figure 9-16 also shows the deposition efficiencies measured by Strong and Swift (in press) in the same nasal casts. Strong and Swift used unattached ^{218}Po for which the measured diffusion coefficient was found to vary between 0.05 cm^2/s for freshly formed ^{218}Po to 0.02 cm^2/s after aging. Figure 9-17 shows that the deposition efficiencies measured by Strong and Swift, represented by $k' = 7.7$, are consistently lower than those obtained by Cheng et al. (1989), which are represented by $k' = 13.2$. The difference in nasal deposition efficiencies measured by Cheng et al. (1989), although relatively small, is significant because it implies approximately a factor of two uncertainty in the fractions of inhaled unattached progeny that are able to penetrate the nose to be deposited in the bronchi. The impact of this uncertainty on the respective predictions of nasal penetration efficiency for unattached ^{218}Po is illustrated in Figure 9-18. The data of George and Breslin (1969) are replotted in Figure 9-18 and compared with the penetration efficiencies predicted as a function of flow rate by Strong and Swift (1990) and Cheng et al. (1989) for an assumed value of the diffusion coefficient of unattached ^{218}Po. Unfortunately, George and Breslin were not able to measure the diffusion coefficient of unattached ^{218}Po at the time of their experiments. For this comparison, it is assumed that the most likely value was 0.035 cm^2/s (E. O. Knutson, Environmental Measurements Laboratory, New York, personal communication, 1990). The diffusion coefficient implied by the data of Cheng et al. (1989) is 0.015 cm^2/s, which is probably too low.

The committee concluded from these data that the nasal penetration efficiency for unattached radon progeny is uncertain by about a factor of two, but that the data of George and Breslin (1969) still provide the best estimate. The impact of this uncertainty on estimates of lung dose in mine and home environments was examined elsewhere in this report.*

To evaluate lung dose in subject's who breathe habitually through their mouth, it has been customary to assume that the filtration efficiency of the oral passageway for unattached radon progeny is negligibly low (NCRP, 1984). However, the recent studies with hollow casts of these airways indicate that oral filtration is substantial. In Figure 9-19, the oral filtration efficiencies measured by Cheng et al. (1989) are compared with their values for the nasal passages. It is seen that oral filtration is approximately 75% of the values for nasal filtration. A similar ratio of oral:nasal efficiencies was observed by Strong and Swift (1990), although the absolute values were lower, as noted above. However,

---

*After the committee completed its work, further experimental studies of the penetration of unattached radon progeny through hollow casts of the human nasal passages, pharynx, and larynx were carried out by J. C. Strong and D. L. Swift (at the Biomedical Research Laboratory, AEA Technology, Harwell, England) and also by P. K. Hopke and D. L. Swift (at Clarkson University, Potsdam, New York). These researchers studied two casts that had been reconstructed from MRI scans of subjects in vivo. One of the casts, for an adult male, had previously been studied by Cheng et al. (1989). The second was taken from a 1.5-year-old infant. The results obtained for both casts were found to support Cheng et al.'s experimental observations. This emerging congruence of experimental data now indicates that George and Breslin's (1969) study of human subjects may not in fact provide the best estimate of nasal penetration efficiency. Further study of this possibility is recommended in the Summary and Recommendations. Tables 3-4 and 9-6 include data for both values of nasal deposition.

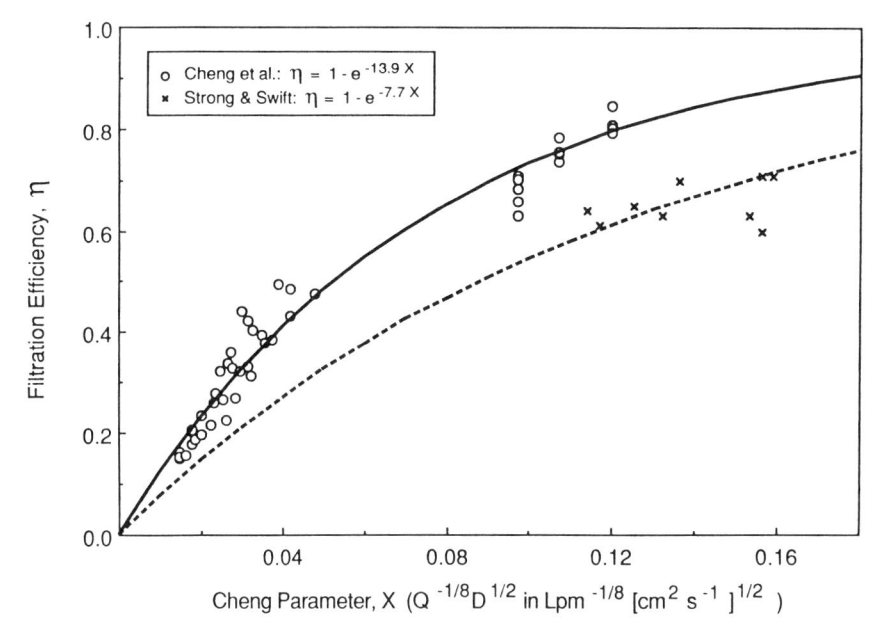

FIGURE 9-17 Filtration efficiencies for submicron particles measured in hollow casts of the human nasal passages. The data are plotted against Cheng et al.'s (1989) turbulent diffusion parameter, X. Cheng et al.'s data for particles in the size range 0.2 μm to 0.05 μm diameter ($X \leq 0.05$), and also their data for 0.0018-μm-diameter particles of unattached ^{212}Pb ($0.10 \leq X \leq 0.12$), are compared with Strong and Swift's (1990) data for unattached ^{218}Po. The curves show the respective efficiency functions fitted to these data.

in view of the preliminary nature of these data and the lack of confirmatory evidence in vivo, the committee took a more conservative approach by assuming that the filtration efficiency of the oral passageway for unattached radon progeny is only 50% of the value for nasal filtration efficiency.

It has been shown that so-called mouth breathers normally inhale partly through the nose (Niinimaa et al., 1980, 1981). Therefore, for the purpose of evaluating dose to the lung from exposure to radon progeny, it is unrealistic to consider the example of a pure mouth breather. On the other hand, at a sufficiently high rate of ventilation in response to heavy work, referred to as the "switch point," a normal nose breather will augment nasal flow by breathing partly through the mouth. The typical pattern of change between nasal and oral breathing found in adult subjects by Niinimaa et al. (1980, 1981) is shown in Figure 9-20. The proportion of the total airflow inhaled nasally by so-called mouth breathers is typically found to decrease from a major fraction at rest to a minor fraction for heavy work. Normal nose breathers were found to switch from 100% nasal breathing to partial mouth breathing at a total respiratory ventilation rate (V_E) of about 2.1 m^3/h. As discussed above, the committee

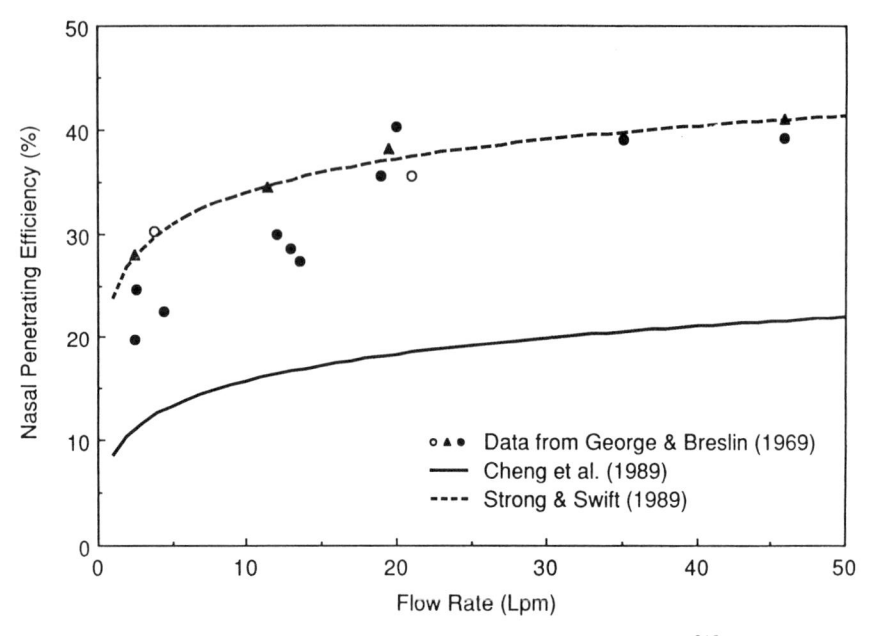

FIGURE 9-18 Comparison of nasal penetration efficiency of unattached ^{218}Po measured in human subjects by George and Breslin (1969) with values predicted by Strong and Swift (1990) and Cheng et al. (1989) on the basis of an assumed particle diffusion coefficient.

assumed that the deposition efficiencies of the nose and mouth for radon progeny particles in the thermodynamic size range are given by Equation 9-14, with k' = 7.7 and k' = 3.9, respectively. This expression is assumed to apply for an adult male subject. In order to scale the nasal and oral deposition efficiencies for other subjects (adult females, children, and infants), the committee adopted the procedure described by Swift (1989). It is assumed for scaling purposes that the flow rate Q in Equation 9-14 can be replaced by the dimensionless Reynolds number Re (Cheng et al., 1988):

$$Re = LV/\nu \quad (9\text{-}15)$$

where L is a hydraulic diameter (in cm), V is the fluid velocity (in cm/s), and ν is the kinematic fluid viscosity (in cm^2/s). Substituting the volumetric flow rate Q for V in Equation 9-15 gives

$$Re = k''LQ/L^2 = k''Q/L \quad (9\text{-}16)$$

Therefore, to scale nasal and oral deposition efficiency for body size, the factor to be applied to Q in Equation 9-14 is L_{ref}/L_s, where L_{ref} is a characteristic airway dimension for the reference adult male and L_s is the

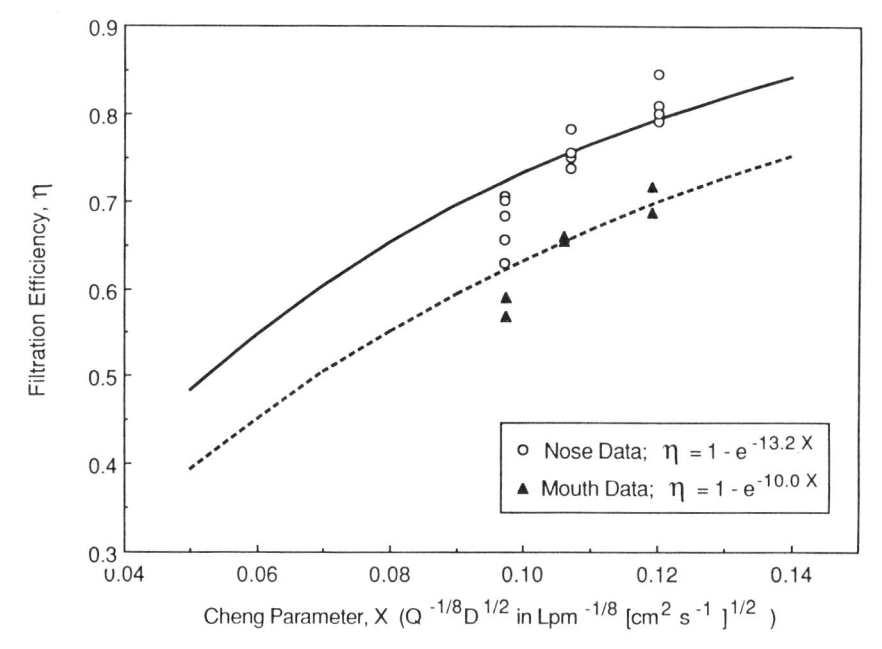

FIGURE 9-19 Comparative filtration efficiencies of the nasal and oral passages for unattached ^{212}Pb measured by Cheng et al. (1989). The measurements were made in hollow casts constructed using in vivo MRI data.

corresponding dimension for the subject considered. Yu and Xu (1987) and Swift (1989) considered that the diameter of the trachea provides an adequate index dimension. The values of tracheal diameter assumed by the committee to represent adults, children, and infants were shown in Figure 9-7.

Finally, it is also necessary to consider the efficiencies of the nose and mouth for removing that part of the radon progeny aerosol spectrum that overlaps the aerodynamic size range. The experimental data on nasal deposition of particles in the size range from about 1 to 10 μm in aerodynamic diameter in human subjects are shown in Figure 9-21 (Stahlhofen et al., 1989). Rudolf et al. (1986) represented these data on the aerodynamic deposition efficiency of the nose, η_{ae} (nose), by

$$\eta_{ae} \text{ (nose)} = 1 - (3.0 \times 10^{-4} d_{ae}^2 Q + 1)^{-1} \qquad (9\text{-}17)$$

The aerodynamic deposition efficiency of the oral cavity and oropharynx is found to be lower than that of the nose. The bulk of deposition in the oral passageway during mouth breathing is considered to occur in the larynx (Rudolf et al., 1986). Rudolf and colleagues represented the experimental data on oral deposition in adult male subjects by:

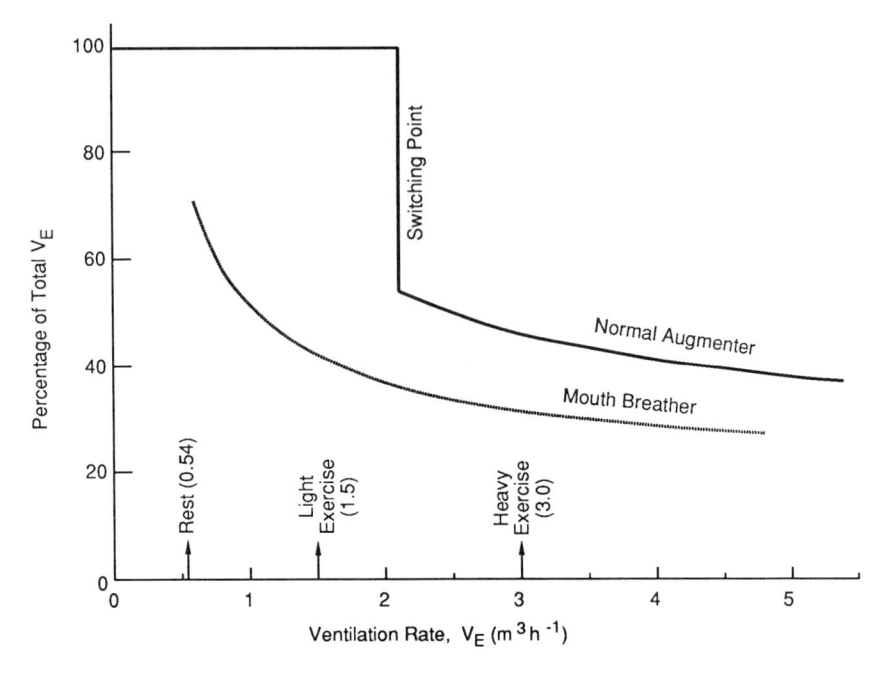

FIGURE 9-20 Percentage of total ventilatory airflow passing through the nasal route in "normal augmenters" and in "mouth breathers." The average ventilation rate, V_E, at which "normal augmenters" switch from 100% nasal breathing to partial breathing through the mouth is shown at $V_E = 2.1$ m^3/h.

$$\eta_{ae} \text{ (mouth)} = 1 - [1.5 \times 10^{-5}(d_{ae}^2 Q^{2/3} V_t^{-1/4})^{1.7} + 1]^{-1} \quad (9\text{-}18)$$

In order to scale these nasal and oral aerodynamic deposition efficiencies for body size, Equations 9-17 and 9-18 can be rewritten in terms of the stokes inertial parameter (Stk) (defined earlier in this chapter), where

$$Stk = d_{ae}^2 V / 18\eta L \quad (9\text{-}19)$$

where V is air velocity (in cm/s), η is dynamic air viscosity (in g cm^{-2} s^{-1}), and L is characteristic airway dimension (in cm).

For a given aerodynamic particle size, the inertial parameter depends on the ratio V/L, and thus on the ratio Q/L^3. The factor $1/L^3$ can therefore be applied to Q to scale inertial effects in the nasal or oral passages for airway dimensions. As assumed above to scale thermodynamic deposition efficiencies, the committee adopted the diameter of the trachea as the reference airway dimension (Yu and Xu, 1987; Swift, 1989).

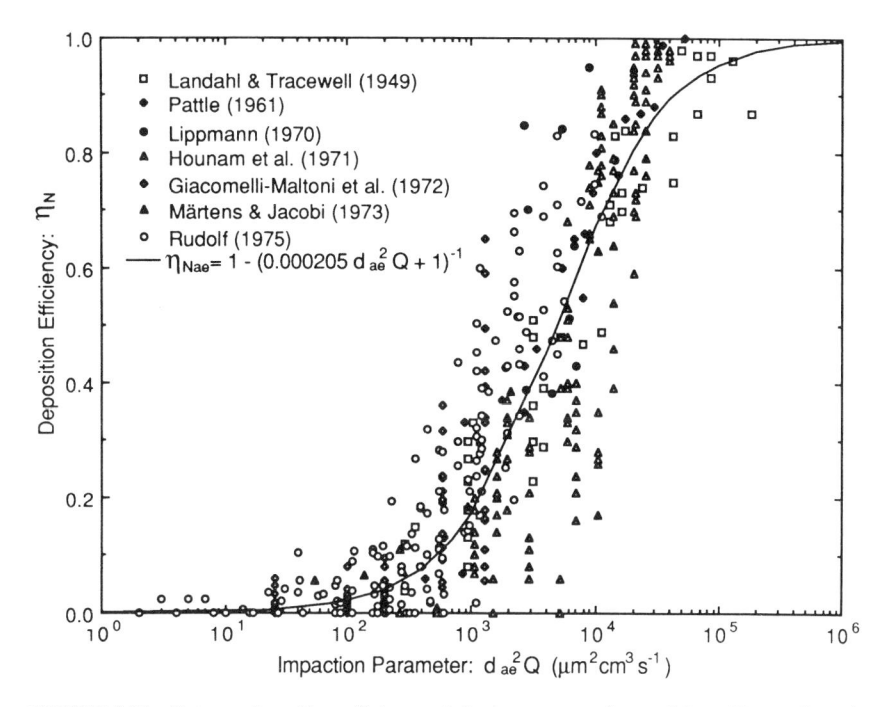

FIGURE 9-21 Data on deposition efficiency of the human nose for particles with aerodynamic diameter greater than 1 μm. The solid curve shows a hyperbolic approximation to the median values fitted as a function of the inertial parameter $d_{ae}^2 \, Q$.

The expressions given above have been used to evaluate nasal and oral filtration of radon progeny aerosols inhaled by adults, children, and infants in all calculations of doses to the lung. The thermodynamic and aerodynamic filtration efficiencies are combined as described earlier (Heyder et al., 1985).

Definition of Exposure and Reference Breathing Rates

In the final part of this chapter, the results of combining all aspects of the panel's dosimetric model for radon and thoron progeny that have been described above are presented. This is done to evaluate the conversion coefficient between "exposure to potential α-energy" and the "dose" received by the tissues of the respiratory tract that are deemed sensitive to bronchogenic cancer. The exposure-dose conversion coefficient is examined as a function of the particular conditions of exposure and determined by the radon progeny aerosol size and characteristics of the exposed subject. Exposures to potential α-energy are expressed here in the familiar unit working level month (WLM), where one working level (WL) is any combination of short-lived radon (or thoron) progeny in 1 liter of air that will result in the emission of 1.3×10^5 MeV (2.08×10^{-8}

J) of potential α-energy, and 1 WLM corresponds to an exposure for 1 working month (of 170 hours) to an airborne concentration of 1 WL. Thus, in terms of physical quantities (SI units), 1 WLM of exposure simply corresponds to 2.08 × 10^{-5} (J m^{-3}) × 170 (h) = 3.5 × 10^{-3} J h m^{-3}.

Contributions of Individual Radon Progeny to the WLM

The amounts (in Bq) of each individual radon progeny that is inhaled by a subject whose exposure to potential α-energy is 3.5 × 10^{-3} J h m^{-3} (1 WLM) are related to the degree of radioactive equilibrium between the progeny, and to the subject's breathing rate. If the breathing rate is denoted by B (in m^3 h^{-1}) and the activity-concentration ratios of ^{218}Po, ^{214}Pb, and ^{214}Bi/^{214}Po with respect to radon gas by F_{RaA}, F_{RaB}, and F_{RaC}, the activities of each progeny inhaled are given by:

$$I_{RaA} = \frac{3.5 \times 10^{-3} B F_{RaA}}{5.8 \times 10^{-10} F_{RaA} + 2.86 \times 10^{-9} F_{RaB} + 2.1 \times 10^{-9} F_{RaC}} \text{Bq/WLM} \quad (9\text{-}20)$$

$$I_{RaB} = \frac{I_{RaA} \times F_{RaB}}{F_{RaA}} \text{Bq/WLM} \quad (9\text{-}21)$$

$$I_{RaC} + \frac{I_{RaA} \times F_{RaC}}{F_{RaA}} \text{Bq/WLM} \quad (9\text{-}22)$$

where 5.8 × 10^{-10} is the potential α-energy (in joule) associated with 1 Bq of the α-emitting progeny ^{218}Po (RaA), 28.6 × 10^{-10} is the potential α-energy (J) associated with 1 Bq of the β-emitting progeny ^{214}Pb (RaB), and 21.0 × 10^{-10} is the potential α-energy (J) associated with 1 Bq of the β-emitting progeny ^{214}Bi (RaC).

The dosimetric model is solved for intake of a given mixture of activities. To represent the activity-intake from 1 WLM exposure to the "unattached" fraction of radon progeny, it is assumed (Reineking and Porstendörfer, 1990) that:

$$F_{RaB}(\text{unattached}) = 0.1 F_{RaA}(\text{unattached})$$

$$F_{RaC}(\text{unattached}) = 0$$

This mixture of ^{218}Po and ^{214}Bi activities is assumed below to calculate dose conversion coefficients for unit exposure to unattached radon progeny (1 WLM) in the size range 0.0006 μm (0.6 nm) to 0.01 μm (10 nm) activity median thermodynamic diameter (denoted by AMTD).

To represent the activity intake from unit exposure to the "attached" fraction of radon progeny, it is assumed that:

$$F_{RaA}(\text{attached}) = 0.8$$
$$F_{RaB}(\text{attached}) = 0.4$$
$$F_{RaC}(\text{attached}) = 0.2.$$

Exposure-dose conversion coefficients are derived below as functions of the attached radon progeny aerosol size over the continuous range of size that may extend from 0.01 μm (10 nm) to 1 μm (1000 nm) AMTD.

In practice, the activity-concentration ratios of ^{218}Po, ^{214}Pb, and ^{214}Bi/^{214}Po will vary with the particular environmental conditions of exposure. However, it can be shown that doses calculated in terms of unit exposure to potential α-energy are insensitive to the actual ratios of the progeny concentrations (Jacobi and Eisfeld, 1980; James, 1988) for both "unattached" and "attached" radon progeny.

Contributions of Individual Thoron Progeny to the WLM

In an analogous manner to Equations 9-20 through 9-22 for radon progeny, the activities of each of the thoron progeny that are inhaled per WLM exposure are given by:

$$I_{ThA} = \frac{3.5 \times 10^{-3} BF_{ThaA}}{5.3 \times 10^{-13} F_{ThA} + 6.91 \times 10^{-8} F_{ThB} + 6.56 \times 10^{-9} F_{ThC}} \text{Bq/WLM} \quad (9\text{-}23)$$

$$I_{ThB} = \frac{I_{ThA} \times F_{ThB}}{F_{ThA}} \text{Bq/WLM} \quad (9\text{-}24)$$

$$I_{ThC} + \frac{I_{ThA} \times F_{ThC}}{F_{ThA}} \text{Bq/WLM} \quad (9\text{-}25)$$

where 5.3×10^{-13} is the potential α-energy (in joule) associated with 1 Bq of the α-emitting progeny ^{216}Po (ThA), 6.91×10^{-8} is the potential α-energy (J) associated with 1 Bq of the β-emitting progeny ^{212}Pb (ThB), and 6.56×10^{-9} is the potential α-energy (J) associated with 1 Bq of the α-emitting progeny ^{212}Bi (ThC).

Since the number of joule per Bq intake of ^{216}Po is five orders of magnitude lower than the corresponding values for ^{212}Pb and ^{212}Bi, exposure to ^{216}Po does not contribute significantly to calculated doses. Therefore, to represent the activity intake from 1 WLM exposure to the "unattached" fraction of thoron progeny, it is assumed that:

$$F_{ThA}(\text{unattached}) = 1.0$$
$$F_{ThB}(\text{unattached}) = 1.0$$
$$F_{ThC}(\text{unattached}) = 0$$

To represent the activity-intake from 1 WLM exposure to the "attached" fraction of thoron progeny, it is assumed that:

$$F_{ThA}(\text{attached}) = 1.0$$
$$F_{ThB}(\text{attached}) = 0.02$$
$$F_{ThC}(\text{attached}) = 0.005$$

Reference Breathing Rates

To estimate the intakes of radon (or thoron) progeny activity by men, women, children, and infants for unit exposure to potential α-energy at various levels of physical exertion, the committee has assumed the breathing rates given in Table 9-5. These values were derived by Roy and Courtay (1990) from a review of Godfrey et al. (1971), Godfrey (1973), ICRP (1975), Taussig et al. (1977), Gaultier (1978), Cotes (1979), Gaultier et al. (1981), Scherrer (1981), Flandrois et al. (1982), Cooper and Weiler-Ravell (1984), and Zapletal (1987). The associated values of tidal volume (in cm^3) and respiratory frequency (in min^{-1}) for each subject are also given in Table 9-5 (Roy and Courtay, 1990), together with the fractions of the inspired airflow that are assumed to pass through the nose and the mouth in "normal" nose-breathers and in subjects who habitually breath through the "mouth" (Niinimaa et al., 1980, 1981). These values have been substituted in the deposition and nasal/oral filtration models (described above) to calculate the amounts of radon (or thoron) progeny deposited in each airway generation, as a function of aerosol size and breathing rate, for each subject.

Exposure-Dose Conversion Coefficients

In this final section, the dependence of the radon progeny exposure-dose conversion coefficient on influencing factors such as aerosol size, breathing rate, the assumed clearance behavior of the progeny, the choice of target cell population for which dose is evaluated, the presence of airway disease, and the gender or age of the exposed subject are examined. In each case, the dose received by the nuclei of target cells is evaluated as a continuous function of the assumed size of radon progeny on entry into the respiratory tract. The calculations relate to the equilibrium size attained by inhaled radon progeny under physiological conditions, where the air is saturated with water vapor. As discussed above, the ambient aerosol particles to which radon progeny become attached are assumed to be unstable in saturated air and to grow rapidly in the nose or pharyngeal airways to double their ambient size.

The curves presented below give values of dose (as the ordinate) for unit exposure (1 WLM or 3.5 J h m^{-3}) of each subject to the particular

TABLE 9-5 Summary of the Respiratory Data Assumed by the Panel to Calculate Exposure-Dose Conversion Coefficients for Various Subjects Exposed to Radon and Thoron Progeny

Level of Exertion/ Respiratory Parameter		Subject					
		Man	Woman	Child (10-yr)	Child (5-yr)	Infant (1-yr)	Infant (1-mo)
FRC	(cm³)	3,300	2,660	1,500	770	240	110
Sleep:	F(normal) = 1.0, F(mouth) = 0.7						
V_E	(m³ h⁻¹)	0.45	0.32	0.31	0.24	0.153	0.079
f	(min⁻¹)	12	12	17	21	34	40
V_t	(cm³)	625	450	305	190	75	33
Rest:	F(normal) = 1.0, F(mouth) = 0.7						
V_E	(m³ h⁻¹)	0.54	0.39	0.38	0.32	0.22	—
f	(min⁻¹)	12	14	19	25	36	—
V_t	(cm³)	750	460	330	210	100	—
Light Exercise:	F(normal 0) = 1.0, F(mouth) = 0.4						
V_E	(m³ h⁻¹)	1.5	1.26	1.11	0.57	0.35	0.12
f	(min⁻¹)	20	21	32	39	46	50
V_t	(cm³)	1,250	1,000	580	245	125	39
Heavy Exercise:	F(normal) = 0.47, F(mouth) = 0.3						
V_E	(m³ h⁻¹)	3	2.7	2.1	—	—	—
f	(min⁻¹)	26	28	45	—	—	—
V_t	(cm³)			1,920	1,610	760	—

radon progeny aerosol size shown on the abscissa. To enable quantitative comparison and application of the dose conversion coefficients evaluated for various conditions, these values are also presented in Table 9-6, which includes the two values of nasal deposition efficiency for the smallest particles based on findings of George and Breslin (1969) and Cheng et al. (1989).

Influence of Aerosol Size, Clearance Behavior, and Target Cells

The calculated dependence on radon progeny aerosol size of doses received by the nuclei of various target cells in an adult male is examined in Figures 9-22 through 9-24. These figures relate to a man undergoing light exercise, who is assumed to breathe through his nose. The three curves shown in Figure 9-22 show doses calculated for secretory cell nuclei in different regions of the respiratory tract, on the assumption that radon progeny, once deposited, remain in mucus and are cleared continuously toward the throat. The highest values are obtained if the dose is averaged over secretory cell nuclei in just the lobar and segmental bronchi (generations 2 through 5 in the lung model). Intermediate values are obtained when doses are averaged for the larger population of

TABLE 9-6 Summary of Coefficients to Convert Exposure to Radon Progeny Potential Alpha-Energy into the Average[a] Dose to Basal Cell Nuclei in the Bronchi of Different Subjects (Normal Nasal Breathers) as a Function of the Radon Progeny Aerosol Size and the Subject's Level of Physical Exertion

Subject	Radon Progeny AMTD (μm) With Assumed Nasal Deposition	Sleep	Rest	Light Exercise	Heavy Exercise
Adult Male	0.0011[b]	48.9	59.6	153.0	321.2
	0.0011[c]	23.4	28.9	80.9	210.7
	0.02	18.3	20.0	31.5	41.6
	0.15	4.66	5.04	7.86	11.8
	0.25	3.35	3.64	6.31	14.9
	0.3	3.03	3.31	6.22	18.4
	0.5	2.63	2.93	7.51	38.6
Adult Female	0.0011[b]	39.9	48.8	152.5	340.9
	0.0011[c]	18.6	23.1	80.0	223.2
	0.02	19.1	21.4	36.2	49.2
	0.15	4.66	5.17	8.62	13.4
	0.25	3.29	3.64	6.75	17.1
	0.3	2.95	3.27	6.59	21.1
	0.5	2.48	2.77	7.77	44.4
Child age 10 yr	0.0011[b]	48.8	60.2	166.5	—
	0.0011[c]	23.0	28.8	87.5	—
	0.02	22.3	24.9	40.1	—
	0.15	5.38	5.98	9.58	—
	0.25	3.81	4.25	7.61	—
	0.3	3.43	3.84	7.47	—
	0.5	2.89	3.30	8.80	—
Child age 5 yr	0.0011[b]	55.7	74.9	129.4	—
	0.0011[c]	26.1	36.0	65.6	—
	0.02	25.8	29.7	38.2	—
	0.15	6.04	6.98	8.94	—
	0.25	4.34	5.05	6.66	—
	0.3	3.93	4.61	6.26	—
	0.5	3.33	4.08	6.42	—
Infant age 1 yr	0.0011[b]	63.8	94.3	148.6	—
	0.0011[c]	29.6	45.2	74.3	—
	0.02	33.2	39.6	48.2	—
	0.15	7.77	9.06	10.9	—
	0.25	5.56	6.65	8.19	—
	0.3	5.02	6.11	7.69	—
	0.5	4.17	5.47	7.67	—
Infant age 1 mo	0.0011[b]	50.1	—	78.8	—
	0.0011[c]	22.4	—	36.7	—
	0.02	36.8	—	45.9	—
	0.15	9.02	—	10.9	—
	0.25	6.25	—	7.35	—
	0.3	5.54	—	6.47	—
	0.5	4.25	—	5.01	—

[a] Average value of the exposure-dose conversion coefficient calculated on the alternative assumptions that radon progeny are i) insoluble or ii) partially soluble in mucus.
[b] Nasal deposition of unattached progeny according to George and Breslin (1969).
[c] Nasal deposition of unattached progeny according to Cheng et al. (1989).

DOSIMETRIC MODEL FOR RADON AND THORON PROGENY

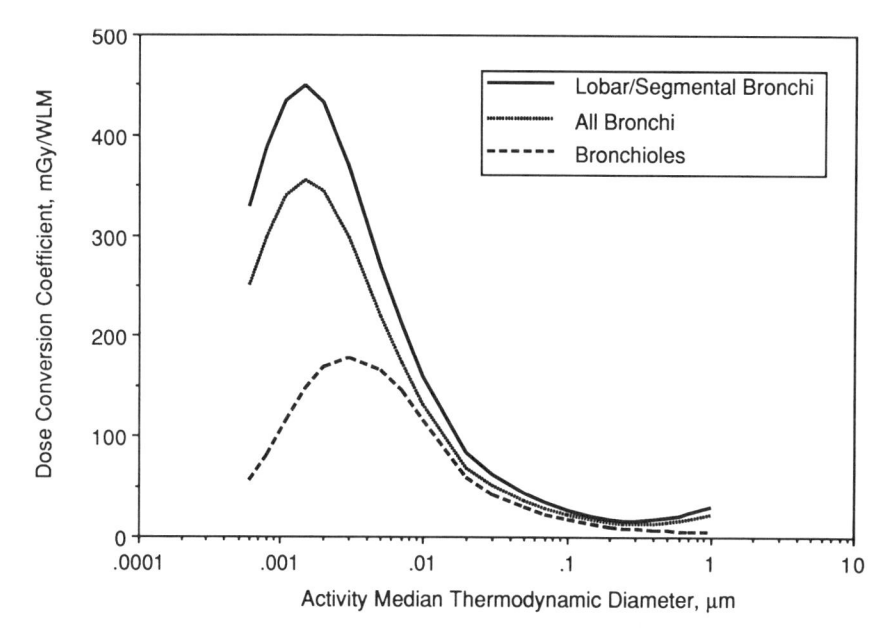

FIGURE 9-22 Effects of radon progeny aerosol size on calculated dose to secretory cell nuclei—adult male (insoluble/light exercise/nose breather).

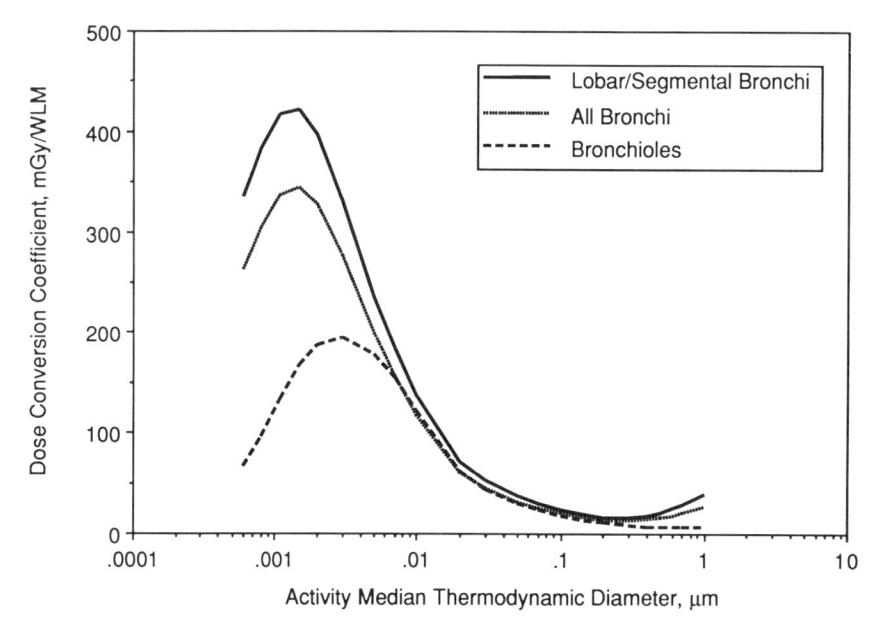

FIGURE 9-23 Effects of radon progeny aerosol size on calculated dose to secretory cell nuclei—adult male (soluble/light exercise/nose breather).

secretory cells that occurs throughout the bronchi (generations 1 through 8 in the lung model). The reference doses are approximately twofold lower if they are averaged for secretory cells in the bronchioles (generations 9 through 15 in the lung model).

It is noted that, for targets in both "lobar/segmental bronchi" and "all bronchi," the dose per unit exposure is approximately 25-fold higher for unattached progeny (with AMTD ≈ 0.001 μm) than it is for attached progeny with equilibrium AMTD in the range 0.3 to 0.5 μm. This ratio is somewhat lower (at approximately 20-fold) for targets in the bronchioles. For ultrafine radon progeny aerosols (AMTD < 0.01 μm), the dose per unit exposure is strongly influenced by the assumed nasal filtration efficiency. Figure 9-23 shows the equivalent exposure-dose conversion coefficients that are calculated if the radon progeny are assumed to be partially (30%) taken up by epithelial tissue at the site of deposition. Comparison with Figure 9-22 shows that this degree of uncertainty in the clearance behavior of radon progeny has a small effect on doses calculated for secretory cell targets.

However, if basal cells are instead assumed to be the principal targets, significantly different values of the exposure-dose conversion coefficient are calculated (Figure 9-24). In this case, dose conversion coefficients are approximately twofold lower for ultrafine radon progeny aerosols than they are for secretory cells, and uncertainty in the clearance behavior of radon progeny has a greater impact.

Influence of Exercise

Figure 9-25 shows exposure-dose conversion coefficients calculated for secretory cell nuclei throughout the bronchi for a man at various levels of physical exertion. It is seen that the calculated doses increase markedly with exercise for both unattached progeny and for large attached aerosols (AMTD ≈ 0.5 μm). These findings are significant for the evaluation of doses for exposure of underground miners.

Influence of Age and Gender

Figure 9-26 shows exposure-dose conversion coefficients calculated for a man, a woman, and children and infants (of either sex) of different age. In this case, the subjects are taken to be resting, and the target cells are taken to be secretory cells throughout the bronchi. It is seen that dose conversion coefficients are calculated to be somewhat lower for a woman than for a man, whereas they are generally higher in children and in the 1-year-old infant.

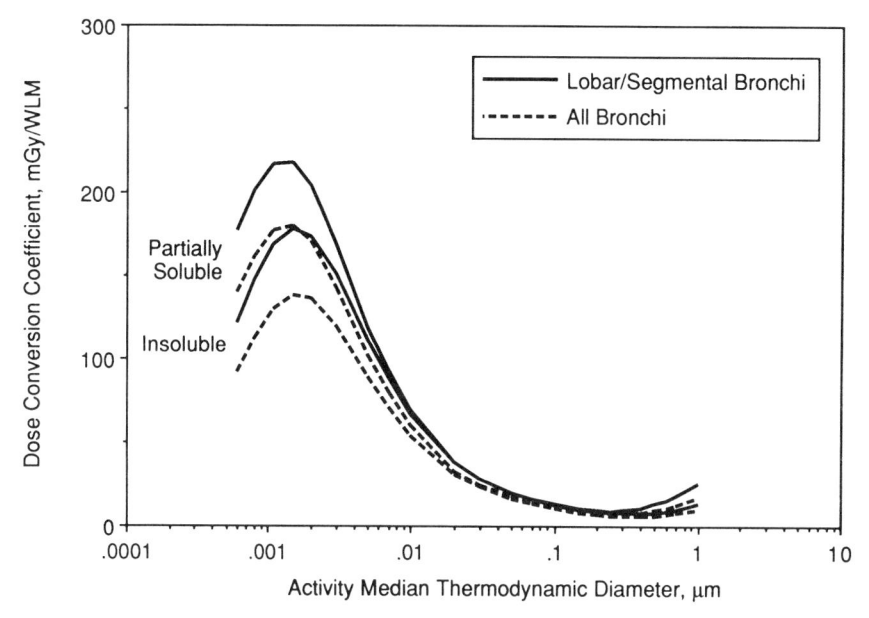

FIGURE 9-24 Effects of radon progeny aerosol size on calculated dose to basal cell nuclei—adult male (light exercise/nose breather).

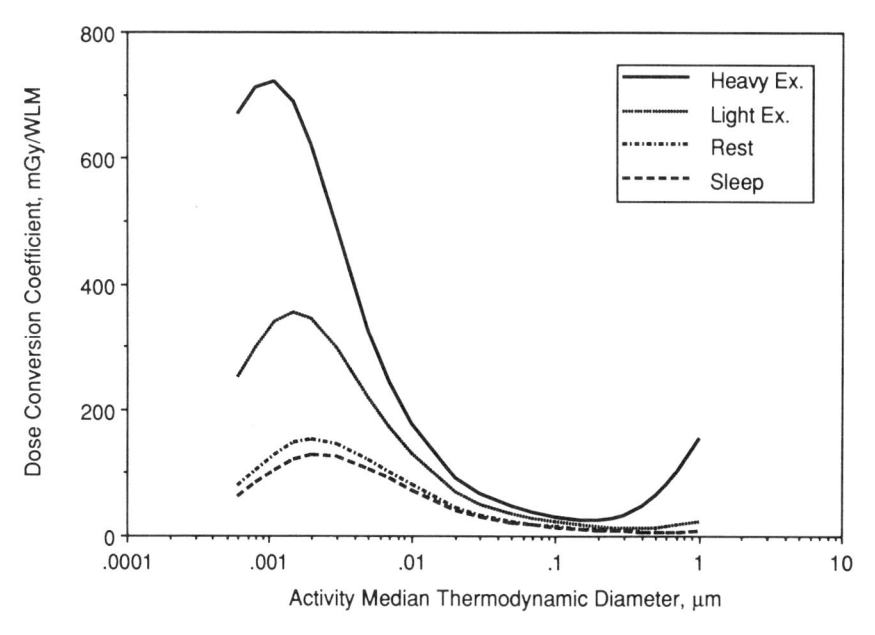

FIGURE 9-25 Effects of radon progeny aerosol size and exercise on calculated dose to secretory cell nuclei in the bronchi—adult male (nose breather).

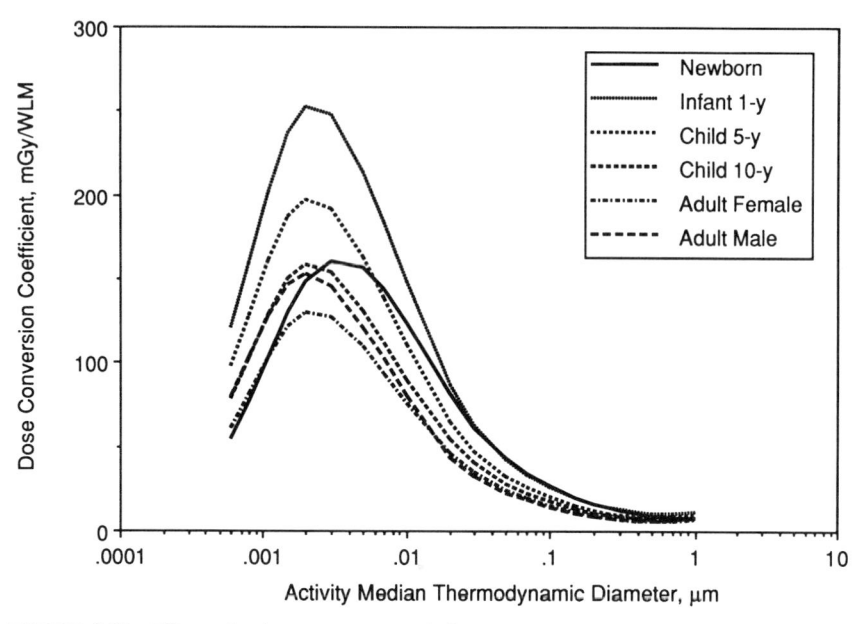

FIGURE 9-26 Effects of radon progeny aerosol size on calculated dose to secretory cell nuclei in bronchi for different subjects (resting/nose breather).

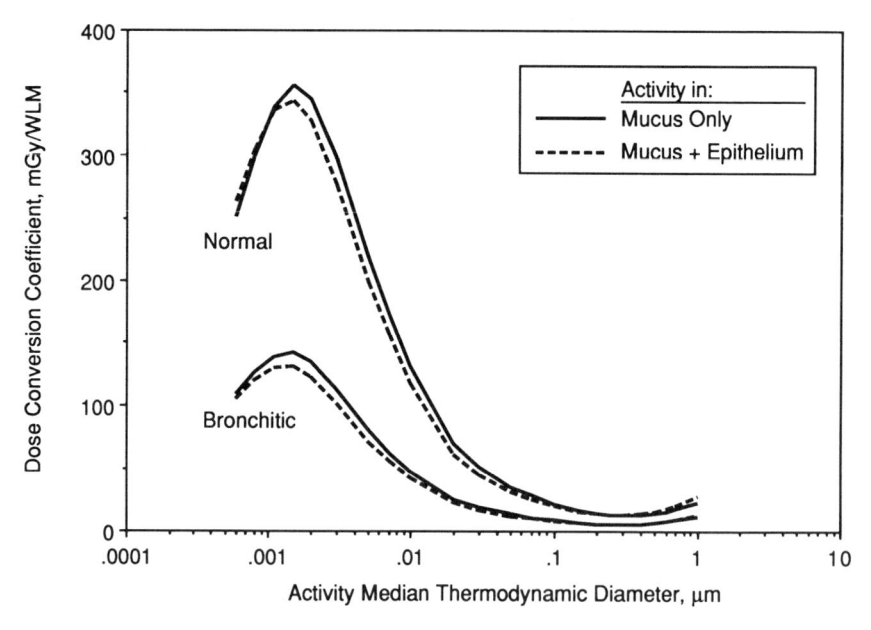

FIGURE 9-27 Modeled effect of bronchitis on calculated dose to secretory cell nuclei in bronchi—adult male (light exercise/nose breather).

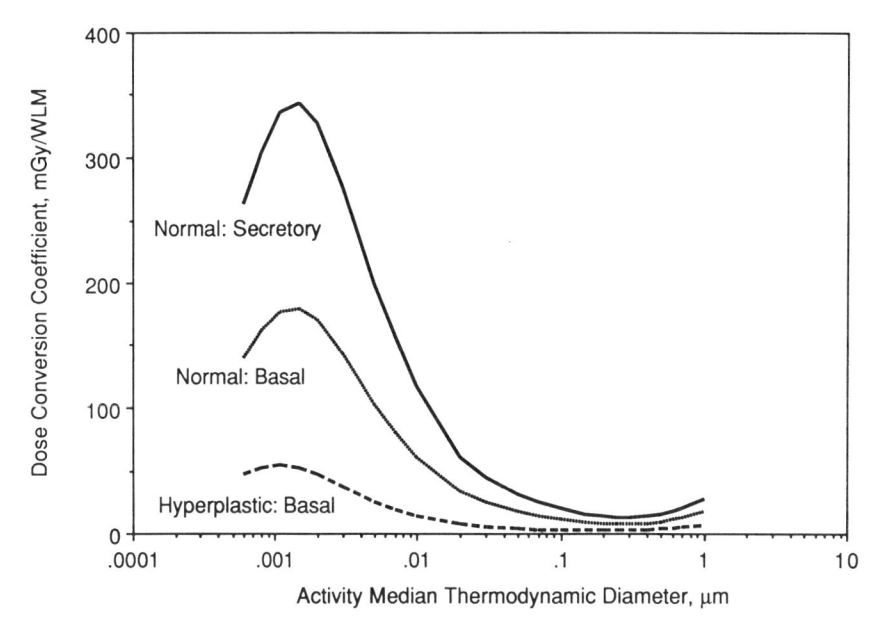

FIGURE 9-28 Effect of epithelial hyperplasia on calculated dose to bronchial target cells—adult male (light exercise/nose breather).

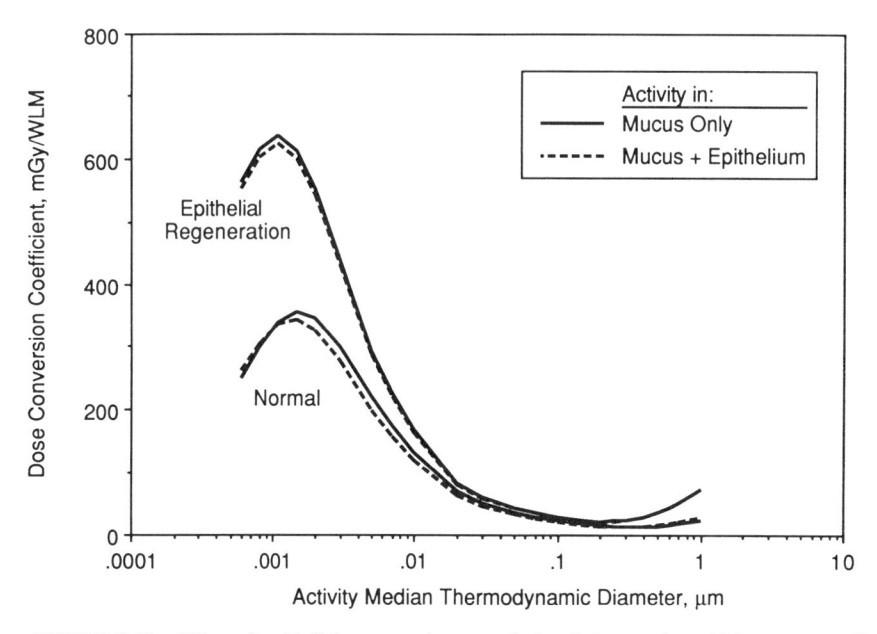

FIGURE 9-29 Effect of epithelial regeneration on calculated dose to bronchial secretory cell nuclei—adult male (light exercise/nose breather).

Influence of Airway Disease

Figures 9-27 through 9-29 illustrate the effects of modeling various disease conditions on the calculated exposure-dose conversion coefficient. In the case of a man with bronchitis (Figure 9-27), the effect of thickened mucus is to reduce the calculated doses by about a factor two. Markedly lower doses are also calculated for target cells in areas of hyperplastic epithelium. Figure 9-28 compares doses calculated for basal cells in hyperplastic epithelium (where no secretory cells are present) with those calculated for both secretory and basal cell targets in normal epithelium. Finally, Figure 9-29 compares doses calculated for secretory cells in thinned epithelium that is undergoing regeneration with those in epithelium of normal thickness. In this case, target cells in the damaged epithelium are calculated to receive about twofold higher doses than those in normal epithelium.

REFERENCES

Armstrong, T. W., and K. C. Chandler. 1973. SPAR, a FORTRAN Program for Computing Stopping Powers and Ranges for Muons, Charged Pions, Protons and Heavy Ions. ORNL-4869. Oak Ridge, Tenn.: Oak Ridge National Laboratory.
Bianco, A., F. R. Gibb, and P. E. Morrow. 1974. Inhalation study of a submicron size lead-212 aerosol. Pp. 1214-1219 in Proceedings of the 3rd International Congress of the International Radiological Protection Association. CONF-730907. Washington, D.C.: U.S. Atomic Energy Commission.
Booker, D. V., A. C. Chamberlain, D. Newton, and A. N. B. Stott. 1969. Uptake of radioactive lead following inhalation and injection. Br. J. Radiol. 42:457-466.
Cheng, Y. S., Y. Yamada, H. C. Yeh, and D. L. Swift. 1988. Diffusional deposition of ultrafine aerosols in a human nasal cast. J. Aerosol Sci. 19:741-752.
Cheng, Y. S., D. L. Swift, Y. F. Su, and H. C. Yeh. 1989. Deposition of radon progeny in human head airways. Proceedings of the DOE Technical Exchange Meeting on Assessing Indoor Radon Health Risks, September 18-19, 1989, Grand Junction, Colo. Department of Energy CONF 8909190. Springfield, Va.: National Technical Information Service.
Cheng, Y. S., Y. Yamada, H. C. Yeh, and D. L. Swift. 1990. Deposition of ultrafine aerosols in a human oral cast. Aerosol Sci. Technol. 12:1075-1081.
Cohen, B. S. 1987. Deposition of ultrafine particles in the human tracheobronchial tree: A determinant of dose from radon daughters. Pp. 475-486 in Radon and Its Decay Products: Occurrence, Properties, and Health Effects, P. K. Hopke, ed. Washington, D.C.: American Chemical Society.
Cohen, B. S., R. G. Sussman, and M. Lippmann. 1990. Ultrafine particle deposition in a human tracheobronchial cast. Aerosol Sci. Technol. 12:1082-1091.
Cook, C. D., R. B. Cherry, D. O'Brien, P. Karlberg, and C. A. Smith. 1955. Studies of respiratory physiology in the newborn infant. I. Observations on normal premature and full-term infants. J. Clin. Invest. 34:975-982.
Cooper, D. M., and D. Weiler-Ravell. 1984. Gas exchange response to exercise in children. Am. Rev. Respir. Dis. 129(Suppl.):S47-S48.
Cotes, J. E. 1979. Lung function assessment and application in medicine. Oxford: Blackwell Scientific Publications.

Cuddihy, R. G., and H. C. Yeh. 1988. Respiratory tract clearance of particles and substances dissociated from particles. Pp. 169-193 in Inhalation Toxicology: The Design and Interpretation of Inhalation Studies and Their Use in Risk Assessment, U. Mohr, ed. New York: Springer-Verlag.

Egan, M. J., and W. Nixon. 1985. A model of aerosol deposition in the lung for use in inhalation dose assessments. Radiat. Prot. Dosim. 11:5-17.

Egan, M. J., and W. Nixon. 1987. Mathematical modelling of fine particle deposition in the respiratory system. Pp. 34-40 in Deposition and Clearance of Aerosols in the Human Respiratory Tract, W. Hofmann, ed. Vienna: Facultas Universitätsverlag.

Egan, M. J., W. Nixon, N. I. Robinson, A. C. James, and R. F. Phalen. 1989. Inhaled aerosol transport and deposition calculations for the ICRP task group. J. Aerosol Sci. 20:1301-1304.

Flandrois, R., H. Grandmontagne, M. H. Mayet, R. Favier, et al. 1982. La consommation d'oxygene maximale chez le jeune français, sa variation avec l'âge et le sexe et l'entranement. J. Physiol. (Paris) 78:186-194.

Gaultier, C. 1978. Développement des volumes pulmonaires pendant les 36 premier mois de la vie chez l'homme. Doctoral thesis in human biology research. Paris: CHU St. Antoine.

Gaultier, C., L. Perret, N. Boule, A. Buvry, and F. Girard. 1981. Occlusion pressure and breathing pattern in healthy children. Respir. Physiol. 46:71-80.

Gehr, P. 1987. Inhalation pathways in relation to infants and children. Pp. 67-78 in Age-Related Factors in Radionuclide Metabolism and Dosimetry, G. B. Gerber, H. Métivier, and H. Smith, eds. Hingham, Maine: Kluwer Academic Publishers.

George, A. C., and A. J. Breslin. 1969. Deposition of radon daughters in humans exposed to uranium mine atmospheres. Health Phys. 17:115-124.

Godfrey, S. 1973. Exercise Testing in Children. London: Saunders.

Godfrey, S., C. T. M. Davies, E. Wozniak, and C. A. Barnes. 1971. Cardio-respiratory response to exercise in normal children. Clin. Sci. 40:419-431.

Greenhalgh, J. R., A. Birchall, A. C. James, H. Smith, and A. Hodgson. 1982. Differential retention of ^{212}Pb ions and insoluble particles in nasal mucosa of the rat. Phys. Med. Biol. 27:837-851.

Gurman, J. L., R. B. Schlesinger, and M. Lippmann. 1980. A variable-opening mechanical larynx for use in aerosol deposition studies. Am. Ind. Hyg. Assoc. J. 41:678-680.

Gurman, J. L., M. Lippmann, and R. B. Schlesinger. 1984. Particle deposition in replicate casts of the human upper tracheobronchial tree under constant and cyclic inspiratory flow. I. Experimental. Aerosol Sci. Technol. 3:245-252.

Hansen, J. E., E. P. Ampaya, G. H. Bryant, and J. J. Navin. 1975. Branching pattern of airways and air spaces of a single human terminal bronchiole. J. Appl. Physiol. 38:983-986.

Harley, N. H. 1971. Spatial distribution of radon daughter and plutonium-239 alpha lung dose based on experimental energy absorption measurements. Ph.D. thesis. New York University, New York.

Harley, N. H., and B. S. Pasternack. 1982. Environmental radon daughter alpha dose factors in a five-lobed human lung. Health Phys. 42:789-799.

Hart, M. D., M. M. Orzalesi, and C. D. Cook. 1963. Relation between anatomic respiratory dead space and body size and lung volume. J. Appl. Physiol. 18:519-522.

Heyder, J., and J. Gebhart. 1977. Gravitational deposition of particles from laminar aerosol flow through inclined circular tubes. Aerosol Sci. 8:289-295.

Heyder, J., J. Gebhart, and G. Scheuch. 1985. Interaction of diffusional and gravitational particle transport in aerosols. Aerosol Sci. Technol. 4:315-326.

Hofmann, W. 1982. Dose calculations for the respiratory tract from inhaled natural radioactive nuclides as a function of age. II. Basal cell dose distributions and associated lung cancer risk. Health Phys. 43:31-44.

Hursh, J. B., and T. T. Mercer. 1970. Measurement of ^{212}Pb loss rate from human lungs. J. Appl. Physiol. 28:268-274.

Ingham, D. B. 1975. Diffusion of aerosols from a stream flowing through a cylindrical tube. Aerosol Sci. 6:125-132.

International Commission on Radiological Protection (ICRP). 1975. Report of the Task Group on Reference Man. International Commission on Radiological Protection, Publication 23. Oxford: Pergamon.

Jacobi, W., and K. Eisfeld. 1980. Dose to tissues and effective dose equivalent by inhalation of radon-222, radon-220 and their short-lived daughters. GSF Report S-626. Munich-Neuherberg, West Germany: Gesellschaft für Strahlen und Umweltforschung.

James, A. C. 1984. Dosimetric approaches to risk assessment for indoor exposure to radon daughters. Radiat. Prot. Dosim. 7:353-366.

James, A. C. 1988. Lung dosimetry. Pp. 259-309 in Radon and Its Decay Products in Indoor Air, W. W. Nazaroff and A. V. Nero, eds. New York: Wiley Interscience.

James, A. C., J. R. Greenhalgh, and H. Smith. 1977. Clearance of lead-212 ions from rabbit bronchial epithelium to blood. Phys. Med. Biol. 22:932-948.

James, A. C. 1990. Experimental data and theoretical models for the dosimetry of human exposure to radon and thoron progeny. DOE/Technical Report Series (in preparation).

Kirichenko, V. N., D. G. Khachirov, S. A. Dubrovin, V. E. Klyuch, and A. V. Bykhovskii. 1970. Experimental study of the distribution of short-lived radon daughters in the respiratory tract. Gig. Sanit. 2:222-227.

Mercer, R. R., M. L. Russell, and J. D. Crapo. 1989. Airway cell and nuclear depth distribution in human and rat lungs. Health Phys., in press.

National Council on Radiation Protection and Measurements (NCRP). 1984. Evaluation of Occupational and Environmental Exposures to Radon and Radon Daughters in the United States. Report No. 78. Bethesda, Md.: National Council on Radiation Protection and Measurements.

Niinimaa, W., P. Cole, S. Mintz, and R. J. Shephard. 1980. The switching point from nasal to oronasal breathing. Respir. Physiol. 42:61-71.

Niinimaa, W., P. Cole, S. Mintz, and R. J. Shephard. 1981. Oronasal distribution of respiratory airflow. Respir. Physiol. 43:69-75.

Nuclear Energy Agency Group of Experts (NEA). 1983. Dosimetry Aspects of Exposure to Radon and Thoron Daughter Products. Paris: Organization for Economic Cooperation and Development (OECD).

Pack, A. I., M. B. Hooper, W. Nixon, and J. C. A. Taylor. 1977. A computational model of pulmonary gas transport incorporating effective diffusion. Respir. Physiol. 29:101-124.

Phalen, R. F., M. J. Oldham, C. B. Beaucage, T. T. Crocker, and J. D. Mortensen. 1985. Postnatal enlargement of human tracheobronchial airways. Anat. Rec. 212:368-380.

Pich, J. 1972. Theory of gravitational deposition of particles from laminar flow in channels. Aerosol Sci. 3:351-361.

Reineking, A., and J. Porstendörfer. 1990. Unattached fractions of short lived Rn decay products in indoor and outdoor environments: an improved single screen method and results. Health Phys. 58:715-728.

Roy, M., and C. Courtay. 1990. Daily activities and breathing parameters for use in respiratory tract dosimetry. Radiat. Prot. Dosim., in press.

Rudolf, G., J. Gebhart, J. Heyder, C. F. Schiller, and W. Stahlhofen. 1986. An empirical formula describing aerosol deposition in man for any particle size. J. Aerosol Sci. 17:350-355.

Scherrer, J. 1981. Précis de physiologie du yravail. In Notions Ergonomie. Masson: Paris.

Scherer, P. W., L. H. Shendalman, N. H. Greene, and A. Bouhuys. 1975. Measurement of axial diffusivities in a model of the bronchial airways. J. Appl. Physiol. 38:719-723.

Sinclair, D. R., R. J. Countess, and G. S. Hoopes. 1974. The effect of relative humidity on the size of atmospheric aerosol particles. Atmos. Environ. 8:1111-1117.

Stahlhofen, W., G. Rudolf, and A. C. James. 1989. Intercomparison of experimental regional aerosol deposition data. J. Aerosol Med. 2:285-308.

Strong, J. C., and D. L. Swift. 1990. Deposition of unattached radon daughters in models of human nasal airways. To be presented at the 29th Hanford Symposium on Health and the Environment Indoor Radon and Lung Cancer: Reality or Myth?, October 16-19, 1990, Richland, Wash.

Swift, D. L. 1989. Age-related scaling for aerosol and vapor deposition in the human respiratory tract. Health Phys. 57:293-297.

Taulbee, D. B., and C. P. Yu. 1975. A theory of aerosol deposition in the human respiratory tract. J. Appl. Physiol. 38:77-85.

Taussig, L. M., R. T. Harris, and M. D. Lebowitz. 1977. Lung function in infants and young children. Am. Rev. Respir. Dis. 116:233-239.

Weibel, E. R. 1963. Morphometry of the Human Lung. New York: Academic Press.

Wood, L. D. H., S. Pritchard, T. R. Weng, K. Kruger, A. C. Bryan, and H. Levison. 1971. Relationship between anatomic dead space and body size in health, asthma and cystic fibrosis. Am. Rev. Respir. Dis. 104:215-222.

Yeh, H. C., and G. M. Schum. 1980. Models of human lung airways and their application to particle deposition. Bull. Math. Biol. 42:461-480.

Yu, C. P., and C. K. Diu. 1982. A probabilistic model for intersubject deposition variability of inhaled particles. Aerosol Sci. Technol. 1:355-362.

Yu, C. P., and G. B. Xu. 1987. Predicted particle deposition of diesel particles in young humans. J. Aerosol Sci. 18:419-429.

Zapletal, A. 1987. Lung Function Testing in Children and Adoloescents, H. Herzog, ed. Basel: Karger.

Zeltner, T. B., J. H. Caduff, P. Gehr, J. Pfenninger, and P. H. Burri. 1987. The postnatal development and growth of the human lung. I. Morphometry. Respir. Physiol. 67:247-267.

Index

A

Actinium, 61, 64
Activity median thermodynamic diameter, 32–33, 40, 41–42, 45, 49, 108, 139, 215, 224
Adenocarcinomas, 172, 173, 179, 184, 187, 189
Aerosols, 2, 21, 32–33, 90–131
  characteristics, 32–33, 72, 142–143
  penetration, 92–93
  see also Clearance; Deposition; Diffusion; Particle size
Age factors, 1, 2–4, 12, 14, 16, 34, 42–43, 53, 54, 63, 72, 113, 148, 204, 230, 233
  cancer and, 53, 55–56, 175, 177, 188
  cell proliferation, 54, 175
  disease effects, 154
  lung volume, 144, 147, 148
  see also Children; Infants
Airway diseases
  allergies, 151
  bronchitis, 2, 45, 46, 49, 152, 153, 154, 155, 183, 196, 232, 233
  chronic obstructive pulmonary disease, 153
  dose effects, 46–49, 180
Algorithms, 118, 122
Allergies, 151
Alpha radiation, 1, 3–4, 6, 9, 15, 26–27, 106, 137, 166–189
  age and susceptibility, 54
  cell-level dose, 56, 166, 186, 194–195
  dosimetry, general, 1, 3–4, 6, 14, 166, 186
  in equations, 22–23, 24
Altitude, 157
Alveoli, 38, 72, 76, 77, 139, 146–147, 148, 149, 152–153, 154, 167–168, 170, 172, 189, 202, 208
AMAD, 108–109
Anatomical factors, 144, 147–148, 157
deposition, 6, 138, 141, 144, 147–148
see also Morphometry; *specific anatomical systems*
Animal studies, 6, 54, 55–56, 137, 154, 155, 157, 189, 223–224
Atom bomb survivors, 53, 57
Automated tape samplers, 115–116

B

Basal cells, 2, 3, 6, 35, 37, 41, 43, 45, 46, 48, 50, 166, 167, 181–184, 194–195, 197, 198, 231
  carcinogenesis, 166, 167, 181–184
  thoron, 50
BEIR IV report, 4, 5, 14, 16, 17, 31, 32, 52, 53, 55, 56, 57
Belgium, 110–113
Benzo[*a*]pyrene, 55
Biological processes, general, 52
Bismuth, 61, 104, 124, 126–128, 130, 205, 224
Breathing patterns, 1, 6, 77, 137, 138, 140, 144, 148–149, 208, 214, 218–219
  altitude factors, 157
  in equations, 208–209, 224
  ethnic differences, 156
  physical activity and, 2, 12, 13, 149–150, 151, 227, 230, 231, 233
  reference rates, 6, 226
  see also Nasal processes; Oral processes
Bronchi, 3, 6, 28, 54, 79, 80, 148, 168, 203, 206
  adult females, 42–43
  cellular pathology of cancer, 167, 172, 180, 181, 184, 185, 186, 189, 223
  epithelium, 35–37, 43, 46–48, 145, 157, 168, 170, 175, 181, 183, 184, 186, 189, 194–197, 199, 234
  hyperplasia, 46–47, 48, 55, 233, 234

239

morphometry, 67–71, 73, 75, 200, 202, 204, 211
particle deposition, 1, 6, 20, 38–39, 47–48, 73–74, 77, 145–146, 148, 152, 195, 200, 206–207, 208, 211, 212–213, 218
stem cells, 6–7
target cells, 35–38, 42, 189, 227–230, 234
Bronchioles, 37, 38, 67, 71, 72, 145, 148, 168, 181, 184, 185, 198–201, 204, 206, 212
Bronchitis, 2, 45, 46, 49, 152, 153, 154, 155, 183, 196, 232, 233

C

Carbon tetrachloride, 56
Carcinogenesis and carcinogens, 2, 5–6, 31, 52
 age factors, 175, 177, 188
 basal cells, 166, 167, 181–184
 bronchial cells, 167, 172, 180, 181, 184, 185, 186, 189, 223
 classification, carcinomas, 172–173, 175–176
 epidemiological data, 1, 11–12, 56, 169, 176, 186–188
 epithelia, 168, 169–170, 175, 181–183, 189, 194, 233
 genetics, 54, 177, 178–180
 hyperplasia, 46–47, 48, 49, 53–56, 233, 234
 liver cancer, 54, 56
 mucosal processes, 168, 181–182, 186
 other than radon, 1, 12, 15, 26
 secretory cells, 166, 167, 172, 181, 183, 184, 186, 189
 skin, 54
 smoking, 169, 170, 174, 175, 176, 177, 185–188
 see also Hyperplasia; Lung cancer; *specific carcinomas*
Cell cultures, 54, 167
Cells and cellular processes, 38, 166–189
 alpha dosimetry, cell level, 56, 166, 186, 194–195
 basal cells, 2, 3, 6, 35, 37, 41, 43, 45, 46, 48, 50, 166, 167, 181–184, 194–195, 197, 198, 231
 bronchi, carcinogenesis, 167, 172, 180, 181, 184, 185, 186, 189, 223
 goblet cells, 168, 181
 hyperplasia, 46–47, 48, 49, 53–56, 233, 234
 Kulchitsky cells, 7, 168
 lung cancer, cells of origin, 56, 166–189
 proliferation, 53–55, 175, 180–182, 189
 smoking and cellular carcinogenesis, 169, 170, 174, 175, 176, 177, 185–188

 stem cells, 6–7
 target cells, 15, 25, 31, 32, 35–38, 39, 42, 43, 49, 166, 189, 194–195, 198, 227–230, 234
 see also Secretory cells and processes
Children, 2, 3–4, 14, 35, 43–44, 45, 46, 148, 201, 202, 204, 228; *see also* Infants
Chemical processes, unattached fraction, 24–25
Chromosomes, 178–180
Chronic obstructive pulmonary disease, 153
Cilia, 46–47, 67, 78, 79, 138, 144, 145, 147, 168, 181, 182, 195, 196, 203
"A Citizen's Guide to Radon," 10
Classification, carcinomas, 172–173, 175–176
Clearance, respiratory tract, 73, 137, 138, 144–147, 148, 151, 227–230
 coughing, 147
 models, 77–81, 202–207
 mucus, 15, 35, 46, 78–80, 145–146, 202–203, 205, 206, 227
 nasal, 78, 145
 preexisting conditions, 152–155
 trachea, 79, 80, 146, 151, 195, 203
Clusters, particles, 24–25, 92, 116, 125, 126
Committee on the Biological Effects of Ionizing Radiation, *see* BEIR IV report
Concentration levels, 15–16, 20, 26, 90–131
 diffusion, 95–96, 101–102, 109–110
 wire screen samplers, 25, 27, 94, 95, 96–100, 102–107, 110–124
Cost factors, mitigation, 11, 14
Coughing, 147

D

Decay, radioactive, 9, 10, 26, 61, 62, 63, 64, 101, 110, 122, 194, 199, 203
 equation, 21–22, 23, 205
Density factors, 139, 142–143
Department of Energy, 194
Deposition, respiratory tract, 3, 6, 72, 137–144, 147–155
 anatomical factors, general, 6, 138, 141, 144, 147–148
 bronchial, 1, 6, 20, 38–39, 47–48, 73–74, 77, 145–146, 148, 152, 195, 200, 206–207, 208, 211, 212–213, 218
 diffusional, 73, 76, 90, 138, 139, 140, 141, 144, 149, 208, 210
 electrostatic, 24, 25, 138, 141
 in equations, 94, 207–212, 215, 216, 220, 222
 geometry of, 66–67, 138, 195
 gravitational sedimentation, 74, 76, 139–140, 144, 149, 209–210
 historical perspectives, research, 137–138

impaction factors, 74, 76, 117, 149, 150, 151, 208, 210
inertial factors, 117, 138, 140, 145, 149, 150, 152
localized, 37–38, 76, 77, 81, 155
in lungs, 6, 137–144, 147–148, 149–157
models, 1, 14, 15, 24, 39, 66–67, 71, 72–77, 200, 207–215
nasal, 1, 3, 38, 39–40, 66–67, 72–73, 74, 149, 150–152, 216, 218–222
oral, 72–73, 74, 216
particle size and, 72–73, 75, 76, 138–143, 157, 212–215
preexisting conditions, 17, 152–155
regional, 37–38, 77
theory, 39, 74, 76, 207, 209–210
tracheal, 6, 20, 73–74, 77, 148, 151, 154, 207, 212
see also Wire mesh screens
Diffusion, 24, 139
Brownian, 139, 210
concentration levels and, 95–96, 101–102, 109–110
conversion coefficients, 24, 33, 90–92
deposition, diffusional, 73, 76, 90, 138, 139, 140, 141, 144, 149, 208, 210
equations, 24, 90–95, 208
particle size and, 25, 90–95
samplers, 95–96, 101–102, 109–110, 118
theoretical models, 91, 92
through tubes, 92–93, 95, 101, 195
Disease, preexisting, 2, 6, 46–49, 148, 151, 152–155, 234
see also specific diseases
DNA, 31, 178, 179, 180–181, 195
Dose and dosimetry, 15–16, 31–57, 60–82
airway disease, general dose effects, 46–49, 180
alpha energy, general, 1, 3–4, 6, 14, 166, 186
cell-level, 56, 166, 186, 194–195
defined, 60
exposure and, 2–4, 5–6, 31–32, 34–35, 137, 225, 226–227, 228
historical perspectives, 24, 60–82
models, 1–4, 14, 15, 31–51, 60–82, 180, 194–202
risk assessment, extrapolation, 1, 4, 5, 12, 14, 31–57, 110, 137, 157
target cells, 194–195
thoron, 49–51, 54, 225–226, 228

E

Einstein-Cunningham constant, 91
Electricity and electrostatics, 24, 25, 32, 109, 113–114, 118–119, 138, 141
Electron microscopes, 72, 175

Emphysema, 153, 154
Environmental Protection Agency, 4, 10
Epidemiological data, 1, 5, 11, 12, 20, 32, 137, 155–157, 169, 173, 176, 177, 184, 186–188
cancer, 1, 11–12, 56, 169, 176, 186–188
historical perspectives, 11, 169, 176
international, 11, 12
smoking, 56, 186–188
see also Risk assessment
Epithelia, 2, 3, 4, 6, 7, 14, 53, 146, 198–199, 206, 234
bronchial, 35–37, 43, 46–48, 145, 157, 168, 170, 175, 181, 183, 184, 186, 189, 194–197, 199, 234
cancer pathology, 168, 169–170, 175, 181–183, 189, 194, 233
lung, 167–169, 174
thoron uptake, 49–50
tracheal, 53, 145, 181, 182, 183
Equations, 51
alpha radiation, 22–23, 24
breathing, 208–209, 224
particle deposition, 94, 207–212, 215, 216, 220, 222
particle diffusion, 24, 90–95, 208
particle growth, 143
radioactive decay, 21–22, 23, 205
smoking, 23
unattached fractions, 104–107, 114–115
working level month, 15, 22, 32, 223–227
Ethnic factors, 148, 156
Evaporation, 143
Exercise, *see* Physical activity
Exposure, 15–16, 32–33
defined, 223–224
dose and, 2–4, 5–6, 31–32, 34–35, 137, 225, 226–227, 228
duration, 12, 13, 14, 15, 16, 55, 155–156, 186
indoor atmospheres, 12, 27–28, 32–33, 109–117, 130
mine atmospheres, 12, 26, 107–109
standards, 26
working level months, 15, 22, 32, 223–227

F

Fibrosis, 153, 154
Formaldehyde, 56
Formulas, *see* Equations
Fractionation, 55

G

Gender differences, 1, 2–3, 4, 5, 12, 13, 15, 34, 42–43, 45, 46, 53, 148, 176, 184, 201–202, 228, 230, 233

Genetics, 54, 177, 178–180, 189
  anatomy, 148
  carcinogenesis, general, 54, 177, 178–180
  DNA, 31, 178, 179, 180–181, 195
  lung cancer, 54, 177, 178–180
  target cell damage, 31
Geometry
  airway, 15, 35, 63, 156, 195, 196, 199–202, 203, 204
  particle deposition, 66–67, 138, 195
  see also Morphometry
Germany, Federal Republic, 11, 111–113, 118
Goblet cells, 168, 181
Gravitational factors
  impaction, 74, 76, 117, 149, 150, 151, 208, 210
  sedimentation, 74, 76, 139–140, 144, 149, 209, 210

H

Historical perspectives, 9–11, 26
  cancer epidemiology, 169, 176
  concentration measurement, 15
  deposition of aerosols, research, 137–138
  dosimetric models, 24, 60–82
  epidemiology, 11, 169, 176
  unattached fractionation, 15, 95, 105–109
Hygroscopicity, 7, 33, 143, 144
Hyperplasia, 46–47, 48, 53–56, 233, 234

I

Impaction, 74, 76, 117, 149, 150, 151, 208, 210
Inertia, 117, 138, 140, 145, 149, 150, 152
Infants, 2, 3–4, 14, 35, 43–44, 45, 46, 201, 204, 228; *see also* Children
Inflammation, 5, 14, 152
Injury, 53, 55–56, 174, 180, 186
Insulation, 33
International activities and perspectives, 11, 172
  epidemiological investigations, 11, 12
  measurement standards, 15
International Commission on Radiological Protection, 3–4, 14, 16, 17, 95–96, 101–102, 103, 105–106

J

Japan, atom bomb survivors, 53, 57
Joule-hours per cubic meter, 15

K

Kinetic theory, 91
Kulchitsky cells, 7, 168

L

Large cell undifferentiated carcinoma, 172, 173, 175–176, 184, 187
Lead, 49, 61, 104, 106, 110, 115, 126–128, 130, 203, 205, 206–207, 224
Liver cancer, 54, 56
Localized deposition, 37–38, 75, 76, 77, 81, 155
Lung cancer, 1, 4, 6, 9, 11, 16, 72, 137
  adenocarcinomas, 172, 173, 179, 184, 187, 189
  age, growth, injury and, 53, 55–56, 175, 177, 188
  cells of origin, 56, 166–189
  gender differences, 53
  genetics, general, 54, 177, 178–180
  large cell undifferentiated carcinoma, 172, 173, 175–176, 184, 187
  plutonium and, 56
  small cell undifferentiated carcinoma, 171–172, 173, 175, 178–179, 183–184, 186–187, 188
  squamous cell carcinoma, 170–171, 173, 174–175, 177, 179, 181–184, 186, 187, 188
Lungs, 215
  age factors, 144, 147, 148
  clearance of aerosols, 79–80, 144–147, 157
  deposition of aerosols, 6, 137–144, 147–148, 149–157
  emphysema, 153, 154
  epithelia, 167–169, 174
  fibrosis, 153, 154
  morphometry, 6, 68–69, 70, 71, 72, 74, 76, 147, 199–200, 201
  volume, 138, 142, 144, 147–148, 149, 150, 151, 153, 156 199–200, 201, 202

M

Magnetic resonance imaging, 217, 218
Mass factors, particles, 142
Measurement and measurement techniques, 11, 24–25
  changes, 20
  diffusion sampler, 95–96, 101–102, 109–110, 118
  international standards, 15
  Joule-hours per cubic meter, 15
  mucous clearance, 77–78
  particle size, 90
  radiolabeling, 78–80, 145, 149, 154
  unattached fraction, 23–25, 26, 95–130
  wire screen samplers, 25, 27, 94, 95, 96–100, 102–107, 110–124
  working level months, 15, 22, 32, 223–227
  see also Dose and dosimetry; Equations; Morphometry

INDEX

Mitigation measures, cost of, 11, 14
Models and modeling
  age factors, 54
  airway, 15, 35, 64, 156, 195, 196, 199–202, 203, 204
  clearance, 77–81, 202–207
  deposition, 1, 14, 15, 24, 39, 66–67, 71, 72–77, 200, 207–215
  diffusion, 91, 92
  dosimetric, 1–4, 14, 15, 24, 31–51, 60–82, 180, 194–202
  historical perspectives, 24, 60–82
  morphometric, 61, 63–82
  radon progeny behavior, indoor, 21
  respiratory tract, 16, 61, 64, 71–82
  time factors, 6, 12, 13, 14, 15, 16, 55, 60, 155–156, 186, 208
  see also Animal studies; Equations; Risk assessment; Theory
Moisture, 143, 151
  atmospheric, 25
  hygroscopicity, 7, 33, 143, 144
Monte Carlo methods, 70, 122
Morphometry, 61–82
  bronchial, 67–71, 73, 75, 200, 202, 204, 211
  head airways, 63–67
  lungs, 6, 68–69, 70, 71, 72, 74, 76, 147, 199–200, 201
  nasal, 63–67, 72–73
  oral processes, 66–67, 72–73, 74
  particle size and, 72–76
  respiratory tract, general, 61, 63, 67, 71–72
  tracheal, 67–71, 73, 75, 204
Mouth, *see* Oral processes
Mucus, 1, 7, 52, 54, 56, 65, 67, 152, 155, 195, 197
  carcinogenesis, 168, 181–182, 186
  clearance, 15, 35, 46, 78–80, 145–146, 202–203, 205, 206, 227
  smoking, 56, 196

N

Nasal processes, 1, 2, 6, 38, 39–40, 43–44, 56, 215–223
  clearance and, 78, 145
  deposition, 1, 3, 38, 39–40, 66, 72–73, 74, 149, 150–152, 216, 218–222
  morphometry, 63–67, 72–73
  unattached fraction, 40, 218, 219
  volume, 151
NCRP, 16, 17
Nucleation theory, 24–25

O

Oral processes, 3, 6, 38, 39–40, 72, 142, 151–152, 215–223

deposition of particles, 72–73, 74, 216
morphometry, 67, 72–73, 74

P

Parenchyma, 150, 155
Particle size, 1, 2, 6, 7, 20, 227–230, 231, 232
  activity median thermodynamic diameter, 32–33, 40, 41–42, 45, 49, 108, 139, 215, 224
  aerodynamic diameter, 139–143
  behavior, aerodynamic, 21
  clearance and, 144
  clusters, particles, 24–25, 92, 116, 125, 126
  deposition and, 72–73, 75, 76, 138–143, 157, 212–215
  diffusion, 25, 90–95, 208
  distributions, 12, 15, 25, 26, 27–28, 110, 117–130, 131, 142, 215
  morphometry, 72–76
  ultrafine particles, 16, 75, 76, 92, 94, 95, 139, 212
  unattached fraction, 1, 2, 11, 12, 15, 23–25, 26, 32, 40, 55, 95–130, 169, 176, 218, 219, 230
Pharynx, 77, 142, 145, 149, 151
Physical activity, 2, 12, 13, 149–150, 151, 227, 230, 231, 233
Polonium, 9, 24, 25, 26, 56, 61, 96, 101–102, 103, 106, 110, 111, 113, 116, 118, 124–125, 128, 194–197, 199, 200, 203, 205, 216, 217, 218, 224
Plutonium, 54, 56
Preexisting conditions, 17, 152–155
  see also Disease, preexisting; Smoking
Pulmonary acinus, 72

R

Racial/ethnic factors, 148, 156
Radiolabeling, 73–74, 78–81, 145, 149, 154
Radium, 26, 54
Reading Prong Formation, 10
Regional deposition, 37–38, 77
Retention, of aerosols, 138, 144, 157; *see also* Clearance
Respiratory tract, 6
  aerosol growth, 2
  morphometry, 61, 63, 67, 71–72
  models, 16, 61, 63, 71–82
  see also specific anatomical subsystems
Risk assessment, 3, 4, 11–12, 16–17, 166, 185
  dosimetric extrapolation, 1, 4, 5, 12, 14, 31–57, 110, 137, 157

S

Sampling and samplers
  diffusion, 95–96, 101–102, 109–110
  wire screen samplers, 25, 27, 94, 95, 96–100, 102–107, 110–124
Secretory cells and processes, 2, 3, 6, 25, 37, 41, 42–43, 45, 46, 47, 77, 163, 194, 195, 196, 197, 198, 227, 229, 230
  carcinogenesis, 166, 167, 172, 181, 183, 184, 186, 189
  hyperplasia and, 46, 48, 233, 234
  thoron, 50
  see also Mucus
Sedimentation, gravitational, 74, 76, 139–140, 144, 149, 209, 210
Sex differences, *see* Gender differences
SIMPLEX algorithms, 118
Skin tumorigenesis, 54
Small cell undifferentiated carcinoma, 171–172, 173, 175, 178–179, 183–184, 186–187, 188
Smoking, 1, 2, 4, 5, 6, 12, 13, 15, 27, 28, 32, 45, 52, 56–57, 79, 128, 129, 130, 155, 196
  cellular carcinogenesis and, 169, 170, 174, 175, 176, 177, 185–188
  in equations, 23
Squamous cell carcinoma, 170–171, 173, 174–175, 177, 179, 181–184, 186, 187, 188
Standards, 11
  carcinoma classification, 172–172, 175–176
  exposure, 26
  international, 15
  wire screen samplers, 96
State-level action, 10
Stem cells, 6–7
Sulfur dioxide, 25

T

Target cells, 15, 25, 31, 32, 35–38, 39, 42, 43, 49, 166, 189, 194–195, 198, 227–230, 234
Temporal factors
  exposure duration, 12, 13, 14, 15, 16, 55, 155–156, 186
  tumor formation, 6
Testing and test methodology, *see* Measurement and measurement techniques
Thermal processes and effects, 212, 223
  activity median thermodynamic diameter, 32–33, 40, 41–42, 45, 49, 108, 139, 215, 224

Brownian diffusion, 139, 210
  cooking, 128, 130
  evaporation, 143
Theory, 194
  deposition, 39, 74, 76, 207, 209–210
  diffusion, 91, 92
  nucleation, 24–25
  wire screen penetration, 96
Thorium, 26, 61, 63
Thoron, 49–51, 54, 225–226, 228
Time factors, *see* Temporal factors
Tissue damage, *see* Injury
Tobacco, *see* Smoking
Toxins, other than radon, 1, 12, 15, 25, 56
Trachea, 53, 67–71
  clearance, 79, 80, 146, 151, 195, 203
  deposition, 6, 20, 73–74, 77, 148, 151, 154, 207, 212
  epithelium, 53, 145, 181, 182, 183
  morphometry, 67–71, 73, 75, 204
  wounding, 55–56
Tubes, diffusion, 92–93, 95, 101, 195

U

Ultrafine particles, 16, 75, 76, 92, 94, 95, 139, 212
Unattached fraction, 1, 2, 12, 23–25, 26, 32, 55, 95–130, 230
  in equations, 104–107, 114–115
  measurement techniques, 23–25, 26, 95–130
  nasal processes, 40, 218, 219
Urethane, 56

V

Volumetrics, 76, 95, 220
  alveolar, 208
  lungs, 138, 142, 144, 147–148, 149, 150, 151, 153, 156, 199–200, 201, 202
  nasal, 151

W

Wire screen samplers, 25, 27, 94, 95, 96–100, 102–107, 110–124
Women, 42–43, 46
  see also Gender differences
Working level month, 15, 223–227
  formula, 22, 32, 225–226
World Health Organization, 172
Wounds, 53, 55–56, 174, 180, 186